The Chemistry of Muscle-based Foods

Edited by

D.E. Johnston

Department of Agriculture for Northern Ireland and The Queen's University of Belfast, UK

M.K. Knight

Griffith Laboratories, Somercotes, Derbyshire, UK

D.A. Ledward

Department of Agriculture for Northern Ireland and The Queen's University of Belfast, UK

The Proceedings of a Symposium organized jointly by the Food Chemistry Group of the Royal Society of Chemistry and the Society of Chemical Industry Meat Panel, The Queen's University of Belfast, 9-11 September 1991.

Special Publication No. 106

ISBN 0-85186-237-3

A catalogue record of this book is available from the British Library

© The Royal Society of Chemistry 1992

All rights Reserved
No part of this book may be reproduced or transmitted in any form or by any means–graphic, electronic, including photocopying, recording, taping or information storage and retrieval systems–without written permission from the Royal Society of Chemistry

Published by The Royal Society of Chemistry.
Thomas Graham House, Science Park, Cambridge CB4 4WF
Typeset by Keytec Typesetting Ltd, Bridport, Dorset
Printed by Redwood Press Ltd, Melksham, Wiltshire

The Chemistry of Muscle-based Foods

Preface

To the majority of consumers muscle-based foods are the centre point and usually the most expensive item of a meal. It is not surprising therefore that these valuable foods have been extensively researched over many generations. Their study, quite naturally, involves a multitude of disciplines but predominant is a full understanding of the chemistry of these complex systems. This book is the proceedings of a symposium held in The Queen's University of Belfast 9-11 September 1991 and organized by the Food Chemistry Group of the Royal Society of Chemistry and the Food Commodities and Ingredients Group of the Society of Chemical Industry at which acknowledged international experts were invited to summarize their views about the chemistry of these foods.

The symposium and thus the book naturally lent itself to four discreet sections: Production Factors, Conversion of Muscle into Food, Chemistry of Raw and Cooked Products, and Chemistry of New and Existing Technologies. The first is concerned with the actual growth and development of the animals which ultimately end up as muscle-based foods. This area has traditionally been the study of physiologists and animal nutritionists but increasingly chemical understanding of how manipulation of the growth of the animal can modify the yield, composition, and quality of the flesh is being applied. These three chapters by J.D. Wood, K.J. McCracken, and J.M. Jones update us with current knowledge as related to pigs, cattle, and poultry. As these chapters demonstrate, it is with the white meats derived from poultry and pigs that the greatest advances have been made, because of their higher growth rates and more integrated systems of production, slaughter, and processing.

The second section, of two chapters, by R.A. Lawrie and B.W. Moss, deals with the conversion of living muscle into a food. Thus the biochemistry of the multitude of changes taking place when living muscle becomes meat, fish, or poultry flesh is discussed and the implications regarding quality are assessed. In this area, of course, animal welfare considerations are of paramount importance as Dr Moss explains and must be taken into account in any review of the topic.

The third section, consisting of nine chapters, sets out our current understanding of the chemistry of muscle-based foods and its relationship to the perceived eating and nutritional quality. Thus in the following two chapters (by N.F.S Gault and A.J. Bailey and T.J. Sims) the chemistry involved in dictating the toughness or texture of raw and cooked meat is described and the intricate relationships between the various muscle structures are evaluated. Colour is perhaps the most important attribute of meat quality since if it is not attractive a consumer will not even attempt to evaluate the toughness and flavour. This important area, as relating to both the raw and the cooked product, is considered in the chapter by D.A. Ledward. Although a few years ago most people would identify texture as the major eating quality attribute of meat it is increasingly being recognized that flavour is as important. There have been developments in recent years which have helped to unravel this very involved area of chemistry and these are summarized and discussed in a chapter by L.J. Farmer. Flavour, however, is inherently affected by post-slaughter factors and a chapter by J.I. Gray is devoted to the chemistry of the oxidative changes that can take place in meat. The following two chapters are concerned with the other very important aspect of any food, namely nutritional quality. The lipid composition of meat and the complex area of bioavailability of iron are explored by J.B. Rossell and R.J. Neale respectively.

The fourth section presents the chemistry underlying both new and existing technologies. For many years surimi has been a most important product of fish processing, especially valued in Japan and the Far East. Increasingly though, the potential of the technique for application to other muscle-based foods has been recognized, as also has its potential for increasing use within the fish area. The chemistry and operations involved in fish and red meat surimi are dealt with in chapters by I.M. Mackie and M.K. Knight. One of the oldest established methods of preserving meat products is curing using a mixture of common salt and nitrate/nitrite. In a chapter by L.H. Skibsted current views on the intricate web of reactions involved are outlined. The last three chapters in the book address what could be deemed relatively new technologies. Thus, the potential of ultrasonics is outlined by R.T. Roberts and high-pressure processing for meat is reviewed by D.E. Johnston. The final chapter covers the irradiation processing or preservation of muscle-based foods. This proven and safe technique is increasingly receiving attention throughout the world and M.H. Stevenson outlines our present knowledge.

This book, though written by different authors, is an integrated treatise which we hope will be a valuable source of reference for researchers, students, and technologists for a good few years.

D.E. Johnston
M.K. Knight
D.A. Ledward

Contents

Production Factors

Effects of Production Factors on Meat Quality in Pigs   3
J.D. Wood, P.D. Warriss, and M.B. Enser

Production Factors: Beef   16
K.J. McCracken

Factors Influencing Poultry Meat Quality   27
J.M. Jones

Conversion of Muscle into Food

Conversion of Muscle into Meat: Biochemistry   43
R.A. Lawrie

Lean Meat, Animal Welfare, and Meat Quality   62
B.W. Moss

Chemistry of Raw and Cooked Products

Structural Aspects of Raw Meat   79
N.F.S. Gault

Structural Aspects of Cooked Meat   106
T.J. Sims and A.J. Bailey

Colour of Raw and Cooked Meat   128
D.A. Ledward

Oxidative Flavour Changes in Meats: Their Origin and Prevention   145
J.I. Gray and R.L. Crackel

Meat Flavour   169
L.J. Farmer

Meat Iron Bioavailability: Chemical and Nutritional Considerations   183
R.J. Neale

Chemistry of Lipids 193
J.B. Rossell

Structural Aspects of Processed Meat 203
D.F. Lewis

Lipids: Nutritional Aspects 204
R.C. Cottrell

Chemistry of New and Existing Technologies

Surimi from Fish 207
I. M. Mackie

Red Meat and Poultry Surimi 222
M.K. Knight

Cured Meat Products and Their Oxidative Stability 266
L.H. Skibsted

High Intensity Ultrasonics 287
R.T. Roberts

High Pressure 298
D.E. Johnston

Irradiation of Meat and Poultry 308
M.H. Stevenson

Subject Index 325

Acknowledgements

The conference organizers wish to record their gratitude to the following organizations for their sponsorship of the conference
Department of Agriculture for Northern Ireland
Meat and Livestock Commission
Moy Park Ltd
Stewarts Supermarkets Ltd
Bernard Matthews
Livestock Marketing Commission for Northern Ireland
Northern Ireland Tourist Board

Production Factors

Effects of Production Factors on Meat Quality in Pigs

J.D. Wood, P.D. Warriss, and M.B. Enser

DEPARTMENT OF MEAT ANIMAL SCIENCE, SCHOOL OF VETERINARY SCIENCE, UNIVERSITY OF BRISTOL, LANGFORD, BRISTOL BS18 7DY, UK

1 Introduction

Until recently, producers of pigmeat were concerned solely with aspects of physical performance in the growing animals, *i.e.* growth rate, feed conversion efficiency, and carcass lean content (as predicted from a measurement of fat thickness). In the future, however, it is likely that the level of meat quality will also be a factor in profitability as consumer and supermarket demands for product quality increase. Particular interest will be attached to influences on meat quality which are independent of lean:fat content since consumers wish to buy lean meat, yet there is evidence that the continuing trend to lower fatness levels has tended to lower some aspects of meat quality.

2 Recent Changes in Carcass Composition and the Fat Content of Pigmeat

For many years, producers in the UK and other countries have received a higher price per kg for carcasses with thinner backfat. Taking account of production and processing costs, these carcasses produce the lean cuts demanded by consumers more economically than preparing defatted cuts from fatter carcasses. As a result of price incentives, producers have made production changes in the following main areas: genetics (use of superior breeding animals selected on the basis of growth and carcass criteria); nutrition (use of high protein–high energy diets to maximize the potential of leaner stock); and the balance between sexes (entire male animals grow faster and are leaner than castrates: as a result only 5% of male pigs are now castrated in the UK). There has also been an increased use of automatic probes for carcass classification so that carcasses are more accurately placed in pricing grades than previously. The advantage of the automatic probe over the intrascope is that it measures muscle as well as fat thickness and provides a more accurate prediction of percentage lean.

The results of these changes in British pigs are illustrated in Table 1.[1] In the past fifteen years, the average pig carcass has decreased in fat thickness by a third, causing reductions in all other measures of fat content. The lipid content of meat consumed by people removing visible fat before cooking or on the plate probably corresponds most closely to 'defatted lean', which is the lean tissue separated from subcutaneous and intermuscular fat and bone by dissection. Other work in pigs of the same fat thickness shows that the joints of the carcass varied in the lipid content of defatted lean from 1.7% in the leg to 5.3% in the collar (shoulder).[2] The important back or loin joint contained 3.9% fat in defatted lean and *ca.* 0.8% lipid in a core taken from the centre of the major *longissimus* muscle.

As the total fat content of the carcass has declined with time, the fatty acid composition of the fat in the various parts of the carcass has also changed towards a more unsaturated fat. Dietary advice is for the British diet as a whole to contain a ratio of polyunsaturated to saturated (P:S) fatty acids of *ca.* 0.4 and it can be seen in Table 1 that this ratio is almost achieved in normal pig meat (0.32). In carcasses with 8 mm P_2 fat thickness the ratio is *ca.* 0.4.[3] In contrast, beef and lamb meat have P:S ratios of *ca.* 0.1.[4]

More detailed figures are given in Table 2 for the fatty acid composition

Table 1 *Changes in the composition of the average British pig carcass (weight 63 kg)*[1]

	1975	1989
P_2 fat thicknes (mm)	18	12
Separable fat in carcass (%)	31	20
Lipid in defatted lean (%)	5.5	3.7
Lipid in subcutaneous fat (%)	85.4	77.2
Marbling fat (%)[a]	1.1	0.8
Fatty acids (%)		
saturated	42.6	40.7
polyunsaturated	9.7	13.2

[a] Ether-extractable lipid in *longissimus*.

Table 2 *Fatty acid composition of lean and intermuscular fat in 38 pig carcasses averaging 12 mm P_2 fat thickness*[2]

	C14:0	C16:0	C16:1	C18:0	C18:1	C18:2	C18:3
Ham (leg)	1.3	22.2	2.4	13.1	34.1	14.5	1.6
Collar	1.6	24.5	2.3	14.6	33.8	14.5	1.3
Hand (shoulder)	1.4	22.6	2.3	13.8	33.9	13.8	1.1
Back	1.4	22.6	2.3	14.9	34.7	13.2	0.7
Streak	1.6	23.2	1.9	13.7	32.6	13.5	1.0
Carcass average	1.5	24.6	2.2	14.2	34.1	13.1	1.4

of lean and intermuscular fat in the average British pig carcass in 1989. There is only minor variation between joints in the concentrations of the major fatty acids.

3 Associations between Carcass Composition and Meat Quality

Most emphasis in this chapter will be given to eating quality which is the most important aspect of pigmeat quality. Within eating quality, tenderness and juiciness are possibly the most crucial with flavour becoming important in certain situations, *e.g.* when taints become apparent.

A study published in 1981[5] provided the first evidence in the UK that eating quality could be detrimentally affected by low levels of fat in the carcass. Before this, the general view was that there was no clear association between fatness and eating quality.[6] The results of the 1981 study (Table 3), on only eight pigs in each of 'lean' and 'very lean' groups, showed that juiciness as assessed by trained taste panellists in roast loin was significantly reduced in the leanest group with tenderness also tending to fall. Toughness measured objectively was higher in the loin muscle of the leanest pigs.

Since 1981, other studies have observed that several aspects of meat quality including eating quality tend to decline as carcass fatness is reduced (lean increased). A comprehensive study published in 1986 (Table 4)[7,8] found that the subcutaneous fat in carcasses with 8 mm P_2 fat thickness was less firm that that in carcasses with 16 mm P_2. Butchers also considered that cuts from the leanest carcasses lacked visual appeal, having more fat separation. Most importantly, both a trained taste panel and a consumer panel gave lower scores for juiciness (and tenderness in the case of consumers) to the leanest cuts.

In the study described in Table 4, the correlations between taste panel scores and measures of meat composition were low, ranging from 0.04 to 0.31 for P_2 fat thickness and from 0.13 to 0.31 for muscle lipid (marbling fat). The correlations were highest for marbling fat and juiciness was the aspect of eating quality most affected. Other studies have also shown that

Table 3 *Eating quality in 'lean' and 'very lean' pork carcasses*[5]

	Lean	Very lean	Significance
P_2 fat thickness (mm)	14	8	
Longissimus depth (mm)	46	52	***
Lean in carcass (%)	57.2	62.9	***
Toughness (J)	0.14	0.16	NS
Tenderness (−7 to +7)	1.8	0.9	NS
Flavour (−7 to +7)	2.4	2.5	NS
Juiciness (0–3)	1.4	0.9	*

Table 4 Eating quality in carcasses of different fat thickness[7,8]

	P_2 fat thickness (mm)		Significance
	8	16	
Trained taste panel[a]			
Tenderness	1.0	1.1	NS
Juiciness	1.1	1.3	**
Consumer panel[b]			
Tenderness	35	37	**
Juiciness	16	23	***
Butcher panel[c]			
Fat softness	32	6	***
Fat separation	46	11	***

[a] Tenderness score −7 to +7, juiciness 0–4.
[b] Percentage of samples rated very tender or very juicy.
[c] Percentage of samples with soft or very soft fat and with excessive tissue separation.

marbling fat is the aspect of fatness most closely related to eating quality characteristics.[9]

4 Marbling Fat and Eating Quality

Marbling fat is the adipose tissue located between muscle fibre bundles in the perimysial connective tissue. It is measured following extraction with a solvent such as diethyl ether. If a more polar solvent such as chloroform: methanol (2:1) is used or if acid hydrolysis precedes diethyl ether extraction, the amount of lipid extracted is increased by *ca.* 20% because the more polar phospholipids are extracted in addition to the neutral lipids, mainly triacylglycerols (Table 5).[10] The use of different extraction pro-

Table 5 Marbling fat and eating quality in 128 Duroc-and Large White-sired crossbred pigs[10]

	Duroc	Large White	Significance
P_2 fat thickness (mm)	14.1	13.0	**
Marbling fat[a]			
Extractable lipid			
thoracic	1.4	1.0	***
lumbar	1.4	0.9	***
Total lipid			
thoracic	1.8	1.4	***
lumbar	1.8	1.3	***
Tenderness (−7 to +7)	−0.3	−0.0	NS
Juiciness (0–4)	1.3	1.3	NS
Flavour (−7 to +7)	2.1	2.0	NS

[a] Cores from the centre of *longissimus* were extracted with diethyl ether (extractable lipid) or diethyl ether following acid hydrolysis (total lipid).

cedures in published studies has contributed to different conclusions being reached about optimal levels of marbling fat for eating quality characteristics.

Although correlations between the concentration of marbling fat and the scores given by taste panellists are generally low, it is accepted that the positive effect of marbling fat on eating quality is important alongside the other factors involved. This is why commercial breeding companies and research workers are currently searching for techniques for measuring marbling fat in live animals and carcasses. Ultrasonic techniques seem promising.

A study conducted at Bristol showed that although marbling fat increased with total carcass fat in eleven pig breeds studied, there was considerable variation about the mean regression line, caused by breed variation.[11] The dark-haired breeds had generally high values in relation to P_2 backfat thickness and among these the Duroc had the highest concentration of marbling fat.

Several European and American studies show that there is a tendency for eating quality characteristics to be higher from Duroc pigs, because of this greater concentration of marbling fat. However, scores for eating quality have not always been higher in pure Durocs or Duroc crosses in individual studies as the results in Table 5 show. In this case the proportion of Duroc genes was 0.5. A recent large-scale British evaluation of the Duroc breed showed that when the proportion of Duroc genes rose to 0.75, tenderness was increased from 4.83 to 5.25 on a scale of 1 to 8 (C.C. Warkup, D.B. Lowe, and P.D. Warriss, unpublished observations).

5 Pale Soft Exudative (PSE) Meat and Eating Quality

One reason for the generally negative association between fatness and meat quality, including eating quality, is that certain breeds or strains with above average conformation (shape) and leanness are also genetically predisposed to producing PSE meat. This meat, possibly because of a lower water-holding capacity in muscle fibres, is tougher and less juicy than would be expected from the lower marbling fat concentration.[12]

It was originally believed that the gene which has a major influence on the development of PSE (halothane gene, n) was totally recessive, *i.e.* that the heterozygotes Nn would not be intermediate between nn and NN but would have normal post mortem muscle characteristics close to NN. However, there is now good evidence that under normal commercial slaughter conditions the incidence of PSE carcasses in Nn pigs is higher than in NN. Recent Canadian results on halothane genotypes derived from the Lacombe breed are shown in Table 6.[13] All indices of PSE obtained in the heterozygotes were intermediate between the two homozygotes and toughness was clearly higher in Nn than in NN. More recent results from the Canadian group show that the expression of the halothane gene depends on age or weight. For toughness as well as other measures of the

Table 6 *Halothane genotype and meat quality (90 kg live weight) in pigs derived from the Lacombe breed*[13]

	NN	Nn	nn
Measurements made in longissimus			
pH_1	6.3^a	5.9^b	5.7^c
Colour a*	8.5^b	9.0^a	9.2^a
Soluble protein ($g\,kg^{-1}$)	160^a	126^b	109^c
Toughness (kg)	5.8^c	7.2^b	7.8^a

Means with different superscripts differ significantly ($P < 0.05$).

PSE condition, the difference between NN and Nn increased as carcass weight increased through the commercial range.[14]

There is evidence that heterozygotes for the halothane gene are more sensitive to preslaughter handling conditions than non-carriers. This may explain the apparent conflict between the early and more recent results on the incidence of PSE in halothane genotypes. For example, Danish research showed that the proportion of PSE carcasses in the Nn genotype increased in abattoirs in which handling was 'less considerate'.[15] However, this interaction between genotype and environment was not shown in a recent study conducted at Bristol.[16] Meat-type and white-type pigs (which would be expected to have relatively high and low incidences of Nn respectively) from four commercial breeding companies were subjected to two transport regimens (1 and 4 hours) and two lairage treatments (2 and 21 hours). There was no evidence to suggest that the meat-type pigs reacted more adversely to the extreme treatments although it is possible that the treatments were not sufficiently extreme to trigger the sequence of biochemical events leading to PSE meat.

6 Factors Affecting Tenderness in Pigmeat

Several factors affect tenderness including preslaughter handling treatment, stunning method, carcass chilling rate, carcass suspension system, and conditioning time. The production factors likely to affect tenderness include breed, carcass weight, fat level, and feeding treatment. Breed effects are probably mainly associated with marbling fat and the genetic susceptibility to stress as already discussed. Feeding treatments include the level of intake and the type of diet.

Effects of Feeding Level on Tenderness

The results of a major study conducted by the Meat and Livestock Commission showed that tenderness was increased in meat from pigs given feed *ad libitum* compared with those given 80% of an *ad libitum* intake (Table 7).[17] The pigs which consumed more feed and grew more quickly

Table 7 *Effects of feeding level on fatness and eating quality in pigs slaughtered at 80 kg live weight*[17]

	Ad libitum	Restricted	Significance
P_2 fat thickness (mm)	12.8	11.1	*
Marbling fat (%)	0.85	0.75	*
Tenderness (1–8)	5.20	4.73	*
Juiciness (1–8)	4.44	4.25	*
Pork flavour intensity (1–8)	4.52	4.57	NS

Means with an asterisk are significantly different at the 95% level or greater.

were fatter and had higher concentrations of marbling fat than those eating at the lower level but it was shown[18] that this was not important in explaining the difference in tenderness. Instead the difference was statistically explained by variation in growth rate (daily liveweight gain).

The influence of growth rate on tenderness could be due to its effect on the pattern of muscle protein deposition. In fast growth, proteins are synthesized and degraded more rapidly, *i.e.* they turn over more quickly, the degradation stage being controlled by proteinase enzymes. It has been suggested that increased activity of the calcium-dependent proteinases could also occur post mortem whilst the pH is still fairly high, resulting in greater tenderization.[18] However, these ideas are speculative at present and have not been properly tested.

Effects of β-Adrenergic Agonist Drugs

Some support for the role of proteinase enzymes in the control of tenderness in pork comes from research which has shown that the administration of β-adrenergic agonists slightly toughens meat. A major effect of these substances is to reduce the rate of muscle protein degradation *in vivo* thus increasing net protein deposition.[19] In lambs a reduced activity of calpain enzymes post mortem has been demonstrated.[20] Evidence for the effect on toughness in pork is shown in Table 8.[21] The effect was observed in *longissimus* and *semimembranosus* but not *supraspinatus*, which has a higher proportion of red oxidative fibres than the other

Table 8 *Effects of the β-adrenergic agonist salbutamol (Sal, 2.7 p.p.m. in the feed between 28 and 92 kg live weight) on toughness (shear force, kg) in three muscles of 80 gilts*[21]

	Control	Sal	Significance
Longissimus	5.06	5.83	*
Semimembranosus	5.08	5.51	*
Supraspinatus	4.69	4.31	NS

muscles and is intrinsically more tender. Previous work by Warriss and co-workers[22] had strongly suggested that another effect of β-adrenergic agonists is to increase the concentration of white glycolytic muscle fibres and this shift is invariably associated with increased toughness.[23]

Effects of Diets with Different Fatty Acid Compositions

Many studies have shown that pigs, being simple-stomached animals, transfer dietary fatty acids to their adipose tissue, the rate of accumulation depending on, amongst other factors, the activities of enzymes which change one fatty acid into another, *e.g.* the desaturase enzymes. The proportion of dietary linoleic acid (C18:2) removed in this way is small and its accumulation is particularly marked in relation to dietary intake.

A recent study of the effects of dietary C18:2 on its incorporation into muscle and backfat lipids is shown in Table 9.[24] Groups of eight pigs were fed a diet containing 6% lipid (4% added fat composed of different proportions of soya oil and coconut oil) between 30 and 90 kg live weight. Lipids were extracted from *longissimus* with chloroform:methanol (2:1) and separated into neutral and phospholipids. Only total lipid (more than

Table 9 *Effects of the concentration of linoleic acid (C18:2) in the diet on its incorporation into muscle neutral lipids and phospholipids and backfat total lipids*[24]

	C18:2 in diet (%)				
	1.2	1.7	2.1	2.6	3.0
Marbling fat (%)[a]	0.71	0.95	0.91	0.73	0.78
Muscle neutral lipids[a]					
C18:0	11.4	12.9	12.3	11.8	11.5
C18:1	42.0	41.5	43.1	40.2	36.6
C18:2	6.9	8.7	8.7	12.1	14.0
C18:3	0.4	0.5	0.7	0.8	0.9
Muscle phospholipids[a]					
C18:0	13.3	14.5	14.4	15.0	16.0
C18:1	13.9	12.3	11.0	10.6	10.5
C18:2	27.7	28.3	33.1	31.4	32.0
C18:3	0.6	0.3	0.8	0.5	0.6
Backfat total lipids[b]					
C18:0	16.2	16.1	17.6	15.3	15.4
C18:1	35.1	34.0	32.2	32.8	31.5
C18:2	6.8	10.0	11.7	16.4	20.4
C18:3	1.3	1.6	1.8	2.1	2.5
Tenderness[c]	0.78	0.47	0.58	0.48	−0.05

[a] From *longissimus*. Percentage of fatty acids.
[b] From inner layer of backfat at last rib. Percentage of fatty acids.
[c] Score −7 (very tough) to +7 (very tender).

99% neutral lipid) was analysed in backfat. Muscle neutral lipids (triacylglycerols) contained a lower concentration of C18:2 than phospholipids but neutral lipid C18:2 was more affected by dietary intake. Diet had an even greater effect on the total lipids from backfat and it was noticeable that as C18:2 increased the concentration of C18:1 (oleic acid) declined. Subsequent analysis of the meat from the pigs described in Table 9 showed no significant effects of diet on eating quality as determined by taste panellists. However, there was an indication that pigs fed the lowest and highest levels of C18:2 had the highest and lowest scores for tenderness respectively. Other work has shown that in general, the concentrations of polyunsaturated fatty acids are negatively correlated with eating quality traits with saturated fatty acids being positively correlated.[25]

Recent work in the USA has shown that diets with high levels of oleic acid (C18:1) produce meat with superior eating characteristics.[26] Diets containing 0 or 12% high-oleic sunflower oil (85% oleic acid) were fed to pigs between 55 days of age and slaughter at 102 kg live weight. Tenderness and juiciness were both significantly higher ($P < 0.05$) in pigs fed the high oleic-acid diet (Table 10).

7 Effects of Diet on Taints

It was formerly believed that unacceptable taints (odours, flavours) in pigmeat were associated almost exclusively with androstenone or boar taint. Early work by Rhodes at Bristol disputed this view, showing that entire male pigs produced meat which was, on average, similar to that from castrates or gilts in terms of odour and flavour.[27] More recent results from the MLC Stotfold trial are shown in Table 11. Meat from the three sexes had similar pork flavour and only abnormal odour differed significantly in these trained taste panel tests. In concurrent consumer panel tests no sex effects were found.

It has also been demonstrated that variation in the concentration of skatole in subcutaneous fat affects odour and flavour, values above *ca.* 0.2

Table 10 *Effect of a high-oleic acid diet on eating quality of pork chops*[26]

	Control[a]	High oleic[a]
Muscle fatty acids (%)		
C18:1	41.9	52.7
C18:2	14.1	12.7
Taste panel scores (1–8)		
Juiciness	4.5	5.4
Tenderness	5.7	6.4
Flavour	5.6	5.8

[a] Diets contained 29.0 and 71.7% C18:1 respectively.

Table 11 *Eating quality of roast loin muscle from 420 carcasses, average weight 65 kg, as assessed by trained taste panellists*[17,a]

	Entire males	Castrated males	Gilts
Tenderness	5.00	4.96	4.94
Juiciness	4.35	4.41	4.26
Pork flavour	4.59	4.55	4.50
Pork odour	3.62[b]	3.78[b]	3.84[c]
Abnormal odour	3.80[b]	3.51[c]	3.30[d]

[a] Scores 1–8 in every case. Means with different superscripts are significant at 95% level or greater.

p.p.m. being termed unacceptable.[28] Entire males have higher concentrations than castrates, especially when fed *ad libitum*.[29] Skatole is produced in the large intestine as the result of microbial fermentation of diets and some evidence suggested that high-fibre diets were likely to produce high levels of skatole and more tainted meat.[28] However, doubt has been cast over the relevance of the intestinal production of skatole in determining skatole concentrations in backfat by the work of Hawe.[30] In a series of studies he found no significant correlation between skatole concentrations in the caecum and in adipose tissue.

In a recent study (Table 12)[31] we found that as the concentration of sugar beet feed in the diet was increased from 0 to 45% in pigs grown from 20 to 77 kg live weight, the concentration of skatole in backfat declined, as did scores for abnormal odour and abnormal flavour as determined by trained taste panellists. The level of overall liking was increased up to the level of 30% sugar beet feed. These results show that highly fermentable fibrous feeds such as sugar beet feed do not increase the incidence of taints and up to a certain level may actually improve eating quality traits.

Table 12 *Effects of different levels of molassed sugar beet feed in the diet on meat quality in pigs*[31]

	Sugar beet feed (%)			
	0	15	30	45
Fat				
Skatole (p.p.m.)	0.069	0.065	0.038	0.036
Abnormal odour[a]	1.96	1.91	1.78	1.71
Lean				
Tenderness[a]	3.83	4.25	4.76	3.94
Abnormal flavour[a]	2.65	2.33	2.23	2.15
Overall liking[a]	4.46	4.70	4.96	4.40

[a] Taste panel scores 1–8.

8 Integrated Production Systems for Pigmeat Production

Because of the many factors affecting eating quality, particularly tenderness, an increasing trend in many countries is towards systems in which the factors involved are specified in contractual arrangements between retailer, processor, feed compounder, and producer. Typical examples of the items likely to be included in a 'quality assurance' scheme are shown in Table 13[32] and it can be seen that considerations of animal welfare and product safety are frequently included in addition to those affecting eating quality.

9 Conclusions

Pig producers in the UK and other countries have been extremely successful in the past fifteen years or so in producing a product which meets the requirements of consumers for low-fat meat. An additional bonus is that lean pigmeat naturally contains a 'healthier' balance of fatty acids.

There is a tendency for some aspects of meat quality to decline as carcass fatness is reduced but it is possible that a combination of low fatness and high visual and eating quality can be obtained by making changes to production and processing factors other than those which affect fatness. Breeds with a low incidence of genetic susceptibility to stress and high concentrations of marbling fat in relation to fat thickness are beneficial as are *ad libitum* feeding systems and certain diets.

Table 13 *Integrated meat production. Examples of factors specified in quality assurance schemes*[32]

Feed	No antibiotics No growth promoters
Breed	Emphasizing meat quality and hardiness (outdoor pigs)
System	No sow tethers No slats/straw
Transportation	Keep rearing groups together No mixing Short distances Recovery in lairage
Quality measurement	Absence of bruising No pale or dark meat
Carcass processing	Use of hip suspension or electrical stimulation Longer conditioning times

Because many factors affect meat quality it is likely that the future will see an increase in the number of 'quality assurance' schemes which specify inputs likely to result in higher quality pigmeat.

10 References

1. J.D. Wood, in 'Animal Biotechnology and The Quality of Meat Production', ed. L.O. Fiems, B.G. Cottyn, and D.I. Demeyer, Elsevier, Amsterdam, 1991, p.69.
2. Meat and Livestock Commission, 'Fat, Fatty Acid and Protein Content of British Pork', Meat and Livestock Commission, Milton Keynes, 1991.
3. J.D. Wood, M.B. Enser, F.M. Whittington, and C.B. Moncrieff, *Livest. Prod. Sci.*, 1989, **22**, 351.
4. J.D. Wood and A.V. Fisher, 'New Developments in Sheep Production', ed. C.F.R. Slade and T.L.J. Lawrence, British Society of Animal Production, Edinburgh, 1990, p.99.
5. J.D. Wood, D.S. Mottram, and A.J. Brown, *Anim. Prod.*, 1981, **32**, 117.
6. D.N. Rhodes, *J. Sci. Food Agric.*, 1970, **21**, 572.
7. J.D. Wood, R.C.D. Jones, M.A. Francombe, and O.P. Whelehan, *Anim. Prod.*, 1986, **43**, 535.
8. A.J. Kempster, A.W. Dilworth, D.G. Evans, and K.D. Fisher, *Anim. Prod.*, 1986, **43**, 517.
9. C. Bejerholm and P.A. Barton-Gade, Proceedings of the 32nd European Meeting of Meat Research Workers, 1986, p.389.
10. S.A. Edwards, J.D. Wood, C.B. Moncrieff, and S.J. Porter, *Anim. Prod.*, 1992, **54**, 289.
11. P.D. Warriss, S.N. Brown, J.G. Franklin, and S.C. Kestin, *Meat Sci.*, 1990, **28**, 21.
12. C. Bejerholm, Proceedings of the 30th European Meeting of Meat Research Workers, Bristol, 1984, p.196.
13. A.C. Murray, S.D.M. Jones, and A.P. Sather, *Can. J. Anim. Sci.*, 1989, **69**, 83.
14. A.P. Sather, S.D.M. Jones, and A.K.W. Tong, *Can. J. Anim. Sci.*, 1991, **71**, 645.
15. P.A. Barton-Gade, Ref.12, p.8.
16. P.D. Warriss, S.N. Brown, E.A. Bevis, and S.C. Kestin, *Anim. Prod.*, 1990, **50**, 165.
17. Meat and Livestock Commission, 'Stotfold Pig Development Unit. First Trial', Meat and Livestock Commission, Milton Keynes, 1989.
18. C.C. Warkup and A.J. Kempster, *Anim. Prod.*, 1991, **52**, 559.
19. R.F. Thornton and R.K. Tume, Proceedings of the 34th International Congress of Meat Science and Technology, 1988, p.6.
20. D.H. Kretchmar, M.R. Hathaway, R.J. Epley, and W.R. Dayton. *J. Anim. Sci.*, 1990, **68**, 1760.
21. P.D. Warriss, S.N. Brown, T.P. Rolph, and S.C. Kestin, *J. Anim. Sci.*, 1990, **68**, 3669.
22. P.D. Warriss, S.C. Kestin, and S.N. Brown, *Anim. Prod.*, 1989, **48**, 385.
23. M.B. Solomon, R.G. Campbell, N.C. Steel, T.J. Caperna, and J.P. McMurty, *J. Anim. Sci.*, 1988, **66**, 3279.

24. M. Enser, F.M. Whittington, J.D. Wood, and D.J.A. Cole, *Anim. Prod.*, 1990, **50**, 572.
25. N.D. Cameron and M.B. Enser, *Meat Sci.*, 1990, **29**, 295.
26. K.S. Rhee, T.L. Davidson, H.R. Cross, and Y.A. Ziprin, *Meat Sci.*, 1990, **27**, 329.
27. D.N. Rhodes, *J. Sci. Food Agric.*, 1972, **23**, 1483.
28. K. Lundstrom, B. Malmfors, G. Malmfors, S. Stern, H. Petersson, A.B. Mortensen, and S.E. Sorensen, *Livest. Prod. Sci.*, 1988, **18**, 55.
29. R.L.S. Patterson, P.K. Elks, D.B. Lowe, and A.J. Kempster, *Anim. Prod.*, 1990, **50**, 551.
30. S.M. Hawe, PhD Thesis, The Queen's University of Belfast, 1990.
31. A.C. Longland, J.D. Wood, M. Enser, J.C. Carruthers, and H.D. Keal, *Anim. Prod.*, 1991, **52**, 559.
32. A.S. Ambler and J.D. Wood, in 'Recent Advances in Animal Nutrition - 1990', ed. W. Haresign and D.J.A. Cole, Butterworth, London, 1990, p.7.

Production Factors: Beef

K.J. McCracken

DEPARTMENT OF AGRICULTURE FOR NORTHERN IRELAND AND THE
QUEEN'S UNIVERSITY OF BELFAST, NEWFORGE LANE, BELFAST BT9 5PX, UK

1 Introduction

In common with other aspects of agriculture, beef production is subject to a wide range of constraints such as climate, topography, disease, finance, feed resources, and market demand. Because of the long production cycle and low reproduction rates it requires a long-term commitment on the part of producers and is particularly susceptible to external pressures. In some parts of the world the most critical factors may be drought or disease though market demand usually plays some role. In developed, over-supplied market economies such as the EC the most critical factors are normally those related to consumer and political pressures. Within the UK the consumption of meat has remained fairly static during the past 20 years.[1] This has been due to a reduction in 'per capita' consumption and increased population. Total food consumption has fallen, presumably due to reduced energy needs as people become less active, but meat as a proportion of food consumption has maintained its position. Poultry (kg/head) has increased by more than 50% whereas beef consumption has decreased by *ca.* 10%.[1]

However, the consumption of beef and veal is still higher than that of any other species and in 1990 beef and veal production accounted for 29% of UK meat production, a value on a par with poultry meat production.[2] It is, therefore, a very significant part of the meat market in terms of both production and consumption.

2 Factors Affecting Beef Production

The factors which must be considered in relation to the production and consumption of beef are growth rate, feed efficiency, fat cover and conformation (which together affect the yield of 'saleable' meat), and quality factors such as colour, flavour, and tenderness.

Whilst price is probably the main factor affecting consumer choice there is growing awareness in the meat industry of consumer concerns over diet and health[3] and the desire for reliability and consistency of eating quality.[4]

However, from the perspective of the primary producer, factors such as growth rate and feed efficiency are of major importance whereas, from the perspective of the processor, yield of 'saleable meat' and leanness are the motivating factors. Unlike the poultry industry, which is highly integrated, or even pig production, where selection for leanness in response to appropriate economic signals has been a priority for many years, the UK beef industry suffers from a lack of integration. There are many reasons for this including the fact that approximately two-thirds of animals for beef come from the dairy herd, that confusing economic signals have been the norm, and that the system of intervention payments has militated against the development of effective marketing.

Animal Factors

As with other animal species, beef production is concerned with maximizing the yield of 'saleable meat' whilst providing a satisfactory return to the producer and processor. The primary component of 'saleable meat' is muscle but it also contains a variable amount of fat depending on a number of animal-related factors such as breed, weight at slaughter, and available market outlets.

The fat associated with meat is in three discrete components (subcutaneous, intermuscular, and intramuscular). Unlike the pig, where the subcutaneous fat is the major component, much of the fat associated with beef is inter- and intra-muscular although the relative proportions tend to change with increasing fatness. Table 1 (based on data from Kempster *et al.*[5]) illustrates this for a typical 300 kg carcass in relation to the EC carcass classification system where code 1 is very lean and code 5 is very fat. Even for fat class 4, which in terms of current consumer demand is too fat, there is twice as much inter- and intra-muscular fat as subcutaneous fat.

The animal-related factors which affect carcass composition, as well as growth and feed efficiency, are breed, sex (and status), nutrition, and slaughter weight. It is impossible to discuss breed, sex, or nutrition independently of slaughter weight and so the latter factor merits some general introduction.

Table 1 *Estimated weights (kg) of fat by fat class at 300 kg carcass weight (based on Kempster* et al.[5])

	\multicolumn{5}{c}{Fat class}				
	2	3	4	4+	5
Lean meat	200	189	181	176	167
Subcutaneous fat	9	18	24.8	28.5	35
Intermuscular fat	30	35.7	39.6	42.0	46.2
Intramuscular lipid	9.2	10.6	11.4	12.0	12.5

Slaughter Weight. It is a general principle that the daily rate of muscle accretion increases during the early phase of growth, tends to plateau between 30 and 60% of mature weight, and then declines. Fat deposition, on the other hand, reflects the energy intake excess to requirements for maintenance and protein deposition, and is therefore a much more variable component of gain. However, the relative excess tends to increase as animals mature and hence the rate of fat deposition, and consequently the proportion of fat in the body, increases with increasing age/weight.[6] For example, the carcass fat content of *ad libitum* fed Friesian steers increased from 220 g kg^{-1} at 420 kg liveweight to 290 g kg^{-1} at 680 kg,[7] and the separable fat content of Simmental crossbred steers increased from 230 g kg^{-1} at 450 kg to 326 g kg^{-1} at 565 kg.[8] In the study of Keane and Drennan,[9] increasing carcass weight under standard feeding conditions from 268 to 375 kg reduced the proportion of lean meat by 10% and increased the separable fat content by 56%. Since the energy content of adipose tissue is much greater than that of muscle such changes are associated with deterioration in feed efficiency.[6,7] Irrespective of consideration of acceptable levels of fat cover there comes a point in the production cycle where the cost of feed inputs becomes the limiting factor to slaughter weight. A further factor relevant to slaughter weight is joint size though the wide range of slaughter weights reported[2] indicates that there is considerable flexibility in market outlets in relation to the latter. It is probable therefore that in future the major limiting factor to slaughter weight will be the degree of fat cover.

Increasing slaughter weight also has implications for the degree of marbling, *i.e.* content of intramuscular fat, and hence for 'quality' characteristics such as tenderness and juiciness. However, in most cases this is unlikely to be of major importance except with very lean animals slaughtered at light weights. For example, in the study of Parrett *et al.*,[8] although marbling score increased by 37% from the low to the high slaughter weight, there were only small effects on 'quality' characteristics and the correlation coefficients for flavour, juiciness, tenderness, and overall acceptability were all less than 0.2. Similarly, Armbruster *et al.*[10] observed that marbling score, intramuscular lipid, and cooking losses were all positively correlated with carcass weight but there was little effect on sensory attributes.

Breed. There are marked breed differences in relation to rate of growth, feed efficiency, and carcass characteristics. The traditional British beef breeds are relatively efficient in terms of gross feed conversion but are early-maturing and tend to become fat at low carcass weights. At the other extreme the Continental breeds have a larger frame size, are late-maturing, and contain a higher proportion of lean meat at similar carcass weights.[11,12] Some of these aspects are illustrated in the results (Table 2) of an MLC breed evaluation.[12] The Charolais × Friesian (F) grew considerably faster

Table 2 MLC breed evaluation: '18 month' beef at fat class 3 (from MLC Beef Yearbook[12])

	Slaughter Age (d)	Wt. (kg)	Carcass wt. (kg)	Feed efficiency[a]	Lean%	Conformation
Holstein	590	523	261	92	58.9	3.2
Friesian	528	468	238	100	60.8	5.5
Hereford × F	467	413	208	114	61.6	6.6
Charolais × F	530	523	275	108	62.9	8.5
Limousin × F	519	463	245	111	63.5	8.0

[a] Relative to Friesian = 100.

than any of the other breeds and the large difference in mature weight of the Charolais × F and the Hereford × F was reflected in the 67 kg difference in carcass weight at equivalent fat class. All three crosses were more efficient than the Friesian and had better conformation and lean yield. The other striking aspect of these results, which has implications in relation to current trends in much of the UK dairy herd, is the poor feed efficiency and low lean yield and conformation of the Holstein.

Results from breed comparisons carried out in Northern Ireland have shown that, with the strains of cattle used for AI, Hereford × F animals produced 27 kg less carcass weight than Friesians while Simmental × F produced carcasses which were 30 kg heavier than Friesians when all animals were slaughtered at the same level of finish.[13]

The breed differences in relation to slaughter weight and fat class are well illustrated by Figure 1.[14] Perhaps the most interesting aspect of this study was the relatively small difference in slaughter weight at the fat class 2/3 interface but the greater range of weight corresponding to fat class 3 in the Charolais × animals compared with the early maturing Herefords. However, there is considerable within-breed variation and the growth potential of individual animals within breeds has a major effect on the relative rates of lean and fat deposition and hence on body composition. For example in a study of the effects of plane of nutrition and slaughter weight on the growth and composition of British Friesian steers from a selected number of AI bulls the differences between the extreme sire types for growth rate, feed efficiency, and carcass fatness were larger and greater than those for the extreme plane of nutrition treatments.[15,16]

There are also marked differences in conformation between breeds (and also within breeds) and this has major implications for the yield of saleable meat as indicated by Figure 2.[14] However, with a few exceptions, the proportion of different joints in carcasses is little affected by breed.[17,18]

It is difficult to make valid comparisons of sensory attributes between breeds due to different rates of growth, slaughter weights, *etc*. Branaman *et al*.[19] and Foster *et al*.[20] concluded that there was no difference in

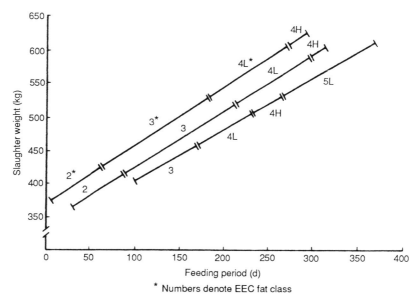

Figure 1 *Breed differences in slaughter weight and feeding period to achieve different levels of fat cover (Kempster)*[14]

tenderness between large and small breeds of cattle. Armbruster et al.[10] found no significant differences in flavour, juiciness, or tenderness of beef from Holstein and Angus steers but Stiffler et al.[18] recorded slightly higher shear force values and reduced tenderness scores from Brahman × steers compared with Holstein or Hereford × Angus steers at similar hot carcass weights.

Sex. There are large differences between males and females in terms of mature weight, growth rate, feed efficiency, and carcass composition.[21–23] On average the difference in carcass weight between steers and heifers at similar levels of subcutaneous fat is *ca.* 80 kg but this value is probably affected by breed and the desired level of subcutaneous fat.

There are also marked differences between bulls and steers.[7,23–25] Feed intakes tend to be similar but the fat content, and hence the energy content of the gain, is lower for bulls. Thus, although the gross energetic efficiency of bulls tends to be lower than for steers, feed efficiency (kg gain per kg feed) is higher. This is illustrated in the results of Fisher et al.[25] who used monozygotic twins (Table 3). Despite the slightly lower feed intake of the steers the carcass contained 22% more separable fat and feed conversion efficiency was reduced by 5%. Similar results have been obtained by Steen[26] in an experiment comparing bulls, steers, and implanted steers in a 15 month beef system (Table 4).

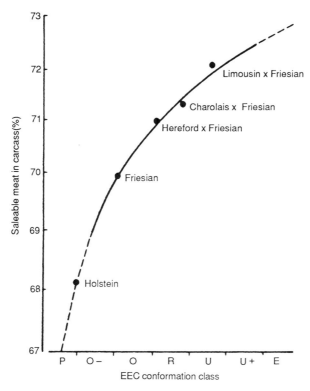

Figure 2 *Relationship between breed, conformation class, and saleable meat yield (Kempster)*[14]

Table 3 *Performance and carcass characteristics of bulls and steers given a complete pelleted diet* ad libitum *and slaughtered at 400 days (Fisher et al.*[25]*)*

	Bull	Steer	Ratio S:B
Final liveweight (kg)	409	379	0.93
FCR (kg feed/kg gain)	7.9	8.3	1.05
Relative feed intake	100	95	0.95
Carcass weight (kg)	220	204	0.93
Weight lean (kg)	136	114	0.84
Weight separable fat (kg)[a]	37	45	1.22

[a] Subcutaneous plus intermuscular fat.

In economic terms therefore, there is a strong case for bull beef rather than steer production. However, there are a number of negative factors in terms of carcass quality associated with bull beef. Firstly they have heavier shoulders and a relatively higher proportion of bone which affects the yield

Table 4 *Growth and carcass characteristics of bulls, steers, and implanted steers slaughtered at 15 months (Steen[12])*

	Bull	Steer	Implant
Carcass weight (kg)	292	261	295
Relative feed intake	100	99	104
Subcutaneous fat depth (mm)	3.8	6.8	6.5

of high-cost joints. On the other hand, because they are leaner, they tend to produce a higher total amount of closely trimmed major retail cuts from the same weight of carcass.[18] Secondly, there is a risk of DFD (dry, firm, dark) carcasses due to inappropriate preslaughter handling. However, it has been shown that this problem does not arise provided that the duration of transport is not excessive and that animals are killed within two hours of leaving the farm of origin.[27] Thirdly, there is a perception that the eating quality of bull beef is inferior to that of steers.[28,29] However, in the study of Stiffler et al.[18] Charolais × bulls compared favourably with Brahman × steers, Holstein steers, and Continental × heifers in terms of shear force, juiciness, tenderness, and overall acceptability. Klastrup et al.[30] found no differences between bulls, steers, and heifers from Hereford/Simmental crosses slaughtered at 16 months of age in regard to a whole range of sensory attributes despite the fact that bulls were leaner and had lower marbling scores.

Dietary Factors

During the past fifty years a considerable literature has been established on the effects of diet composition and plane of nutrition on the growth rate and feed efficiency of bulls, steers, and heifers of a variety of breeds and breed crosses.[6] However, relatively few studies involved detailed examination of carcass composition. This situation has improved in the past 15 years[7,9,23,31–33] and particularly in relation to information on the late-maturing Continental breeds.[34–36]

Type of Diet. With forage-only diets, silage invariably produces more fat and less lean than dried forage, at similar intakes of energy or at similar rates of carcass gain.[37–39] Keane and Drennan[9] also obtained fatter carcasses when finishing steers were given a silage-based diet supplemented with concentrates rather than a dry-forage based diet. However, in this case there were major differences in protein content and in the forage to concentrate ratio between the two diets. On the other hand in a series of three experiments[40,41] carcass gain and characteristics were similar for steers finished on isonitrogenous and isoenergetic diets based on silage or dry forage supplemented with concentrates.

Implanted steers offered high-digestibility silage supplemented with a low input of concentrates produced fatter carcasses than those offered medium-digestibility silage supplemented with a high input of concentrates, but in this case the problem of greater fat deposition with the high-digestibility silage was overcome by offering additional protein in the diet.[41] This suggests that in the earlier experiment[9] and with the low input of concentrates the highly digestible silage may have caused a reduction in protein deposition due to a shortfall in undegradable nitrogen in the rumen.

Plane of Nutrition. Certain general principles are well established; thus restricted, as compared with *ad libitum*, feeding reduces growth rate, dressing percentage, conformation score, and fat content at a given carcass weight. The effects of restriction on feed conversion efficiency are more variable and depend on the degree of restriction, stage of growth, and quality of the diet consumed.

The results (Table 5) of Andersen and Ingvartsen[24] who used dry roughage plus concentrate at four feeding levels (*ad libitum*; 85%; 70%; 70% to 125 kg below slaughter liveweight then *ad libitum*) show that feed efficiency improved in both bulls and steers when the feeding level was 85% compared with *ad libitum*. However, it is unlikely that restriction would ever be of practical significance with bulls. On the other hand restriction of steers, whilst increasing the age at slaughter, provides benefits in terms of feed efficiency and carcass composition. In the Andersen and Ingvartsen[24] study the feed conversion ratio of restricted steers growing at 700–800 g d^{-1} was similar to that of *ad libitum* fed bulls gaining 1100–1200 g d^{-1} to the same slaughter weight. This result appears surprising at first sight but can be explained by the higher maintenance requirement of bulls[6,42] and the low efficiency of protein deposition compared with fat deposition.[11]

Another interesting aspect of the Andersen and Ingvartsen[24] study is the group fed 70% followed by *ad libitum*, somewhat akin to the situation occurring with overwintered stores. The overall growth, feed efficiency, and carcass composition was comparable to that observed in the group given 85% of *ad libitum*. This concurs with other observations that the

Table 5 *Effect of plane of nutrition on liveweight gain* (kg d^{-1}) *and feed efficiency (Scandinavian Feed Units* kg gain^{-1}) *of bulls and steers from 90 to 550 kg (after Andersen and Ingvartsen*[24])

		Bulls			Steers	
	Ad lib.	85%	70%	Ad lib.	85%	70%
Liveweight gain	1.13	1.06	0.84	0.93	0.88	0.76
Feed efficiency	5.92	4.45	4.68	6.96	5.69	5.36

phenomenon of compensatory growth has little effect on carcass composition of animals taken to the same liveweight.[43]

There is considerable information on the interaction between plane of nutrition and stage of growth. Generally speaking restriction during early growth, *i.e.* the period of rapid protein gain, causes a greater reduction in liveweight gain and tends to reduce feed efficiency whereas restriction in the later stages of growth has relatively less effect on growth rate and usually improves feed efficiency.[9,11,24,44,47]

The qualitative effects of feed restriction on the energy composition of the gain and on carcass composition have been well documented.[6,7,9,11,23,33] However, when one begins to try to establish quantitative relationships between plane of nutrition, carcass weight, and carcass composition the information from different experiments is quite variable and the use of different methods of dissection and chemical analysis of different bulked components of the empty body and carcass cause considerable problems of interpretation.

One of the most useful published studies is that of Andersen *et al.*[7] These results can be interpolated to suggest that, with restriction to 85% of *ad libitum*, the liveweight consistent with equivalent degrees of fatness is *ca.* 80 kg higher than that for *ad libitum* fed animals. A study on the interaction of plane of nutrition and slaughter weight on growth and carcass composition of Friesian steers[48-50] is nearing completion in the author's laboratory. The proportion of separable fat in the carcasses of animals fed *ad libitum* and slaughtered at 500 kg was similar to that of those restricted to 70% of *ad libitum* and slaughtered at 600 kg. At the same slaughter weight, restriction (to 70%) reduced separable fat, abdominal fat, and intramuscular fat by *ca.* 20%. Not surprisingly, in view of the previous discussion on sensory attributes, there were no significant effects of plane of nutrition or slaughter weight on shear force, colour, or cooking loss and it is unlikely that there would be any measurable effects on eating quality.

3 Conclusions

This chapter has highlighted the need for strategies which maximize the yield of lean, wholesome beef. These include the use of late-maturing breeds, genetic selection within breeds for high rates of lean deposition, use of entire males instead of castrates, appropriate nutrition, and slaughter at appropriate weights. The first three of these strategies are consistent with rapid growth rates, and hence throughput, and with efficient feed use. Hence, there is no major conflict between the objectives of producers, processors, and consumers. Recent research in the author's laboratory, indicating that appropriate nutrition may be restriction of feed intake, represents a partial conflict, in that this scenario involves reduced rates of gain and hence additional overheads, though the good feed efficiency observed with restriction is a mitigating factor. The economic basis for such an approach therefore depends on an appropriate pricing structure to encourage production of lean meat. It is nevertheless probable that restriction of feed intake will play a significant role in production of steers and heifer beef in the forseeable future. The alternative strategy of slaughtering at lighter weights, although effective in terms of providing a lean product, is economically unacceptable in an industry with such a long production cycle. Strategies such as the use of exogenous hormones or other chemicals such as β-agonists have been deliberately excluded from the discussion in that it is unlikely that these will be acceptable to consumers who are increasingly concerned about residues in food products.

It is clear, however, that there is the potential to meet consumer demand for lean, wholesome beef with acceptable sensory attributes and there is every reason why beef should maintain its position as a significant contributor to nutrition, health, and delectation of consumers in the future.

4 References

1. A.J. Kempster (personal communication).
2. MLC, Beef Yearbook, Meat and Livestock Commission, White Friars Press Ltd., Tonbridge, 1990, p.11.
3. S. Fallows and H. Gorden, Food Policy Research Briefing Paper, University of Bradford, 1985.
4. MLC Corporate Plan 1991-94, Milton Keynes, 1991.
5. A.J. Kempster, G.L. Cook, and M. Grantly-Smith, *Meat Sci.*, 1986, **17**, 107.
6. ARC, The Nutrient Requirements of Ruminant Livestock: Technical Review by an ARC Working Party, CAB, Slough, 1980.
7. H.R. Andersen, K.L. Ingvartsen, and S. Klastrup, *Livest. Prodn. Sci.*, 1984, **11**, 571.
8. D.F. Parrett, J.R. Romans, P.J. Bechtel, B.A. Weichenthal, and L.L. Berger, *J. Anim. Sci.*, 1985, **61**, 436.
9. M.G. Keane and M.J. Drennan, *Irish J. Agric. Res.*, 1980, **19**, 53.
10. G. Armbruster, A.Y.M. Nour, M.L. Thonney, and J.R. Stouffer, *J. Food. Sci.*, 1983, **48**, 835.
11. Y. Geay and J. Robelin, *Livest. Prodn. Sci.*, 1979, **6**, 263.
12. MLC Beef Yearbook, Meat and Livestock Commission, White Friars Press Ltd., Tonbridge, 1984.
13. R.W.J. Steen and D.M Chestnutt, 'Beef Production', Occas. Publ. No. 14, Agricultural Research Institute of Northern Ireland 1986.
14. A.J. Kempster (personal communication).
15. K.J. McCracken, C.A. Moore, E.F. Unsworth, F.J. Gordon, R.W.J. Steen, and D.J. Kilpatrick, *Anim. Prod.*, 1991, **52**, 566.
16. K.J. McCracken, C.A. Moore, R.W.J. Steen, E.F. Unsworth, D.J. Kilpatrick, and F.J. Gordon, *Proc. Nutr. Soc.*, 1991, **50**, 84A.
17. M.A. Carroll and D. Conniffe, in 'Growth and Development of Mammals', ed. G.A. Lodge and G.E. Lamming, Butterworth, London, 1967, p.389.
18. D.M. Stiffler, C.L. Griffin, C.E. Murphey, G.C. Smith, and J.W. Savell, *Meat Sci.*, 1985, **13**, 167.
19. G.A. Branaman, A.M. Pearson, W.T. Magee, R.M. Griswold, and G.A. Brown, *J. Anim. Sci.*, 1962, **21**, 321.

20. P.E. Foster, S.M. McCurdy, E.L. Martin, and M.M. Hard, *Home Econ. Res. J.*, 1979, **8**, 127.
21. R.T. Berg and R.H. Butterfield, *J. Anim. Sci.*, 1968, **27**, 611.
22. J. Robelin, *Ann. Zootech.*, 1979, **28**, 209.
23. A. Fortin, S. Simpendorfer, J.T. Reid, H.J. Ayala, R. Anrique, and A.F. Kertz, *J. Anim. Sci.*, 1980, **51**, 604.
24. H.R. Andersen and K.L. Ingvartsen, *Livest. Prodn. Sci.*, 1984, **11**, 559.
25. A.V. Fisher, J.D. Wood, and O.P. Whelehan, *Anim. Prod.*, 1986, **42**, 203.
26. R.W.J. Steen, *Anim. Prod.*, 1985, **41**, 301.
27. A.B. Mohan Raj, Ph.D. Thesis, The Queen's University of Belfast, 1988.
28. R.A. Field, *J. Anim. Sci.*, 1971, **32**, 849.
29. S.C. Seideman, H.R. Cross, R.R. Oltjen, and B.D. Schanbacher, *J. Anim. Sci.*, 1982, **55**, 826.
30. S. Klastrup, H.R. Cross, B.D. Schanbacher, and R.W. Mandigo, *J. Anim. Sci.*, 1984, **58**, 75.
31. J.J.M. Frood, Ph.D. Thesis, University of Reading, 1976.
32. T.G. Truscott, J.D. Wood, and H.J.H. MacFie, *J. Agric. Sci. (Camb.).*, 1983, **100**, 257.
33. J.F.D. Greenhalgh, *Proc. Nutr. Soc.*, 1986, **45**, 119.
34. Y. Geay, *J. Anim. Sci.*, 1984, **58**, 766.
35. J. Robelin and Y. Geay, in 'Herbivore Nutrition in the Subtropics and Tropics', ed. F.M.C. Gilchrist and R.J. Mackis, Science Press, Pretoria, 1984, p.525.
36. K. Rohr and R. Daenicke, *J. Anim. Sci.*, 1984, **58**, 753.
37. R.B. McCarrick, Proceedings of the 10th International Grassland Congress, Helsinki, 1966, p.575.
38. R.B. McCarrick, Proceedings of the 3rd Occasional Symposium, British Grassland Society, 1967, p.721.
39. A.V. Flynn, *Anim. Prod. Res. Rep. An. Forais Taluntais*, 1976, 23.
40. R.W.J. Steen and C.A. Moore, *Anim. Prod.*, 1988, **47**, 29.
41. R.W.J. Steen and C.A. Moore, *Anim. Prod.*, 1988, **49**, 233.
42. R.W.J. Steen, *Anim. Prod.*, 1986, **42**, 29.
43. A.J.F. Webster, *Proc. Nutr. Soc.*, 1977, **36**, 53.
44. R.L. Henrickson, L.S. Pope, and R.F. Hendrickson, *J. Anim. Sci.*, 1965, **24**, 507.
45. M. Sorensen, J. Lykkeaa, and H.R. Andersen, *Arbog. Landokon Forsogslab., København*, 1972, 370.
46. D. Levy, Z. Holzer, H. Newmark, and S. Amir, *Anim. Prod.*, 1974, **18**, 67.
47. H.R. Andersen, *Livest. Prodn. Sci.*, 1975, **2**, 341.
48. C.A. Moore, K.J. McCracken, E.F. Unsworth, R.W.J. Steen, D.J. Gordon, and D.J. Kilpatrick, *Anim. Prod.*, 1991, **52**, 572.
49. K.J. McCracken, C.A. Moore, E.F. Unsworth, F.J. Gordon, R.W.J. Steen, and D.J. Kilpatrick, *Anim. Prod.*, 1991, **52**, 592.
50. K.J. McCracken, C.A. Moore, E.F. Unsworth, F.J. Gordon, R.W.J. Steen, and D.J. Kilpatrick, Proceedings of the 12th Symposium on Energy Metabolism, Zürich, EAAP Publ. No. 58, 1991, p.186.

Factors Influencing Poultry Meat Quality

J.M. Jones

CONSULTANT, 18 SYWELL CLOSE, OLD CATTON, NORWICH NR6 7EW, UK

1 Introduction

Poultry, in the form of chickens, turkeys, ducks, geese, guinea fowl, and quail, is accepted as a dietary component in many countries and worldwide total slaughterings have increased by 4% each year since 1981. However, five countries, namely the USA, China, the former USSR, Brazil, and France, account for over half the world's total output, which in 1991 was expected to exceed 40 million tonnes.[1] In 1990 chicken was the only meat to increase sales in Great Britain, accounting for 33% of the carcass market.[2,3] Output of products such as Traditional Farm Fresh turkey and 'Label' poultry aimed at specific markets continues to expand.[4] For instance, in France, 'Label' chickens produced from slow growing breeds fed high cereal diets constitute some 20% of the total consumption of fresh (chilled) chicken carcasses.[5]

Extensive breeding and nutritional programmes have ensured that a shorter time is required to grow birds to a marketable weight, this being particularly noticeable in the case of broiler chickens where liveweights of 2 kg may be reached at 6 weeks of age compared with the 8 or more weeks required some two decades ago. In the case of broilers with a live weight of 2 kg or so, the feed conversion rate (FCR), *i.e.* the kg of diet required to produce one kg of bird, is generally less than 2.0, while in turkeys aged between 8 and 14 weeks FCR values of 2.1–3.7 have been reported,[6,7] although there may be marked differences between strains.[8,9] Growing conditions will affect the FCR, as was shown by Ricard *et al.* who found that pheasants reared in houses required a greater quantity of food to reach the same weight as their counterparts raised as game birds in an outside aviary.[10] FCR values of over 7.0 were observed in the case of guinea fowl raised in the tropics.[11]

Except in the case of products attracting a price premium, prolongation of the fattening period is generally uneconomical because of the increased FCR, and in the case of ducks and geese may be limited by the onset of moulting which results in a higher incidence of pin feathers which are hard to remove. However, it is interesting to note that in some European countries the same geese may be used for both meat and feather production, with the birds being plucked three or four times, starting at about 8 weeks of age and continuing at intervals of 6–8 weeks until slaughter at 28–32 weeks of age.[12,13]

With the possible exception of guinea fowl and quail, the males of commercially important species are heavier than females of the same strain and age and hence may be used to prepare portions of value-added products whereas females will be sold as whole carcasses.

As with other meats, the quality of poultry may be thought of in terms of nutritive, compositional, or sensory attributes and these are influenced by a number of factors which may be divided into three categories, namely the bird, its housing and management, and the effect of preslaughter operations.

2 Yield

The yield, especially of edible parts, is of primary concern to the producer, the processor, and the consumer, but it is clear that there is a wide divergence of opinion as how to calculate yield, *i.e.* whether one bases the calculation on (*a*) live weight before or after fasting, (*b*) carcass weight after evisceration (hot dressing), (*c*) after chilling - here variation may occur depending on method of cooling, or (*d*) presence or absence of giblets and/or neck. Also, difficulties arise regarding the determination of the yields of parts, since these may be determined on meat alone; meat, skin, and bone; meat plus skin; or meat plus bone, while breast yield may be given on a 'bone-out' basis whereas leg yield from the same carcass includes bone. In addition, different terms are used to describe the same muscles in the case of breast meat: *e.g.* the small breast muscle may be known as 'tenderloin' or 'fillet', with the last term also being applied to the large breast muscle by some workers in the poultry industry.

Thus there is considerable difficulty in comparing the results from different laboratories, and in an effort try to bring a degree of uniformity to the situation a standard method of dissection has been developed and is steadily being adopted in the countries of the European Community.[14] In the USA Benoff and colleagues have produced a reference for the dissection of broilers.[15]

The need for uniformity between laboratories at least is emphasized by the fact that the reported eviscerated yields of poultry carcasses (expressed as percentage of live weight) ranges from 60 to 80% and it is to be expected that much of this variation results from the use of varying fasting periods prior to slaughter or from different methods of handling the carcass. However, other factors have been implicated: bird strain;[16,17] sex;[18] age;[8,9,19–21] diet;[21] weight and sex at a particular age.[17,22] In addition, Bilgili *et al.* found that chicken carcass yield increased as bird stocking density increased,[23] although the converse seemed to be true for turkeys.[8,24] Tawfik et al. showed that broilers raised at high temperatures (30–32 °C) for 9 weeks had a greater carcass yield than chickens held under temperatures which declined in a controlled manner from 30 °C in the first week of the experiment to 17–19 °C at the end, although the weekly temperatures were not specified.[19]

In many cases processing plants are producing parts and deboned meat, with the breast being regarded as the most valuable product, and so it is perhaps not too surprising to find that the quantity of breast meat in chickens and ducks has increased over the past decade, often at the expense of fat.[22,25,26]

As with the eviscerated carcass, reported yields of parts can vary widely. For instance, although there seems to be general agreement among laboratories in North America that broiler breast (meat, skin, and bone) represents some 32–36% of the eviscerated weight, yields reported for leg (thighs plus drumsticks) have ranged from 32 to 47%.[15–17,27,28] Studies in France showed the breast muscles to represent some 20% of the eviscerated chicken carcass and the legs ca. 38%.[29] Application of the equation and constants given by Veerkamp suggests that in the case of 49 day old broilers with carcass weights between 1.0 and 1.4 kg, the expected yield of legs would be in the region of 35–36%.[22]

In the case of turkeys, legs represent ca. 30% and the breast portion 33–41% of the eviscerated carcass,[7,8,30,31] while in pheasants breast and leg portions each make up one-third of the eviscerated carcass.[29]

The proportions of breast and leg may be affected by a number of factors. For instance the proportion of breast generally increases with bird age, while the quantities of thigh and drumstick remain constant or change only slightly.[8,13,30] The proportions of breast and leg in chicken and turkey carcasses are markedly affected by reducing levels of dietary protein, amino acids, and energy, as well as by the form (meal or pellets) in which the feed is presented.[31–35] The type of coccidiostat included in the diet may also be important, since Izat et al. found that inclusion of salinomycin resulted in a higher chicken breast meat yield than did halofuginone.[28]

Design of the broiler rearing house may alter the weights of various muscles,[36,37] and the lighting regime employed during the rearing of turkeys can also influence the proportions of the various carcass parts,[24,30] although the environmental factor most likely to influence the financial return on chicken or turkey parts would seem to be the temperature at which the birds are raised, since greater yields of breast meat are obtained at lower temperatures.[24,38,39] Although body weight is depressed at higher temperatures, this is not necessarily the cause of reduced breast yield, since the effect was still noted in broilers grown to the same weights at 21 and 31 °C.[39] Here the total meat yield of the carcass was not altered by temperature, but birds grown at 31 °C had significantly less breast meat.

The influence of the above variables may go some way to explain the significant yield differences found between chickens raised under different production systems in a number of European countries.[40]

3 Composition of Muscle and Its Influence on Poultry Meat Quality

In contrast to the situation with waterfowl, where breast and leg meats contain almost equal amounts of protein,[41,42] the skinless breast meat of chickens and turkeys contains 3 or 4% more protein, as well as less fat, than the thigh and drumstick meats.[4] In the chicken both protein and fat levels increase with bird age with higher levels being found in females.[43,44] In the case of turkeys the levels of protein in breast meat may be influenced by bird strain and by the level of dietary protein.[45]

At concentrations of 0.5 and 0.6 $mg\,g^{-1}$ wet tissue respectively the levels of haem pigments in the breast meat of chickens and turkeys are much lower than in the corresponding thigh meats (1.7 and 2.2 $mg\,g^{-1}$ respectively),[46] although the levels of pigment are influenced by bird age and strain.[47-49] Niewiarowicz and Pikul found ratios of myoglobin: haemoglobin in broiler and turkey breast to be over 3:1 compared with the 2:1 for leg meat.[41] In the case of ducks, where more haem pigments were to be found in breast meat than in leg, the respective ratios were 2:1 and 1.5:1.

Traditionally the breast meat from chicken and turkey has been classified as 'white' or 'light' and the leg meat as 'red' or 'dark', with the two types exhibiting differences in protein extractability (greater with breast) and ultimate pH values (higher in leg), factors which may be expected to influence the meats' performance in products.[4,50]

Recent studies show that 'white' breast muscles (*pectoralis major*, also reported as *superficalis* or *thoracicus*, and *pectoralis minor*, *supracoracoideus*, *profundus*, deep pectoral) do not contain Type I (βR) red fibres, but are composed almost entirely of Type II-B (αW) fibres.[51-54] On the other hand, the 'red' meat of the thigh and drumstick is composed of a number of muscles which may contain substantial numbers of both Type I and Type II-A (αR, intermediate) as well as Type II-B fibres; for example, the *sartorius* of the chicken thigh is composed of 32% βR, 53% αR, and 15% αW.[53] In the turkey the corresponding average values for the *sartorius* were 50, 20, and 30% respectively, while in the *biceps femoris* which is frequently used as a typical 'red' muscle for reference purposes, less than 1% of the fibres were βR and 70% were αR.[55]

A feature of poultry meat is the speed at which it enters *rigor*. For instance, in the case of the major breast muscle from commercially processed chickens, this may occur approximately one hour after slaughter although there may be a wide variation within a batch of birds.[56] It is clear that within the chicken and turkey strains currently used by industry there are populations whose metabolic processes such as rate of glycolysis differ. Indeed chicken muscle has been classified as 'PSE' (pH value below 5.7 at 15 min post mortem) and 'DFD' (above pH 6.4).[57] However, such measurements may not necessarily be good indicators of meat quality, since post mortem changes such as ATP catabolism and lactic acid accumulation in the muscle do not appear to be directly related to *rigor mortis* development in broiler muscles, in contradiction to red meat species.[58]

Not only do birds within a group vary, but it is clear that biochemical and physical variations occur within the major pectoral muscle. Jones noted that the pH values of the cranial (anterior) and caudal (posterior) ends of the turkey muscle could differ by up to 0.3 pH units - the posterior having the higher value.[59] Subsequently Papa and Fletcher reported that the posterior part of the chicken breast muscle took considerably longer to enter *rigor*.[60] Muscle fibre diameters were greater in the caudal end.[51] In the case of major breast muscles from a group of 10 turkeys, Hillebrand *et al.* found that fibres in the anterior part of the muscle, which contained 1.6% intermediate fibres, had a greater diameter (104 μm) than in the posterior part (94 μm) containing less than 1% intermediate fibres.[54] However, the anterior part of muscle from fast glycolysing birds (pH < 6.0 at 15 min) contained 5–7% intermediate fibres. Overall, fast glycolysing muscles (pH < 6.0 at 15 min post mortem) had 2.4% intermediate fibres and the average fibre diameter at 94 μm was less than that of the moderate (pH 6.0–6.7) or slow glycolysing muscles' (pH > 6.7) average of 100 μm. These findings may go some way to explaining the wide variation noted in the tenderness of breast meat from commercially processed poultry.

In addition to the above variations within muscles, there are indications that today's rapidly growing turkeys selected for increased musculature may be susceptible to a myopathy which can affect a number of muscles. Sosnicki *et al.* carried out histological examinations on two breast and three leg muscles from male turkeys aged up to 18 weeks and found that all muscles showed some degree of degradation with the incidence and intensity of degeneration increasing with age.[61,62] Additional studies revealed that the changes were compatible with an ischaemic condition and the authors surmised that lack of exercise might be a major causative factor of muscle damage, the effect being particularly noticeable with the older, less mobile birds.[63,64] Wilson *et al.* found fast growing lines of turkey to be more affected than slow growing birds.[65]

Although none of the foregoing investigations on myopathy included studies on meat quality, some of the histological changes noted, such as the presence of variable numbers of both 'giant' and necrotic cells, were also noted in British studies on the quality of turkey breast meat.[45,66] Here more affected cells were present in tough meat. Clearly more work is required in this area.

At least one other degenerative condition appears to have arisen as a result of selection for breast size without an accompanying increase in muscle vascularity. The degenerative myopathy, first noted in turkey breeder hens but also found in young turkeys and broiler chickens, is characterized by an encapsulated green lesion in the deep breast muscle and atrophy of the caudal end of the muscle. The lesion size appears to be related to the dimensions of the pectoral caudal artery.[67] Management, particularly the avoidance of any stimulus which initiates wing flapping, appears to play an important part in deciding the degree of incidence in genetically susceptible birds.

4 Effect of Diet and Bird Strain on Meat Tenderness

Many studies concerned with poultry meat tenderness have concentrated on the effect of preslaughter handling and/or processing conditions and ignored the fact that the bird and its husbandry may influence the ultimate tenderness of the cooked product. However, there does seem to be a growing need for this area of research to be expanded.

A number of investigations have suggested that significant differences do exist between modern strains of both chicken and turkey,[9,68,69] although Touraille et al. found no effect of bird strain in the case of 16 week old chickens of fast and slow growing lines.[47]

While no effect of bird sex on tenderness was reported for turkeys,[7,70] the situation with chicken breast meat is less clear. For instance, whereas Ricard and Touraille reported the breast meat of females to be more tender,[71] several investigations indicated that the meat from females was tougher than that from males.[5,68,72] The tougher or firmer meat of 'Label' poultry compared with the standard broiler is considered to be a positive feature of this type of production.[5]

Ngoka and Froning, on measuring the shear values from the cooked *pectoralis major* of both male and female turkeys, found the meat of 20 week old turkeys to be more tender than that from 16 week old birds.[70] On the other hand, Seeman and Bohn reported the *pectoralis minor* of 10 week old females to be more tender than that of birds *ca.* 14 weeks of age.[69] Leg meat *(biceps femoris)* was more tender in the older birds. No age effects were noted in male turkeys aged 20 and 23 weeks. It is worth noting that the shear values obtained in these studies were low, and the differences numerically small, and so it is not clear whether the age differences would have been detected by a taste panel. Sensory analysis of chicken breast and thigh meat showed that tenderness decreased between 8 and 16 weeks of age.[47,48]

Moran et al. found that the tenderness of turkey breast meat was unaffected when birds were fed either a typical North American or European diet, where the latter had a lower energy level and energy:protein ratio than the North American diet.[9] In contrast, similar investigations in England on five strains of turkey showed that, in general, slower growing birds, *i.e.* those fed the North American type diet (high energy/low protein) produced a more tender meat.[45] The apparent influence of energy:protein ratio noted in the latter study might help explain the observation of Savage et al. that a diet containing triticale as a protein source produced a more tender breast meat than one where the protein was derived from corn-soy meal, since the metabolizable energy:protein ratio in the latter case was 23 compared with 36 for the triticale-based diet.[73]

Poor control of the environment in the rearing house may influence the tenderness of chicken meat in several ways. Sackett *et al.* reported that exposure to ammonia (75 p.p.m.) or carbon dioxide (5000 p.p.m. or greater) for 7 days prior to slaughter significantly increased breast shear values, as did the carbon monoxide which may result from incomplete combustion of fuel in burners used in the houses.[74] A more indirect effect on carcass quality might result from poor ventilation control in the house, since ammonia has been implicated in the formation of breast blisters, while litter conditions can cause lesions which result in downgrading:[75,76] the on-line trimming of these blemishes may result in a tough meat.[77] A similar situation would hold with birds damaged during catching and transport or in transit along the hanging-on line at the processing plant.

Prolonged holding of the birds in crates during the summer may also create problems since it has been shown that the breast meat of ducks and turkeys exposed to temperatures of 38–42 °C for several hours prior to slaughter is tougher than that from control birds or those which have been cold-stressed.[26,78,79]

5 Product Colour

Poultry differs from red meat in that it may be sold with or without the skin attached and thus skin colour can influence consumer acceptability. Indeed, growing conditions (bird strain and diet) and processing procedures (scalding regime) are manipulated to produce a range of skin colours for use in different locations.[27,80]

Many chicken and turkey portions are being marketed minus the skin and hence the colour of the packaged meat may play a part in consumer acceptability. A problem of concern to some European processing companies is the wide between-muscle variation in meat colour, especially in the case of raw chicken breasts, but which is also noted with turkey meat.

The fact that the variation in chicken meat colour does not appear to be a problem in the USA probably explains why most of the published studies on poultry meat colour have been concerned with turkeys. Metz found that the breast meat of female turkeys became paler as the distance (up to 80 km) and duration of transport (up to 4 h) to the processing plant increased.[81] When the fasting time was extended to between 10 and 21 h there was a positive correlation with muscle pH value at 15 min post mortem, but no correlation with meat colour.

Pingel *et al.* found that meat from ducks exposed to 40–44 °C for two hours prior to slaughter was redder than that of controls and also generally had higher pH values.[26] Froning *et al.*, comparing male turkeys exposed to 42 °C for one hour prior to slaughter with birds cold-stressed at 4.3 °C for 20 min, found that the major breast muscle of heat-stressed birds was darker and redder.[78] Although there was a slightly lower pH value with these birds, it is worth noting that the pH measurements were determined on the minor muscle. Subsequent experiments in which birds were held at

38, 21, or 4.4 °C for four hours prior to slaughter produced different results in that the breast muscle (unspecified) from the heat-stressed birds was paler and had a significantly lower pH value than that from other groups (5.99 *versus* 6.06).[79] Total pigments and myoglobin were significantly reduced in the heat-stressed meat. The reasons for this are not clear, nor is the role of haem pigments in the colour variation established. Several authors have suggested that it is the cytochrome *c* content of the muscle which influences meat colour in both the raw and cooked state.[46,49,82]

Fletcher was of the opinion that breast meat colour variation probably is not due to variation in muscle pH value, rate of post mortem glycolysis, or effect of stress on these, but that stress may have an independent effect on muscle vascularity and subsequent haem content of the muscle.[83]

Recent work from Poland showed that that inclusion of low levels of rapeseed meal or ground rapeseed in broiler diets increased the haem pigment content of the breast meat.[84] In this instance the increase in colour would not necessarily be a disadvantage, since in Poland breast meat is frequently used in comminuted products.

6 Influence of Lipids on Poultry Meat Quality

Many aspects of lipid composition and oxidation are dealt with in subsequent chapters and so the following discussion will be limited to the possible improvement of poultry carcass quality by dietary manipulation.

It is well documented that in the case of poultry species the fatty acids composition (in particular the neutral lipid fraction) of the meat and adipose tissue generally reflects the level and composition of the dietary fat, but that the type of fat used in the diet has little effect on the total lipid content of the meat.[85-88]

Although the leg meat of chicken and turkey generally contains 3-4 times as much lipid as breast meat, the ratios of unsaturated to saturated fatty acids in the two types of tissue are very similar. For instance, Lin *et al.* reported ratios of 1.71 and 1.61 in the phospholipids of dark and white meats respectively from chickens fed a diet containing 55 g kg^{-1} coconut oil.[86] In the case of the neutral fraction the ratios were in the region of 1.2. When olive or linseed oils were used as dietary components the ratios in the neutral fraction of each tissue were over 3.0, but the ratios in the phospholipid had risen to only 2.0.

In the case of chickens of commercial slaughter age the fat content of the skin ranges between 30 and 40% and according to Hulan is composed exclusively of triglycerides.[89]

The apparent ease with which dietary lipid enters poultry tissue means that it should be possible to improve carcass acceptability by manipulating the lipid composition of the bird. For example, in Israel where poultry is plucked without scalding and the number of female carcasses with torn skins may reach 80% in summer months, it was found that if a diet in which 18:1 and 18:2 acids accounted for 34% of the total dietary fatty acids was substituted for a control diet where these acids made up 58% of the total, the resulting carcasses were less greasy and the number of skin tears reduced.[90]

The current interest in the apparent beneficial effects of fish oils, or rather the ($\omega - 3$) acids contained therein, on human health has prompted studies into the possibility of increasing the intake of such acids by incorporating them into poultry meat, the consumption of which is increasing at a time when that of fatty fish seems to be declining in England and North America.

Hulan fed chicken diets containing 0, 4, 8, or 12% redfish meal (10.1% lipid) and found that the contents of ($\omega - 3$) polyunsaturated fatty acids such as eicosapentaenoic and docosahexaenoic in the meat increased at the expense of ($\omega - 6$) acids such as linoleic and arachidonic, with levels in the breast meat being approximately twice those in the thigh.[89] He concluded that the edible (skinless) meat from chickens fed a diet containing 8% redfish meal would contain levels of ($\omega - 3$) acids comparable to those provided by white fish.

It is perhaps unfortunate that the above work did not include a report of sensory analyses of the chicken meat, since investigations in the laboratory where the above work was performed suggested that when cooked meat samples from chickens fed 0, 4, 8, or 12% fishmeal (species unspecified) were analysed directly after cooking there were no flavour differences between the treatments, but if cooked samples were analysed after overnight storage at 4 °C the increased fish 'off-flavour' meant that levels of fishmeal above 4% would give an undesirable product.[91]

One of the problems facing the poultry industry is the need to maintain product stability during fresh or frozen storage: leg meat is more susceptible to oxidation than breast meat and oxidation is more extensive in the turkey than in the chicken.[92] While the variation in oxidation might be due to different levels of phospholipids and polyunsaturated fatty acids in the various tissues,[86] a contributing factor was considered to be the quantity of catalytic 'free' iron in tissues, higher levels being found in dark meat and more in the turkey (2.5 $\mu g\,g^{-1}$ tissue *versus* 0.55 $\mu g\,g^{-1}$ for chicken leg). In breast meat the corresponding values were 0.92 and 0.24 $\mu g\,g^{-1}$ tissue.[92]

Subsequent investigation showed that, while withdrawal of the customary iron supplement from the diet during the last 3–7 weeks of the growing period did not significantly alter the haem iron or total iron content of the meat, the level of lipid peroxidation in turkey leg meat was substantially reduced.[93]

Another - more widely utilized - approach to improving poultry meat stability is to include α-tocopherol in the diets. Several studies have shown that the amount of tocopherol deposited in the tissue depends on both its concentration and the length of time it is fed to the birds.[94,95] In the case of the chicken one finds 50% more tocopherol in leg meat than in breast,[95] whereas in the turkey, where higher levels of tocopherol are fed, presumably to compensate for the poorer ability of the bird to deposit the compound, the levels in thigh meat were generally 2–4 times higher than in breast meat, and it has been suggested that the greater concentration in dark meat is probably due to the increased vascularity and lipid content of the leg.[94,95]

Feeding α-tocopherol at a level of 200 mg kg^{-1} diet significantly reduced lipid oxidation, as judged by the level of thiobarbituric acid-reactive substances (TBARS, estimated as mg malonaldehyde/kg meat), during both chilled and frozen storage of chicken meat, with the protective effect being more noticeable when tocopherol had been fed for 6 weeks prior to slaughter than when short-term supplementation (10 days) was applied.[95] For instance, after 9 days storage at 4 °C, leg meat from chicken given the short-term supplementation had an average TBARS value of 0.92 and those fed tocopherol for 6 weeks prior to slaughter had an average value of 0.62. In comparison, the meat from birds which had received diets with no added tocopherol gave a value of 1.89. No sensory analyses were reported and so it is difficult to correlate the measured changes with consumer acceptance of the product.

7 Conclusions

Breeding programmes are constantly increasing the growth rate and yield of commercially important poultry species, although in the case of turkeys there are indications that such progress may adversely affect meat quality in that muscle vascularity development does not appear to be keeping pace with muscle growth.

It is becoming clear that the sensory attributes of poultry meat are not entirely dependent on the slaughtering and processing conditions and there seems to be a case for further work on the influence of diet on meat texture. Further studies into the preslaughter factors influencing the colour of chicken and turkey meat are also required.

8 References

1. Anon, *Poult. Internat.*, 1991, **30** (8), 22.
2. H. Darrington, *Food Manuf.*, 1991, **66** (7), 29.
3. K. Grikitis, *Food Processing*, 1991, **60** (8), 15.
4. J.M. Jones, in 'Developments in Food Proteins', ed. B.J.F. Hudson, Elsevier Applied Science, London, 1988, Vol. 6, p. 35.
5. J. Culioli, C. Touraille, P. Bordes, and J.P. Girard, *Arch. Geflügelk.*, 1990, **54**, 237.
6. R.E. Salmon, *Br. Poult. Sci.*, 1986, **27**, 629.
7. E.T. Moran, Jr., L.M. Poste, E. McMillan, C. Patterson, and W.H. Revington, *Can. J. Anim. Sci.*, 1987, **67**, 705.
8. J.E. Larsen, R.L. Adams, I.C. Peng, and W.J. Stadelman, *Poult. Sci.*, 1986, **65**, 2076.
9. E.T. Moran, Jr., L.M. Poste, P.R. Ferket, and V. Agar, *Poult. Sci.*, 1984, **63**, 1778.

10. F.H. Ricard, M.J. Petitjean, J.M. Melin, G. Marché, and G. Malineau, *INRA Prod. Anim.*, 1991, **4**, 117.
11. L.N. Agwunobi and T.E. Ekpenyong, *J. Sci. Food Agric.*, 1991, **55**, 207.
12. J. Borge, Proceedings of the Eighth European Symposium on Poultry Meat Quality, Budapest, 1987, p. 40.
13. J. Borge and F. Bogenfurst, Ref.12, p.44.
14. W.P.S.A. 'Method of Dissection of Broiler Carcasses and Description of Parts', ed. J. Fris Jensen, World's Poultry Science Association, Danish Branch, Copenhagen, 1984.
15. F.H. Benoff, J.P. Hudspeth, and C.E. Lyon, 'Reference Guide to Broiler Processing Yields', Cooperative Extension Service Special Bulletin 12, University of Georgia, Athens, 1981.
16. H.L. Orr, E.C. Hunt, and C.J. Randall, *Poult. Sci.*, 1984, **63**, 2197.
17. E.T. Moran, Jr., N. Acar, W.H. Revington, and S.F. Bilgili, Proceedings of the Tenth European Symposium on Poultry Meat Quality, Doorwerth, ed. T.G. Uijttenboogaart and C.H. Veerkamp, Spelderholt Centre for Poultry Research and Information Services, The Netherlands, 1991, Vol.1, p.303.
18. G. Seeman, Proceedings of the Sixth European Symposium on Poultry Meat Quality, Ploufragan, ed. C. Lahellec, F.H. Ricard, and P. Colin, Station Experimentale d'Aviculture, Ploufragan, 1983, p. 341.
19. E.S. Tawfik, A.M.A. Osman, F.W. Klein, and W. Hebeler, *Arch. Geflügelk.*, 1989, **53**, 235.
20. F.H. Ricard, G. Marché, and M.J. Petitjean, Ref.17, p.333.
21. R.E. Salmon and V.I. Stevens, *Br. Poult. Sci.*, 1989, **30**, 283.
22. C.H. Veerkamp, Ref.17, p.355.
23. S.F. Bilgili, W.H. Revington, E.T. Moran, Jr., and R.D. Bushong, Ref.17, p.263.
24. J.C. Halvorsen, P.E. Waibel, E.M. Oju, S.L. Noll, and M.E. El Halawani, *Poult. Sci.*, 1991, **70**, 935.
25. Anon, 'Product Profile - Yield', Cobb Breeding Company, Chelmsford, 1991.
26. H. Pingel, R. Klemm, and U. Knust, Ref.17, p.325.
27. J.E. Marion and R.A. Peterson, *Poult. Sci.*, 1987, **66**, 1174.
28. A.L. Izat, M. Colberg, M.A. Reiber, M.H. Adams, J.T. Skinner, M.C. Cable, H.L. Stilborn, and P.W. Waldroup, *Poult. Sci.*, 1991, **70**, 1419.
29. F.H. Ricard and M.J. Petitjean, *Ann. Zootech.*, 1989, **38**, 11.
30. I.C. Peng, R.L. Adams, E.J. Furumoto, P.Y. Hester, J.E. Larsen, O.A. Pike, and W.J. Stadelman, *Poult. Sci.*, 1985, **64**, 871.
31. P.Y. Hester, K.K. Krueger, and M. Jackson, *Poult. Sci.*, 1990, **69**, 1743.
32. P.R. Ferket and J.L. Sell, *Poult. Sci.*, 1990, **69**, 1982.
33. M. Ristic, F.X. Roth, M. Kirchgessener, and E. Maurus-Kukral, *Mitteilungsbl. Bundesanst. Fleischforsch. Kulmbach*, 1990, **29** (107), 14.
34. J.D. Summers, S. Leeson, and D. Spratt, *Can. J. Anim. Sci.*, 1988, **68**, 241.
35. E.M. Oju, P.E. Waibel, and S.L. Noll, *Poult. Sci.*, 1988, **67**, 1760.
36. C.L. Sandusky and J.L. Heath, *Poult. Sci.*, 1988, **67**, 1557.
37. C.L. Sandusky and J.L. Heath, *Poult. Sci.*, 1988, **67**, 1708.
38. A.M.A. Osman, E.S. Tawfik, M. Ristic, W. Hebeler, and F.W. Klein, *Arch. Geflügelk.*, 1989, **53**, 244.
39. M.A.R. Howlider and S.P. Rose, *Br. Poult. Sci.*, 1989, **30**, 61.
40. M. Ristic, Ref.17, p.339.
41. A. Niewiarowicz and J. Pikul, Ref.18, p.527.

42. H. Pingel and K.-H. Schneider, Ref.12, p.13.
43. T.C. Grey, D. Robinson, J.M. Jones, S.W. Stock, and N.L. Thomas, *Br. Poult. Sci.,* 1983, **24**, 219.
44. A.M.A. Osman, E.S. Tawfik, M. Ristic, W. Hebeler, and F.W. Klein, *Arch. Geflügelk.,* 1990, **54**, 20.
45. T.C. Grey, N.M. Griffiths, J.M. Jones, and D. Robinson, *Lebensm.-Wiss. Technol.,* 1986, **19**, 412.
46. J. Pikul, A. Niewiarowicz, and H. Kupijaj, *J. Sci. Food Agric.,* 1986, **37**, 1236.
47. C. Touraille, F.H. Ricard, J. Kopp, C. Valin, and B. Leclerq, *Arch. Geflügelk.,* 1981, **45**, 97.
48. C. Touraille, J. Kopp, C. Valin, and F.H. Ricard, *Arch. Geflügelk.,* 1981, **45**, 69.
49. B. Girard, J. Vanderstoep, and J.F. Richards, *J. Food Sci.,* 1990, **55**, 1249.
50. J.M. Jones, in 'Progress in Food Proteins - New and Developing Sources', ed. B.J.F. Hudson, Elsevier Applied Science, London, 1991 (in press).
51. D.P. Smith and D.L. Fletcher, *Poult. Sci.* 1988, **67**, 1702.
52. G. Seeman and G. Kozlowski, *Arch. Gelfügelk.,* 1982, **46**, 228.
53. A.R. Sams and D. Janky, *Poult. Sci.,* 1990, **69**, 1433.
54. S.J.W. Hillebrand, M. van der Leun, F.J.M. Smulders, and P.A. Koolmees, Ref.17, p.45.
55. K.J. Wiskus, P.B. Addis, and R.T.-I. Ma, *Poult. Sci.,* 1976, **55**, 562.
56. J.M. Jones and T.C. Grey, in 'Processing of Poultry', ed. G.C. Mead, Elsevier Applied Science, London, 1989, p.127.
57. J. Kijowski and A. Niewiarowicz, *J. Food Technol.,* 1978, **13**, 451.
58. A.R. Sams and D. Janky, *Poult. Sci.,* 1990, **70**, 1003.
59. J.M. Jones, *Turkeys,* 1986, **34** (6), 25.
60. C.M. Papa and D.L. Fletcher, *Poult. Sci.,* 1988, **67**, 635.
61. A. Sosnicki, R.G. Cassens, D.R. McIntyre, R.J. Vimini, and M.L. Greaser, *Food Microstruct.,* 1988, **7**, 147.
62. A. Sosnicki, R.G. Cassens, D.R. McIntyre, R.J. Vimini, and M.L. Greaser, *Br. Poult. Sci.,* 1989, **30**, 69.
63. A. Sosnicki, R.G. Cassens, R.J. Vimini, and M.L. Greaser, *Poult. Sci.,* 1991, **70**, 343.
64. A. Sosnicki, R.G. Cassens, R.J. Vimini, and M.L. Greaser, *Poult. Sci.,* 1991, **70**, 349.
65. B.W. Wilson, P.S. Nieberg, R.J. Buhr, B.J. Kelly, and F.T. Schultz, *Poult. Sci.,* 1990, **69**, 1553.
66. G. Seeman, J.M. Jones, N.M. Griffiths, and T.C. Grey, *Arch. Geflügelk.,* 1986, **50**, 149.
67. M. Swash, K.J. Henrichs, C.L. Berry, and J.M. Jones, in 'Animal Models of Neurological Disease', ed. F.C. Rose and P.O. Behan, Pitman Medical, Tunbridge Wells, 1980, p.3.
68. F. Ehinger and B. Schwindt, *Arch. Geflügelk.,* 1981, **45**, 260.
69. G. Seeman and M. Bohn, Ref.18, p.78.
70. D.A. Ngoka, G.W. Froning, S.R. Lowry, and A.S. Babji, *Poult. Sci.,* 1982, **61**, 1996.
71. F.H. Ricard and C. Touraille, *Arch. Geflügelk.,* 1988, **52**, 27.
72. C.E. Lyon and R.L. Wilson, *Poult. Sci.,* 1986, **65**, 907.
73. T.F. Savage, Z.A. Holmes, A.H. Nilipour, and H.S. Nakaue, *Poult. Sci.,* 1987, **66**, 450.

74. B.A.M. Sackett, G.W. Froning, J.A. Deshazer, and F.J. Struwe, *Poult. Sci.,* 1986, **65**, 511.
75. N.J. Lynn, S.A. Tucker, and T.S. Bray, Ref.17, p.251.
76. W.D. Weaver and R. Meyerhof, *Poult. Sci.,* 1991, **70**, 746.
77. J.M. Jones, *Poult. Internat.,* 1988, **27** (11), 110.
78. G.W. Froning, A.S. Babji, and F.B. Mather, *Poult. Sci.,* 1978, **57**, 630.
79. A.S. Babji, G.W. Froning, and D.A. Ngoka, *Poult. Sci.,* 1982, **61**, 2385.
80. J.M. Jones, in 'Meat Science, Milk Science and Technology', ed. H.R. Cross and A.J. Overby, Elsevier, Amsterdam, 1988, World Animal Science Series, Vol.B3, p.141.
81. M.-H. Metz, Ref.18, p.443.
82. D.A. Ngoka and G.W. Froning, *Poult. Sci.,* 1982, **61**, 2291.
83. D.L. Fletcher, Ref.17, p.11.
84. E. Swierczewska, J. Niemiec, and J. Mroczek, Ref.17, p.347.
85. H.W. Hulan, F.G. Proudfoot, and D.M. Nash, *Poult. Sci.,* 1984, **63**, 324.
86. C.F. Lin, J.I. Gray, A. Ashgar, D.J. Buckley, A.M. Booren, and C.J. Flegal, *J. Food Sci.,* 1989, **54**, 1457.
87. J.C. Yau, J.H. Denton, C.A. Bailey, and A.R. Sams, *Poult. Sci.,* 1991, **70**, 167.
88. J.P. Holsheimer, Ref.17, p.273.
89. H.W. Hulan, Ref.17, p.289.
90. D. Sklan and A. Ayal, *Br. Poult. Sci.,* 1989, **30**, 407.
91. L.M. Poste, *J. Anim. Sci.,* 1990, **68**, 4414.
92. J. Kanner, B. Hazan, and L. Doll, *J. Agric. Food Chem.,* 1988, **36**, 412.
93. J. Kanner, I. Bartov, M.O. Salan, and L. Doll, *J. Agric. Food Chem.,* 1990, **38**, 601.
94. B.W. Sheldon, *Poult. Sci.,* 1984, **63**, 673.
95. C.F. Lin, A. Ashgar, J.I. Gray, D.J. Buckley, A.M. Booren, R.L. Crackel, and C.J. Flegal, *Br. Poult. Sci.,* 1989, **30**, 855.

Conversion of Muscle into Food

Conversion of Muscle into Meat: Biochemistry

R.A. Lawrie

DEPARTMENT OF APPLIED BIOCHEMISTRY AND FOOD SCIENCE, UNIVERSITY OF NOTTINGHAM, SUTTON BONINGTON, LOUGHBOROUGH, LEICESTERSHIRE LE12 5RD, UK

1 Introduction: Main Events Post Mortem

By far the greater part of the commodity we refer to as 'meat' consists of the post mortem aspect of an animal's muscular tissue. The qualification 'post mortem' is necessary because death of the animal initiates metabolic processes in muscle which alter its *in vivo* nature.

When the circulation stops, muscles can no longer obtain energy by respiration, *i.e.* by the complete oxidation of substrates to carbon dioxide and water; instead, in a local endeavour to maintain structural and functional integrity and temperature, energy is obtained by the conversion of glycogen (the principal mode of storing calories in muscle) to lactic acid. The biochemical steps in this post mortem glycolysis, and the enzymes responsible, have long been established. They have been confirmed by modern techniques such as NMR.[1] More recently it has been shown that the enzymes involved are attached to the contractile proteins, whose action they service, and to each other, in a logical spatial sequence which reflects the progressive metabolism of their substrates.[2,3] The extent or binding is determined directly by the degree of activity of the muscle.[4]

Post mortem glycolysis is much less efficient than respiration in providing energy - which means, in essence, the resynthesis of adenosine triphosphate (ATP), the universal energy currency of cells, from adenosine diphosphate (ADP). Although there is a short-term store of energy-rich phosphate, creatine phosphate (CP), in most muscles and this can regenerate ATP from ADP, neither CP nor post mortem glycolysis itself can maintain the *in vivo* ATP level for long. It falls to a value which is insufficient to keep the principal contractile proteins, actin and myosin, apart and in their flexible resting state *in vivo*. They unite irreversibly to form inextensible actomyosin, an event manifested superficially as the stiffness of *rigor mortis*.[5,6] This happens, even if the store of glycogen in the muscle immediately pre mortem is ample, because the production of lactic acid brings the pH down from its *in vivo* level of *ca.* 7.2 to *ca.* 5.5, at which the enzymes which effect post mortem glycolysis, being near their isoelectric point, are inactivated. Other events, such as the deamination of AMP, remove essential cofactors from the system.

Absence of energy prevents resynthesis of complex protein molecules, and these begin to denature, making them susceptible to attack by endogenous proteolytic enzymes, including the calpains and the cathepsins. This autolytic action, which probably begins before post mortem glycolysis ceases, and continues for a considerable number of days thereafter, breaks strategic bonds in the structural proteins - the actomyosin system and connective tissue - whereby the meat becomes more tender, and degrades the more soluble muscle proteins to their constituent peptides and amino acids. The intracellular osmotic pressure thus rises; but, although this tends to increase the water-holding capacity of the meat after *rigor mortis*, it cannot offset the predominant effect of the attainment of the isoelectric point of the muscle proteins, and exudation occurs.

The principal pigment responsible for the bright red colour of fresh meat, myoglobin, oxidizes and denatures. Such oxidation is synergistic with that of the fat, especially unsaturated fat. Rancidity thus tends to develop, although unless storage is prolonged such organoleptically undesirable changes are more than masked by the increasing concentration of the precursors of cooked meat flavour, *viz.* the small molecules derived from proteins, fats, and carbohydrates during post mortem glycolysis and autolysis.

Post mortem glycolysis and proteolysis are the predominant biochemical events in the conversion of muscle into meat. The changes they effect make the meat more liable to spoilage by micro-organisms and by non-biological agents and necessitate preservative procedures (Table 1). The factors which influence their rate and extent will now be considered.

Table 1 *Conversion of muscle into meat: main events*

Blood circulation ceases
Local metabolism continues anaerobically
Post mortem glycolysis commences (glycogen → lactic acid)
pH and energy levels fall
Rigor mortis develops
Intracellular osmotic pressure rises
Proteins denature: fats oxidize
Lysis greatly predominates over synthesis
Exudation: discoloration: rancidity
Microbial growth stimulated
Conditioning changes

2 Factors Affecting the Rate of Post Mortem Glycolysis

The factors which affect the rate of post mortem glycolysis are both intrinsic and extrinsic. The former include species, genotype, animal age, the type of muscle, and the location within it. The latter are more amenable to deliberate manipulation and include the preslaughter administration of drugs or hormones, the circumstances during slaughter (*e.g.* pithing, shackling, electrical stimulation), the temperature to which the muscles are exposed during post mortem glycolysis, the degree of comminution of the muscle, the addition of salts, the application of pressure, and oxygen tension (Table 2).

Intrinsic

For a given muscle (*e.g. longissimus dorsi*), under controlled conditions (*e.g.* under nitrogen at 37 °C), the rate of post mortem glycolysis is generally faster in the average pig than in ox or sheep;[7,8] but, especially in the pig, the breed and the genotype within the breed are important determinants. Overintensive selection of pigs for the ability to elaborate protein and a diminished capacity to lay down fat, in such breeds as Danish Landrace, Hampshire, and Piétrain, led to an increased incidence of pale, soft, and exudative (PSE) pork.[9,10] These undesirable features in the meat reflect denaturation of the muscle proteins either because the pH has fallen to a normal ultimate value (*ca.* 5.5), whilst a near *in vivo* temperature still prevails, owing to rapid post mortem glycolysis[9] or because post mortem glycolysis proceeds to an unusually low ultimate pH

Table 2 *Factors affecting rate of post mortem glycolysis*

Intrinsic
    Species
    Genotype
    Animal age
    Type of muscle
    Intramuscular location

Extrinsic
    Preslaughter drug administration
    Post mortem pithing and shackling
    Post mortem temperature
    Post mortem electrical stimulation
    Post mortem comminution
    Post mortem salting
    Post mortem pressure
    Post mortem oxygen tension

(*e.g.* 4.8).[11] Offer[8] has shown that an increased rate of pH fall and a decreased level of ultimate pH both cause an increased fraction of myosin to denature - and an increased severity of exudation.

At least two distinct genes are believed to be responsible for the PSE condition. One promotes a high rate of post mortem glycolysis and is associated with sensitivity to the anaesthetic halothane (causing malignant hyperthermia). The other is associated with abnormally low ultimate pH, although the rate of pH fall may not be great.[12,13] The high glycolytic potential which the latter signifies is common in pigs of the Hampshire breed, from which halothane sensitivity is virtually absent.[14] On the other hand, pigs of the Piétrain breed tend to be halothane-sensitive and exhibit a high rate of pH fall post mortem. Muscles of the latter have a high proportion of fast-twitch, glycolytic 'white' fibres in comparison with other breeds. An excessive anaerobic release of Ca^{2+} ions from mitochondria and from the sarcoplasmic reticulum may be the trigger for the accelerated post mortem glycolysis in halothane-sensitive pigs.[15]

Apart from the genes responsible for PSE, however, it is evident that there may be other genotypic differences between pigs, since, under identical conditions, the rates of post mortem glycolysis in a given muscle (*e.g. l. dorsi*) of a given breed (Large White) can vary considerably.[7] These data were obtained from animals of approximately the same age; but data from bovine muscles demonstrate that higher rates of post mortem glycolysis may occur in older animals.

The type of muscle is also an important determinant of the rate of pH fall post mortem.[7] It has long been apparent that this rate is greater in so-called 'white' than in 'red' muscles.[16] 'White' muscles have been differentiated in the course of evolution to undertake fast, intermittent, and largely anaerobic action, whereas 'red' muscles are adapted to carry out slower, more continuous action which depends on the provision of oxygen *in vivo*. The red hue is due to their relatively high content of the pigment myoglobin, which acts as a short-term oxygen store in muscle. 'Red' muscles also contain a higher proportion of respiratory enzymes than 'white', whereas the latter have the prerequisites for fast and effective anaerobic metabolism, namely a high ATP-ase activity, a high content of CP to replenish ATP from ADP, a greater store of glycogen to sustain anaerobic glycolysis, and a more developed sarcoplasmic reticulum to effect rapid release and recapture of Ca^{2+} ions by which the glycolytic process can be switched on and off.[17-19] Reflecting their overall hue, the ratio of 'white' to 'red' fibres within a muscle is similarly correlated with the rate of post mortem glycolysis in the location concerned,[20] although biochemical differentiation between fibres is somewhat more complex than their 'red' or 'white' appearance signifies.

Intrinsic differences in the rates of post mortem pH fall between 'red' and 'white' muscles may be confounded by the location of the muscles in the intact side or carcass; for example those which are more superficial to the cooling environment will have their metabolism affected more than those which are more deeply located both physically and physiologically, as will be considered below.[21]

Extrinsic

It is possible, by preslaughter injection of drugs or hormones, to alter the rate of post mortem glycolysis. Thus, calcium salts speed up the process, whereas doses of magnesium sulphate sufficient to induce relaxation in the animal cause a marked slowing.[22] Insofar as the preslaughter administration of anabolic agents tends to increase the proportion of 'white' fibres in the musculature,[23] an intrinsic increase in the rate of post mortem pH fall would be anticipated.

There is evidence that the process of pithing (to destroy the CNS during slaughter) accelerates the rate of post mortem glycolysis in those muscles where movement is induced.[24] The mode of suspension of the carcass has been shown to have an indirect effect if the shackled leg is also used to conduct the current used to stun the animal.[25] Although the duration of the current used to stun animals at slaughter is brief (1–4 s), it can cause some acceleration of post mortem glycolysis.[26]

Of the extrinsic factors, probably the most important is the temperature at which post mortem glycolysis takes place. It will be evident, therefore, that the relative proximity of muscles in the carcass to the exterior could modify their intrinsic rate of post mortem glycolysis. As would be expected, the rate is relatively rapid at *in vivo* temperatures (*ca.* 37–39 °C) and decreases as the temperature falls to ambient values; but, most surprisingly, it increases again as the temperature approaches 0 °C.[27] In unrestrained muscles, shortening has long been known to occur when *rigor mortis* develops at *ca.* 37 °C ('high temperature rigor'); but with the high rate of post mortem glycolysis at chill temperatures, shortening also occurs - the phenomenon of 'cold-shortening' - and it may be as great as at body temperature.[28] Shortening during post mortem glycolysis causes toughness in the meat and it was this feature which first drew attention to 'cold-shortening'. The problem had become prevalent in the early 1960s when hot lamb caracasses were increasingly exposed to very fast refrigeration. Since thermodynamic considerations must still apply, the fact that low temperatures cause an increase in biochemical processes must signify the involvement of an additional factor. It appears that, when muscles are exposed to temperatures below *ca.* 10 °C before post mortem glycolysis has brought the pH below *ca.* 6, they are still physiologically reactive and 'shiver' in an attempt to maintain local temperature. As the temperature falls from *ca.* 15 to 0 °C, Ca^{2+} ions are released from the tubes of the sarcoplasmic reticulum whereby their concentration in the sarcoplasm rises 30–40-fold, causing a massive stimulation of the contractile ATP-ase of the myofibrils and thus their contraction.[29] Such a discharge can be reversed more readily where the sarcoplasmic reticulum is well elaborated, as in 'white' muscles: this structure is not so developed in 'red' muscles[30] and the latter are more susceptible to 'cold-shortening'.[31] The whiter musculature of pork is thus less likely to toughen under 'cold-shortening' than the redder muscles of beef and lamb: nevertheless, if the rate of cooling is exceedingly rapid certain muscles of the pork carcass will undergo accelerated post mortem glycolysis and toughen.[32,33]

'Cold-shortening' and the organoleptic disadvantages it causes can be avoided by deliberately accelerating the rate of post mortem glycolysis until the pH has reached *ca.* 6.2, at which the muscles are no longer reactive to cold and refrigeration can be rapidly applied without toughening. Such acceleration is achieved by electrical stimulation which is applied immediately post mortem - as opposed to electrical stunning which is applied immediately pre mortem. During the 2-4 min of stimulation the rate of post mortem glycolysis is increased 100-150 fold, the pH falling by *ca.* 0.7 units:[34] when the current ceases the rate of pH fall is still approximately twice that of unstimulated muscles at the temperature concerned.

The efficacy of electrical stimulation depends on various factors such as the nature of the current (frequency of pulses, pulse duration, voltage: the higher the latter the more the effect), the mode of application (via the nervous system or directly to the muscle), the time after death, and the type of muscle.[35] Thus with a voltage of 2 V cm^{-1}, a pH fall of 0.5 units can occur in 2 minutes in a typical 'white' muscle compared with 0.3 units in a typical 'red' muscle.[36]

It appears anomalous that, by accelerating pH fall, electrical stimulation does not produce toughness similar to that caused by the 'cold-shortening' which it is designed to prevent. However, although electrical stimulation is associated with sarcomere shortening, and with rapid breakdown of ATP, the effect is transitory and, at the (still) *in vivo* temperature, the sarcoplasmic reticulum can readily recapture the Ca^{2+} ions liberated by the current, thus removing the stimulus to contract whilst the overall ATP level is high enough to effect relaxation of the sarcomeres.[35] On the other hand, at the low temperatures which cause 'cold-shortening', the sarcoplasmic reticulum is unable to operate efficiently, the Ca^{2+} level remains high, stimulation of the ATP-ase continues, and the consequent contraction is irreversible before *rigor mortis* ensues.

The early attainment of the ultimate pH after electrical stimulation is associated with shrinkage of the muscle fibres whereby oxygen from the atmosphere has ready access to some depth into the meat surface causing the myoglobin to transform into bright-red oxymyoglobin. In comparison, a paired non-stimulated portion will take longer to attain the same pH level, and if examined relatively soon after death will appear purplish red, since the pH will still be at more alkaline values at which the structure is swollen, absorbs incident light, and fails to take up oxygen. If examination is delayed until *ca.* 48 hours post mortem, however, both stimulated and non-stimulated samples will brighten in colour to the same extent.

As an alternative to the use of electrical stimulation to avoid 'cold-shortening', altering the mode of carcass suspension can redistribute the tensions within the muscles so that shortening is prevented in various commercially important portions of the carcase,[36] even with early exposure to fast refrigeration.

Because of the relatively small bulk of lamb carcasses in comparison with those of beef, their post mortem temperature can be more speedily and accurately controlled by the refrigeration regime; and it was suggested that it might be possible to chill pre rigor lamb carcasses, without using electrical stimulation to avoid 'cold-shortening'.[37] This was confirmed by Sheridan.[38] He demonstrated that, by swiftly cooling lamb carcasses in air at -20 °C, the meat was as tender as that conventionally chilled at 4 °C for 24 hours. He attributed the absence of 'cold-shortening' to the skeletal restraint of the carcass and to 'hardening' (but not freezing) of its surface - a factor which Davey and Garnett[37] had envisaged as being sufficient to avoid toughening. By a combination of various rates of chilling and modes of electrical stimulation, Marsh et al.[39] confirmed that intermediate rates of post mortem glycolysis, at which neither the high rates of high-temperature *rigor mortis* nor of 'cold-shortening' were involved (with their attendant toughening) and a pH of *ca.* 6.1 was attained in approximately three hours, yielded meat of optimal tenderness.

Offer[8] has demonstrated a further complexity in the effect of temperature on post mortem glycolysis. Thus, although increasing the intensity of prerigor chilling (in the absence of 'cold-shortening' circumstances) generally reduces the denaturation of myosin, it has little protective effect when the intrinsic rate of post mortem glycolysis is high - and the muscle enters *rigor mortis* rapidly. Moreover, when muscles are not chilled, the denaturation of myosin is greater in more slowly-glycolysing muscles in which *rigor mortis* is relatively delayed. Offer[8] explains these unexpected observations by postulating that actin markedly inhibits the denaturation of myosin when the latter combines with it in forming actomyosin.

A number of other contrived environmental factors also affect the rate of post mortem glycolysis. Thus, if muscles are comminuted whilst prerigor the damaged structure increases contact between ions, enzymes, and substrates, causing the rate of pH fall to accelerate markedly. Although comminution of muscles in the presence of sodium chloride also enhances the rate, and the concomitant breakdown of ATP, the water-holding capacity of the tissues nevertheless remains at the *in vivo* level.[40]

Subjection of prerigor muscles to pressures of *ca.* 100 MPa (*ca.* 1000 atm) for 2-4 minutes at room temperature disrupts the sarcoplasmic reticulum whereby Ca^{2+} ions are released and the rate of post mortem glycolysis greatly increases.[41] Concomitantly such pressure causes increased proteolysis, solubilizes myofibrillar proteins, and enhances tenderness.[42] If the process could be applied commercially, it would permit the production of reformed meats with the use of minimal quantities of salt.

If high oxygen pressure is applied to prerigor muscle it can re-establish respiration locally, causing removal of already formed lactic acid and the resynthesis of ATP and CP. Post mortem glycolysis is thus reversed

(Lawrie, 1950, unpublished). In practice this can happen, even under normal atmospheric conditions on the extreme periphery of exposed meat surfaces.

3 Factors Affecting the Extent of Post Mortem Glycolysis

As with its rate, the extent of post mortem glycolysis is also affected by both intrinsic and extrinsic factors (Table 3). Thus, the ultimate pH differs from the normal value. This is *ca*. 5.5, but if the amount of glycogen remaining in the muscles at the moment of death is less than *ca*. 0.7%, if the type of glycogen present resists post mortem breakdown, or if the enzyme system involved is disrupted, higher values will arise. The most dramatic effect of a high ultimate pH is the failure of myoglobin to take up oxygen (when exposed to the atmosphere) and form bright-red oxymyoglobin: instead it remains dull, purplish red ('dark-cutting'). It is also more tender (often excessively so), lacks flavour, and is more susceptible to subsequent microbial spoilage.

Intrinsic

Under comparable conditions, the ultimate pH of a given muscle of the three common meat species, cattle, sheep, and pigs, is of the same order, *ca*. 5.4–5.6; but, insofar as the muscles of pigs tend to have relatively little glycogen at the moment of death, the ultimate pH of the majority is higher than that of corresponding muscles in the other two species.[43]

The muscles of certain breeds of pig are found to have higher ultimate pH values than those of others. Thus, for a given muscle, the ultimate pH is greater in the *l. dorsi* of the Large White than in that of the Landrace.[44] In pigs of the Hampshire breed, however,[12] and in some Landrace pigs,[11]

Table 3 *Factors affecting extent of post mortem glycolysis*

Intrinsic
  Species
  Genotype
  Sex
  Animal age
  Animal temperament
  Type of muscle
  Pathology

Extrinsic
  Stress
  Preslaughter drug administration
  Post mortem temperature

the ultimate pH may be abnormally low - a circumstance associated with the PSE condition.

In examining the muscles of small numbers of animals, it is difficult to identify a consistent effect of sex on the ultimate pH. From a study of a group of ca. 500 bulls, 1000 heifers, and 2000 steers, however,[45] it has been reported that the ultimate pH in *l. dorsi* of bulls (5.84) was significantly higher than that in steers and heifers (5.62).[44] It must be acknowledged, however, that this difference may reflect the greater stress susceptibility of entire males.

Although the ultimate pH of porcine muscles has been reported to be lower in older animals,[46] most investigators have failed to note any significant effect of age on this parameter.

Stress is a major factor in depleting muscle glycogen reserves but most aspects are extrinsic, as will be considered below. It is now appreciated, however, that excitability of temperament (or stress susceptibility) is a significant determinant of ultimate pH, and in the case of cattle can be more dramatic than either fasting or enforced exercise immediately pre-slaughter (Table 4).[22] Close examination of stress-susceptible steers shows that they are frequently trembling and thus causing a low equilibrium level of glycogen in their muscles. The mixing of groups of animals of different origin at abattoirs can elicit such a response.[47,48] The importance of excitability of temperament in causing a high ultimate pH and dark-cutting beef was amply confirmed in a survey of its incidence in 20 countries, which revealed that young bulls were especially involved.[49]

Since the PSE condition can be induced by emotional stress immediately preslaughter,[9] it must be presumed that a very low pH and a high rate of pH fall in the muscles of pigs are further examples[12,50] of the influence of this factor, albeit that the biochemical mechanisms are probably different.

As already indicated, 'red' muscles are biochemically specialized to undertake respiratory metabolism *in vivo*; and they are relatively deficient in the enzymes which convert glycogen into lactic acid,[19] and in glycogen itself. These features cause their ultimate pH to be higher than that of 'white' muscles.[3,17,22] In a detailed analysis of 12 bovine muscles, it was shown that a high proportion of 'red' to 'white' fibres in a muscle (by

Table 4 *Comparative data on initial glycogen* and ultimate pH in* l. dorsi *of steers*

Treatment	Initial glycogen (mg/%)	Ultimate pH
Control (normal)	1126	5.50
Fasted 28 days	768	5.44
Forcibly exercised	1158	5.50
Control (excitable)	390	6.38

* Corrected to a resting *in vivo* pH of 7.2.

which its superficial colour is largely determined) is associated with a high ultimate pH.[20] Similarly the relative numbers of 'red' and 'white' fibres at different locations no doubt explains the differences in ultimate pH which are found within muscles.[44,51] The stress of mixing groups of animals preslaughter depletes the glycogen from 'white' fibres more severely than from 'red'.[52]

Not surprisingly, the concentration of residual glycogen (*i.e.* that which has not been converted into lactic acid during post mortem glycolysis) is usually high in 'white' muscles;[22] but this is not invariably so. Thus, in the relatively 'red' *psoas* and diaphragm of the horse and in bovine *sternocephalicus*, for example, substantial quantities (> 1%) of residual glycogen are found.[53] However, it has a somewhat different structure from that in other muscles,[54] and this may signify why it is more resistant to breakdown during post mortem glycolysis. Using model systems, it was shown that the extent of post mortem glycolysis depends on the proportion of the enzyme phosphorylase which is in the *a* form;[55] and this is less effective in 'red' muscles.[3]

Abnormal accumulation of glycogen in muscle occurs in a number of pathological conditions which reflect genetically determined deficiencies of certain enzymes of the glycolytic pathway[56] but these are unlikely to be significant in the domestic meat animals. Similarly, the very high glycogen content of the muscles of new-born pigs (*ca.* 7%) is not relevant in relation to the conversion of muscle into meat.

Extrinsic

Although inherent stress-susceptibility is now acknowledged as an important factor in lowering *in vivo* reserves of muscle glycogen, the writings of Daniel Defoe and Thomas Hardy indicate that extrinsic circumstances, such as baiting and enforced preslaughter activity, were known to cause stress in animals, and affect the meat thereby. This was many years before scientific investigations demonstrated that depletion of glycogen was a major biochemical result, leading to a high ultimate pH.[58,59] Apart from exhausting activity, extrinsic stressors include fasting, high and low ambient temperatures, anoxia, and conditions in transit and at the abattoir causing anxiety or rage. According to Selye[60] all such stressors involve a common physiological complex arising in the hypothalamus, and proceeding via the pituitary and adrenal glands to the level of the affected tissue. At the latter various protective proteins and polypeptides are induced, including ubiquitin.[61] Insofar as meat quality is concerned, the most significant effect is the lowering of the stored glycogen in muscle: as already indicated, if this falls below *ca.* 0.7% the ultimate pH will be elevated above the normal, *i.e. ca.* 5.5.

Of the domestic species, pigs are particularly liable to suffer glycogen depletion from their muscles in the immediate pre mortem period, as exemplified by the incidence of 'glazy bacon' in pigs which had walked for a short distance to abattoirs in Northern Ireland, instead of being killed on the farms of origin.[43] Nevertheless, although less commonly, both cattle and sheep can react to stress by producing meat of high ultimate pH. Thus, exposure to a sudden incidence of cold damp weather can cause widespread onset of dark-cutting beef in cattle[49] and, although the effect is less dramatic, it can also occur when cattle are slaughtered immediately after a short journey.[62] Preslaughter stressors may have a cumulative effect. Thus, although underfeeding, shearing, and preslaughter washing each tend to lower the level of glycogen in the musculature of lambs, their simultaneous operation has been shown to cause a significantly greater elevation of ultimate pH than the sum of their individual effects.[63] In certain pig genotypes, as has been seen, stress causes a very low ultimate pH (rather than a fast pH fall) to occur, leading to PSE - equally undesirable although opposite in its external appearance.

The inherently greater susceptibility of 'red' muscles to preslaughter stress - a reflection of the proportion of 'red' fibres in their structure - has already been mentioned, and this difference is also shown in their greater sensitivity to extrinsic stress.

Attempts to avoid a high ultimate pH, by feeding an easily assimilable sugar to pigs to build reserves of muscle glycogen, have been made.[64,65] Although the effect tends to be limited, a depression of even 0.3 pH units could be beneficial microbiologically.

The deliberate induction of a high ultimate pH can be achieved by the administration of various drugs preslaughter. Thus, insulin causes blood sugar to be deposited as liver glycogen at the expense of that in muscle, adrenalin causes aerobic glycolysis,[66] and neopyrithamin does so by preventing the metabolism of fat.[67] The effect of insulin tetany on the ultimate pH of various muscles is exemplified in Table 5.

Exposure to a high temperature during post mortem glycolysis will inactivate the enzymes effecting the process and fix the pH at the value it had attained at the time of cooking.[68] Tenderness is greater than in corresponding muscles which are cooked after the attainment of their ultimate pH. On the other hand, prerigor freezing (*i.e.* early in post

Table 5 *Ultimate pH in muscles from normal and insulin-tetanized steers*

Muscle	Normal	Insulin-tetany
Biceps femoris	5.40	6.53
Semimembranosus	5.48	7.09
Adductor	5.59	7.16
Gluteus medius	5.45	6.85
Psoas major	5.56	7.24
Longissimus dorsi(T)*	5.45	7.24
Longissimus dorsi(L)*	5.48	7.13

*T = thoracic, L = lumbar.

mortem glycolysis), when followed by rapid thawing, causes massive shortening, toughness, and exudation ('thaw rigor') and promotes the process of glycolysis, when it resumes, to be rapid and to proceed to an unusually low value (Table 6).

In comminuted meat, the addition of glucose (*ca.* 1.5%) appears to inhibit phosphorylase and yields ultimate pH values above 6.[69]

4 Post Mortem Proteolysis: Factors Affecting the Rate or Extent

Before considering the factors which affect the rate or extent of proteolysis, the nature of the general structural changes which accompany the tenderization of meat during a period of post mortem holding (ageing or conditioning) should be addressed. Although it has been known for nearly 100 years that there is a concomitant increase in soluble nitrogen, it is now clear that this is mainly due to the breakdown of the soluble proteins of the sarcoplasm to amino acids and peptides: neither a reversal of *rigor mortis* (*i.e.* the dissociation of actin from myosin)[70] nor extensive proteolysis of the contractile proteins[71] or of those of the connective tissue[72] occurs. More subtle and specific rupture of peptide bonds in the contractile and connective tissue proteins is now believed to explain the tenderization (Table 7). The general absence of energy post mortem makes it increasingly difficult for the ATP pump of the sarcoplasmic reticulum to maintain a low level of ionic calcium in the vicinity of the contractile proteins: Ca^{2+} ions are also released from the mitochondria under such conditions. The intracellular concentration of Ca^{2+} ions increases and these stimulate the calcium-activated enzymes, calpains I and II.[73] These operate optimally above pH 6, before post mortem glycolysis has been completed, attacking bonds in the contractile proteins (Table 7). The rate of tenderizing appears to be proportional to the concentration of active calpain I, until autolysis of the latter causes the rate to fall.* Ca^{2+} ions, *per se*, may also contribute to loosen the structure non-enzymically.[74]

Table 6 *pH and shear values for rabbit l. dorsi after cooking*

Time post mortem	Cooked immediately		Cooked after prerigor freezing and thawing	
	pH	Shear force (kg)	pH	Shear force (kg)
10 min	6.59	11	5.42	>27
3 h	6.03	15	5.92	10

* Indeed Dransfield[106] has postulated that the activity of calpain I could be enhanced *below* pH 6 insofar as a greater concentration of calcium ions would then have been released from the sarcotubular system and the inhibitory action of calpastatin would diminish as the pH fell from 6 to 5.5.

Table 7 *Sites of action of proteolytic enzymes during ageing*

Calpains (Calcium-activated proteinases)
    (i) Troponin T (pH>6)
    (ii) Z-lines (desmin)
    (iii) Connectin ('gap filaments')
    (iv) M-line proteins, tropomysin

Lysosomal enzymes (including Cathepsins B, D, and L)
    (i) Troponins T and I (pH<6), C-protein (relatively rapidly)
    (ii) Myosin (light and heavy), actin, tripomyosin, nebulin, titin, α-actinin (relatively slowly)
    (iii) Collagen (cross-links of non-helical telopeptides)
    (iv) Mucopolysaccharides (of ground substance)

The fall of pH to acid levels during post mortem glycolysis weakens walls of organelles such as the lysosomes.[75] The latter liberate proteolytic enzymes, the cathepsins, with pH optima below 6.[75,76] They attack contractile proteins and collagen at various strategic points (Table 7). Cathepsin L is probably the most important lysosomal enzyme in effecting tenderizing. It degrades troponins T and I and C-protein rapidly, and myosin, actin, tropomyosin, nebulin, titin, and α-actinin slowly. Its action is faster at pH 5.5 than at 6.[77]

The collagen of the epimysium appears to be much more labile during ageing than that of the perimysium, and, within the endomysium, type III collagen is preferentially attacked in comparison with type I.[78]

Specific factors affecting proteolysis post mortem will now be considered (Table 8).

Intrinsic

Rates of tenderizing during conditioning differ between species. Thus, in beef, lamb, rabbit, and pork the relative rates were shown to be 0.17, 0.21, 0.25, and 0.33% per day, respectively.[79] Subsequently it has been shown

Table 8 *Factors affecting the rate or extent of proteolysis*

Intrinsic
  Species
  Animal age
  Type of muscle

Extrinsic
  Post mortem temperature
  Post mortem time
  Post mortem electrical stimulation
  Post mortem pressure

that interspecies rates of proteolysis are reflected by the relative concentrations of of cathepsins B and L[80] and by the ratio of calpain II:calpastatin[81] in corresponding muscles.

The rate of proteolysis is greater in young animals. Thus it is greater in lamb than in sheep[82] and in vealer calves than in cull cows.[83] Such differences may reflect the biochemical differences responsible for the concomitant decrease in the speed of muscle contraction. The *extent* of tenderizing tends to be less in older animals of a given species, coinciding with the increasing number of stable cross-links in collagen as animal age increases.[84]

Twenty years ago, Goll[85] observed that, during post mortem ageing, the Z-lines in 'white' muscles were more liable than those in 'red' muscles. Thus, in bovine *psoas*, which is more 'red' than the *semitendinosus* in this species, tenderness changed little during holding for 4 days at 2°C, whereas it increased markedly in the latter.[86] Similar differences were found between the 'red' and 'white' muscles of pork and between 'red' and 'white' locations within a given porcine muscle.[87] Dransfield *et al.*[79] demonstrated that the effect of muscle type on the rate and extent of proteolysis, although ten-fold lower than that of temperature, was three-fold greater than those due to the type of animal.

Extensive studies[21,81] have confirmed the reality of such intermuscular differences and established their cause. The rate of ageing is greater in 'white' than in 'red' muscles since, in the former, (*a*) the ratio of calpain II to its inhibitor, calpastatin, is higher (Koohmaraie,[88] however, concluded that calpain I activity was the more important determinant of ageing, its concentration being higher in 'white' muscles), (*b*) the ratio of cathepsins B and L to their inhibitors is higher, and (*c*) the susceptibility to proteolysis of both myofibrillar proteins[21] and of connective tissue protein is greater. In respect of the latter, the extractabilities during ageing of both perimysial[89] and endomysial[78] collagens increase to a greater extent in 'white' muscles. Concomitantly, it is noted that the post mortem increase in osmotic pressure is greater in muscles of this latter type, reflecting the greater increase in small molecular species intracellularly.[21]

Extrinsic

As would be anticipated, the extent of tenderizing during post mortem ageing increases with increase in the environmental temperature (and with the passage of time). Thus, the tenderness increment occurring over 14 days at 0 °C takes place in 2 days at 20 °C,[90] in 1 day at 43 °C, and in even less time at 49 °C.[91] Nevertheless the *rate* of ageing decreases as holding temperatures rise from 40 to 60 °C. Above this temperature it falls rapidly, ceasing at *ca*. 75 °C.[92] Evidently the increase in enzymic activity with increasing temperature is eventually overtaken by heat denaturation of the enzymes involved. It is known that calpains lose activity above 40 °C,[93] and cathepsins above 50 °C.[94]

At a given temperature, the extent of proteolysis increases with time. Thus, under sterile conditions at 37 °C, the nitrogen soluble in trichloroacetic acid, as a percentage of total protein, increased from 11% to 17%, 23%, and 31.5%, respectively, after 20, 40, and 172 days.[95]

The use of electrical stimulation to lower the pH swiftly to *ca.* 6, whereby 'cold-shortening' and toughening are avoided, has already been discussed, but the procedure also interacts with temperature to affect proteolysis. Although most researchers agree that, when muscles are exposed to temperatures capable of causing 'cold-shortening', electrical stimulation will prevent toughening, the mechanism whereby it induces the tenderness when they are not so exposed - an effect first noted by Franklin in 1749 - is more controversial. Many workers have shown that, by speeding the attainment of an ultimate pH of *ca.* 5.5, electrical stimulation promotes the action of the lysosomal proteinases which have pH optima in this region, especially if the temperature is still at *in vivo* levels. George *et al.*[96] calculated that ageing changes proceed at about twice the rate in electrically stimulated muscles during the first 24–30 hours post mortem as in non-stimulated controls. In such circumstances, the effect of electrical stimulation in enhancing tenderness above that in controls diminishes as the period of ageing increases,[97] presumably because tenderness approaches an asymptotic value, beyond which it cannot increase.

On the other hand, Marsh *et al.*[98] believed that electrical stimulation, at least at the frequency of 50 Hz used commercially, increases tenderness by causing disruption of the muscular structure (when the avoidance of 'cold-shortening' is not an issue). They showed that, when the frequency of the current was only 2 Hz, and there was no disruption, electrical stimulation still accelerated the rate of post mortem glycolysis and the attainment of a pH *ca.* 5.5. But it was associated with greater *toughness* than that of non-stimulated controls when the meat was held at 37 °C (and heat shortening was prevented).[99] In such circumstances it was inferred that the delay in pH fall in the non-stimulated muscles enhanced the action of the calpains, since they have pH optima above 6. Others showed, however, that electrical stimulation at 50–60 Hz had no tenderizing effect in meat of high ultimate pH, despite disruption of the structure.[100] At this juncture, it must be concluded that electrical stimulation enhances the action of either calpains or cathepsins or both.

Horgan[101] and McFarlane[102] demonstrated that pressures of *ca.* 150 Mpa destroyed that ATP-ase which provides energy for pumping Ca^{2+} ions into the sarcoplasmic reticulum. The concentration of free Ca^{2+} in the cell thus increases and, by enhancing the action of calpains, promotes tenderness.

5 Other Catabolic Changes

Apart from post mortem glycolysis and proteolysis, the conversion of muscle to meat involves other catabolic changes. Although most of the glycogen present in the muscles at death is converted into lactic acid, a relatively small portion is converted into glucose by α-amylase action. Particularly in dehydrated and freeze-dried meats, the CO group of glucose subsequently unites with NH groups of proteins and amino acids by the Maillard reaction, producing brown pigments, bitter flavours, and brittle texture.[103] Since the rate of glucose formation is more rapid in the muscles of pork and rabbit than in those of beef, and the flesh of the former is relatively pale, Maillard browning is more serious with these species.[104] The production of glucose from glycogen is more marked in 'red' than in 'white' muscles.[105]

The breakdown of proteins to peptides and amino acids and the production of purine and pyrimidine residues from ATP and other nucleotides yields a reservoir of precursors for flavour volatiles on subsequent cooking. Hence the progression of ageing is associated with increased flavour in the meat.

6 References

1. D.G. Gadian, in 'Developments in Meat Science', ed. R.A. Lawrie, Elsevier Applied Science, London, 1980, Vol.1, p.89.
2. F.M. Clarke and D.J. Morton, *Biochem. J.*, 1976, **186**, 104.
3. D.J. Morton, J.F. Weidemann, and F.M. Clarke, Ref.1, Vol.4, p.37.
4. T.P. Walshe, C.J. Masters, D.J. Morton, and F.M. Clarke, *Biochim. Biophys. Acta,* 1981, **381**, 37.
5. E.C. Bate-Smith and J.R. Bendall, *J. Physiol.*, 1947, **106**, 177.
6. R.A. Lawrie, *J. Physiol.*, 1953, **121**, 275.
7. R.A. Lawrie, 'Meat Science', 1st Edn. Pergamon, Oxford, 1966.
8. G. Offer, *Meat Sci.*, 1991, **30**, 157.
9. L. Ludvigsen, *Beretn. Forsøgslab. Kbh.*, 1954, No. 272.
10. E.J. Briskey, *Adv. Food Res.*, 1964, **13**, 90.
11. R.A. Lawrie, D.P. Gatherum, and H.P. Hale, *Nature (London)*, 1958, **182**, 807.
12. G. Eikelenboom and L. Nanni-Costa, *Meat Sci.*, 1988, **23**, 9.
13. P. Sellier, in 'Qualita della Carcassa e della Carne Suine', University of Bologna, 1988, Chapter 9, p.163.
14. G. Monin and P. Sellier, *Meat Sci.*, 1985, **13**, 49.
15. K.S. Cheah, A.M. Cheah, and J.C. Waring, *Meat Sci.*, 1986, **17**, 37.
16. D.M. Needham, *Phys. Rev.*, 1926, **6**, 1.
17. R.A. Lawrie, *Biochem. J.*, 1953, **55**, 298, 305.
18. J. Gergely, D. Pragey, A.F. Scholtz, J.C. Seidel, F. Sréter, and M.M. Thompson, in 'Molecular Basis of Muscular Contraction', ed. S. Ebashi, Igaku Shoin Ltd., Tokyo, 1965, Chapter 12, p.145.
19. C.H. Beatty and R.M. Bocek, in 'The Physiology and Biochemistry of Muscle as a Food', ed. E.J. Briskey, R.G. Cassens, and B.B. Marsh, University of Wisconsin Press, Madison, 1970, Vol.2, Chapter 9, p.155.
20. M.V. Rao and N.F.S. Gault, *Meat Sci.*, 1989, **26**, 5.
21. G. Monin and A. Ouali, Ref.1, Vol.5, p.89.
22. A. Howard and R.A. Lawrie, Special Report, Food Investigation Board,

London, No. 63, 1956.
23. M.J. Clancey, J.M. Lester, and J.F. Roche, *J. Anim. Sci.*, 1986, **63**, 83.
24. A. Watanabe, E. Tsuneishi, and Y. Takimoto, *Meat Sci.*, 1991, **29**, 221.
25. J.L. Aalhus, C. Gariepy, A.C. Murray, A.D.M. Jones, and K.W. Tong, *Meat Sci.*, 1991, **29**, 321.
26. E. Lambooy, Proceedings of the 27th Meeting of European Meat Research Workers, Vienna, 1988, B3.
27. R.P. Newbold and R.K. Scopes, *Biochem. J.*, 1967, **105**, 127.
28. R.H. Locker and C.J. Hagyard, *J. Sci. Food. Agric.*, 1963, **14**, 787.
29. C.L. Davey and K.V. Gilbert, *J. Food Technol.*, 1974, **9**, 51.
30. D.N. Fawcett and J.P. Revel, *J. Biophys. Biochem. Cytol.*, 1961, **10**, Suppl., 89.
31. J.R. Bendall, *J. Sci. Food Agric.*, 1975, **26**, 55.
32. J.R. Bendall, *Meat Sci.*, 1978, **2**, 91.
33. A.J. Gigiel and S.J. James, *Meat Sci.*, 1984, **11**, 1.
34. B.B. Chrystall and C.J. Hagyard, *NZ Agric. Res.*, 1976, **19**, 7.
35. J.R. Bendall, Ref.1, Vol.1, p.37.
36. C.L. Davey, K.V. Gilbert, and P. Curson, Annual Report, Meat Industry Research Institute, Incorporated, New Zealand, 1971, p.39.
37. C.L. Davey and K.J. Garnett, *Meat Sci.*, 1980, **1**, 319.
38. J.J. Sheridan, *Meat Sci.*, 1990, **28**, 31.
39. B.B. Marsh, T.P. Ringkob, D.L. Russell, D.R. Swartz, and L.A. Pagel, *Meat Sci.*, 1987, **21**, 241.
40. R. Hamm, *Z. Lebensm.-Unters. Forsch.*, 1957, **106**, 281.
41. D.J. Horgan and R. Kuypers, *Meat Sci.*, 1983, **8**, 65.
42. J.J. McFarlane, I.J. McKenzie, R.H. Turner, and R.M. Jones, *Meat Sci.*, 1984, **10**, 307.
43. E.H. Callow, Annual Report, Food Investigation Board, London, 1939, p.29.
44. R.A. Lawrie and D.P. Gatherum, *J. Agric. Sci.*, 1962, **58**, 97.
45. L.E. Jeremiah, A.K.W. Tong, and L.L. Gibson, *Meat Sci.*, 1991, **30**, 97.
46. J.D. Sink and M.D. Judge, *Growth*, 1971, **35**, 349.
47. T.W. Jorgensen, Proceedings of the 9th Meeting of European Meat Research Workers, Budapest, 1963, p.3.
48. C.R. Ashmore, F. Carroll, L. Doerr, G. Tompkins, H. Stokes, and W. Parker, *J. Anim. Sci.*, 1973, **36**, 33.
49. P.V. Tarrant, in 'The Problem of Dark-Cutting in Beef', ed. D.E. Hood and P.V. Tarrant, Martinus Nijhoff, The Hague, 1981, Chapter 1, p.3.
50. P. Henckel, P.F. Jorgensen, and P. Jensen, *Meat Sci.*, 1992, **32**, 131.
51. G.R. Beecher, L.L. Kastenschmidt, R.G. Cassens, W.G. Hoekstra, and E.J. Briskey, *J. Food. Sci.*, 1968, **33**, 84.
52. A. Lacourt and P.V. Tarrant, *Meat Sci.*, 1985, **15**, 85.
53. R.A. Lawrie, *Biochim. Biophys. Acta*, 1955, **17**, 282.
54. R.A. Lawrie, D.J. Manners, and A. Wright, *Biochem. J.*, 1959, **73**, 485.
55. R.K. Scopes, *Biochem. J.*, 1974, **142**, 79.
56. G.T. Cori, *Mod. Probl. Paediatr.*, 1957, 344.
57. R.A. McCance and E.M. Widdowson, *J. Physiol.*, 1959, **147** 124.
58. C. Bernard, 'Leçons sur la Diabète et le Glycogénèse animale', Baillière, Paris, 1877, p.426.
59. E.H. Callow, Annual Report, Food Investigation Board, London, 1936, pp.75, 81.

60. H. Selye, *J. Clin. Endocrin.*, 1946, **6**, 117.
61. S. Lindquist and E.A. Craig, *Ann. Rev. Genet.*, 1988, **22**, 631.
62. S.N. Brown, E.A. Bevis, and P.D. Warriss, *Meat Sci.*, 1990, **27**, 249.
63. A.R. Bray, A.E. Grafhuis, and B.B. Chrystall, *Meat Sci.*, 1989, **25**, 59.
64. N.E. Gibbons and D. Rose, *Can. J. Res.*, 1950, **28**, 438.
65. J. Wismer-Pedersen, *Acta Agric. Scand.*, 1959, **9**, 67, 91.
66. C.F. Cori and G.T. Cori, *J. Biol. Chem.*, 1928, **79**, 309.
67. A. Howard and R.A. Lawrie, Special Report, Food Investigation Board, London, No.65, 1957.
68. C.L. Miles and R.A. Lawrie, *J. Food. Technol.*, 1970, **5**, 325.
69. O.A. Young, S.M. Humphrey, and D.J.C. Wild, *Meat Sci.*, 1988, **23**, 211.
70. B.B. Marsh, *J. Sci. Food Agric.*, 1954, **5**, 70.
71. R.H. Locker, *J. Sci. Food Agric.*, 1960, **11**, 520.
72. J.G. Sharp, Proceedings of the 5th Meeting of European Meat Research Workers, Paris, 1959, paper 17.
73. T. Marachi, K. Tanaka, K. Hatanaka, and T. Murakami, *Adv. Enzym. Regul.*, 1981, **19**, 405.
74. M.A.J. Taylor and D.J. Etherington, *Meat Sci.*, 1991, **29**, 211.
75. D.J. Etherington, *J. Anim. Sci.*, 1984, **59**, 1644.
76. I.F. Penny and E. Dransfield, *Meat Sci.*, 1979, **3**, 135.
77. M. Mikami, A.H. Whiting, M.A.J. Taylor, R.A. Maciewicz, and D.J. Etherington, *Meat Sci.*, 1987, **21**, 81.
78. C. Stanton and N.D. Light, *Meat Sci.*, 1990, **27**, 41.
79. E. Dransfield, R.C.D. Jones, and H.J.H. McFie, *Meat Sci.*, 1980–81, **5**, 131.
80. D.J. Etherington, M.A.J. Taylor, and E. Dransfield, *Meat Sci.*, 1987, **20**, 1.
81. A. Ouali and A. Talmant, *Meat Sci.*, 1990, **28**, 33.
82. C. Radouco-Thomas, C. Lataste-Dorolle, R. Zender, R. Busset, H.M. Meyer, and R.F. Mouton, *Food Res.*, 1959, **24**, 453.
83. L. Buchter, in 'Meat Chilling: How and Why?', ed. C.L. Cutting, Symposium, ARC Meat Research Institute, Bristol, 1972, Vol.2, p.4.51.
84. A.J. Bailey, *J. Sci. Food Agric.*, 1972, **23**, 995.
85. D.E. Goll, in 'The Physiology and Biochemistry of Muscle as a Food', ed. E.J. Briskey, R.G. Cassens, and B.B. Marsh, University of Wisconsin Press, Madison, 1970, Vol.2, Chapter 36, p.255.
86. D.E. Goll, M.H. Stromer, R.N. Robson, J. Temple, B.A. Eason, and W.H. Busch, *J. Anim. Sci.*, 1974, **33**, 963.
87. C. Chin-Sheng and F.C. Parrish, *J. Food Sci.*, 1978, **43**, 17.
88. M. Koohmaraie, Proceedings of the 41st Annual Reciprocal Meat Conference, Laramie, 1989, p.89.
89. A.J. Bailey and N.D. Light, 'Connective Tissue in Meat and Meat Products', Elsevier Applied Science, London, 1989.
90. P.E. Bouton, A. Howard, and R.A. Lawrie, Special Report, Food Investigation Board, London, No.67, 1958.
91. G.D. Wilson, P.D. Brown, W.F. Chesbro, B. Ginger, and C.E. Weir, *Food Technol.*, 1960, **14**, 143, 186.
92. C.L. Davey and K.V. Gilbert, *J. Sci. Food Agric.*, 1976, **27**, 244.
93. W.R. Dayton, D.E. Goll, M.G. Zeece, R.M. Robson, and W.J. Reville, *Biochemistry*, 1970, **15**, 2150.
94. A.A. Swanson, B.J. Martin, and S.S. Spicer, *Biochem. J.*, 1974, **137**, 223.
95. J.G. Sharp, *J. Sci. Food Agric.*, 1963, **14**, 468.

96. A.R. George, J.R. Bendall, and R.C.D. Jones, *Meat Sci.*, 1980, **4**, 51.
97. P.V. Harris and R. Shorthose, Ref.1, Vol.4, p.1.
98. B.B. Marsh, J.V. Lochner, G. Takahashi, and D.D. Kragness, *Meat Sci.*, 1981, **5**, 479.
99. G. Takahashi, J.V. Lochner, and B.B. Marsh, *Meat Sci.*, 1984, **11**, 207.
100. T.R. Dutson, J.W. Savell, and G.C. Smith, *Meat Sci.*, 1982, **6**, 159.
101. D.J. Horgan, *Meat Sci.*, 1981, **5**, 297.
102. J.J. McFarlane, Ref.1, Vol.3, p.155.
103. C.H. Lea and R.S. Hannan, *Nature (London)*, 1950, **65**, 438.
104. J.G. Sharp, Annual Report, Food Investigation Board, London, 1958, p.7.
105. R.M. Bocek, G.M. Basinger, and C.H. Beatty, *Am. J. Physiol.*, 1966, **210**, 1108.
106. E. Dransfield, *Meat Sci.*, 1992, **31**, 85.

Lean Meat, Animal Welfare, and Meat Quality

B.W. Moss

DEPARTMENT OF AGRICULTURE FOR NORTHERN IRELAND AND THE
QUEEN'S UNIVERSITY OF BELFAST, NEWFORGE LANE, BELFAST BT9 5PX, UK.

1 Introduction

The quality attributes of meat can be considered under four main headings: leanness, appearance/colour, texture/tenderness, and flavour. In this classification aspects such as succulence and juiciness would be considered under the texture/tenderness heading. At point of sale the consumer is influenced by three aspects: price or, more correctly, value for money, appearance, and their previous experience of the product (Figure 1). The appearance of the product includes both the leanness, *i.e.* the ratio of fat to lean, and the colour of the lean meat in the joint, or meat product. Previous experience includes not only their previous sensory experiences of that particular product/brand but also advertising and media influences, *e.g.* reported implication of animal fat in coronary heart disease. Vacuum packed beef cannot be sold at retail level since the dark colour due to myoglobin formation in the vacuum packs is associated in the consumer's mind with the dark colour of aged spoiled meat even though the latter is due to metmyoglobin formation and is dark brown rather than purplish.

The relationship between leanness and colour of lean, and some other meat quality aspects, will be discussed in depth later. At this point it is worth considering whether in selection for lean meat consumers would choose one which had less fat cover or more muscle. This is important for two reasons; firstly many payments from slaughterer to producer have been based on fat cover as an indicator of lean rather than lean content *per se*, and secondly the physiological mechanisms to achieve less fat are likely to be different from those which increase muscle mass within the animal. The implications of different physiological profiles needs to be considered when discussing the relationship between lean meat, animal welfare, and meat quality.

In the relationship between lean meat, animal welfare, and meat quality there are two questions to be discussed. The first question is 'does the production of leaner animals have implications for meat quality?', and the

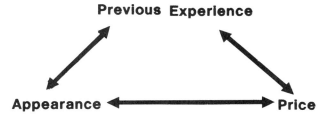

Figure 1 *Factors influencing consumers' choice at point of sale*

second is 'how is animal welfare implicated?'. There are three main practical ways forward to producing leaner animals; these are by nutritional changes, genetic selection, and sex type. The influence of nutritional changes on lean meat and meat quality has been reviewed[1] and this chapter will concentrate mainly on genetic means and sex type. A fourth method of producing leaner animals by the use of exogenous substances (drugs, hormones) will not be considered, given the present climate and consumer attitudes, as a practical proposal. The evidence from previous research studies in this area may shed some light on the physiological mechanisms that might control both lean meat and meat quality.

2 Animal Welfare - the Stress Cycle

In the context of this chapter the aspect of animal welfare to be considered will be 'stress' particularly in relation to preslaughter handling and meat quality. 'Stress' is a human emotive feeling which we apply to animals and as such is subjective. It has been argued that given the physiological similarities of farm animals there is no reason to assume that this should not apply.[2] To be more objective we need to define the term stress, One of the early treatises on stress considered stress to be composed of two components,[3] the stressor and the response. Various definitions used later for stress have led to some confusion.[4-7] One approach using stressor and response as integral parts of stress also incorporated the concept of animals being understressed or overstressed,[8] when the animals' responses were outside the acceptable or normal range. The stressor is the noxious stimulus applied to the animal and the response is the physiological and behavioural response of the animal to the stressor. Mathematically we could write this as: stress = stressor × response. This allows a more objective approach because we can attempt to quantify the stressor by measuring the environmental and social interactions on a group of animals, and the responses can be quantifiable physiological changes. It is still difficult, however, to interpret these without applying the concepts of human feelings to those of animals.

Behavioural studies can be very useful in assessing stress but are not without difficulties in interpretation. Aggressive behaviour may result as a response to some stressor such as stocking density, confinement, or another aggressive act, and may itself be a stressor. Adaptation to the environment may occur, such that physiological stress responses change. For example, the effects of handling on animals in early life, which at the time is a stressor, can subsequently alter responses to further stressors in later life.[9-14] At all times during an animal's life it experiences stress from a range of stressors. For a meat-producing animal stress may be maximum at birth and at slaughter. Such a stress cycle is shown diagrammatically in Figure 2. Of particular relevance to this chapter is the increase in stress in the period prior to slaughter and its influence on meat quality.

3 Preslaughter Handling and Meat Quality

The extrinsic and intrinsic factors influencing meat quality have been reviewed previously.[15,16] The preslaughter factors influencing meat quality are genetics, sex/gender, husbandry, nutrition, transport, lairage, and stunning. The aspect of meat quality concerned is that which is influenced by the rate and extent of post mortem glycolysis. A scheme showing the main biochemical links in the aerobic and anaerobic pathways in the muscle is shown in Figure 3. The detailed intermediates and consequences of slaughter have been discussed in detail previously.[16] After slaughter the muscle attempts to maintain its cellular integrity without supply of substrates to the muscle and removal of waste products; thus glycogen is used up and lactate accumulates in the muscle. If the rate of these glycolytic processes is rapid such that there is a combination of low pH and high

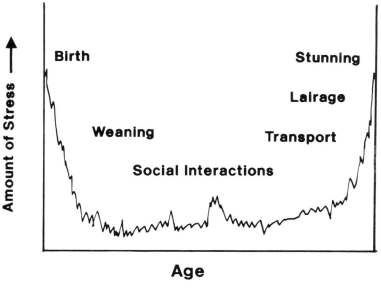

Figure 2 *The stress cycle*

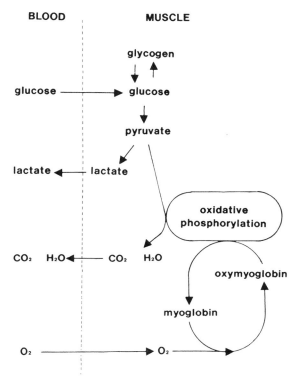

Figure 3 *Pathways of glucose metabolism in muscle (key compounds at branch points shown only)*

temperature (*i.e.* near body temperature) the resulting meat is pale, soft, and exudative (PSE) and is of poor quality. If, however, lactate production is limited by glycogen reserves the ultimate pH is high and the meat is dark, firm, and dry (DFD), and has poor keeping quality (Figure 4). Both abnormal conditions, PSE and DFD, are influenced by stress in the preslaughter handling period.[17] PSE increases in situations where the stress immediately prior to slaughter, *e.g.* movement to stunning pen and stunning, is high, or the animal is more sensitive to stress (stress sensitive). DFD arises when the animal has been subjected to stress of longer duration which depleted muscle glycogen or to lack of food in a period prior to slaughter. Both PSE and DFD occur in pigs, whereas only the DFD condition occurs in cattle and is often referred to as the 'dark cutting' condition. A survey of pig producers delivering to one factory in Northern Ireland indicated that on the basis of time of last feed *ca.* 20% of the pigs slaughtered may be prone to the DFD condition.[18] Studies on cattle indicate that stress, particularly that involved in mixing strange animals in the lairage, may have greater influence on the incidence of DFD condition than duration of time without food.[18]

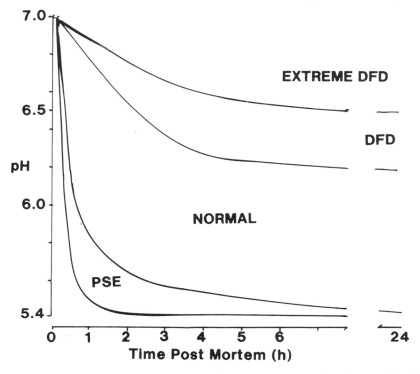

Figure 4 *Typical rates of post mortem decline in porcine muscle showing classification into PSE, normal, and DFD*

4 Classification and Identification of PSE Carcasses

PSE carcasses are those which show a rapid post mortem glycolysis and thus the pH measurement at 45 min post slaughter (pH_1) is used to identify carcasses. Different workers used different cut-off points and PSE carcasses have been defined as having a pH_1 of 6.0 of less,[19] or pH_1 of 5.9 or less.[20] These cut-off points are based on *in vitro* studies of muscle protein denaturation at different pH and temperature combinations.[21] Recently a mathematical approach has been used to describe the changes in protein denaturation in pig carcasses in relation to PSE.[22] When the carcasses of progeny test pigs were classified as PSE on the basis of a pH_1 of 5.9 or less and then their appearance was examined it was found that not all of these carcasses appeared pale. In the Landrace breed 50% were considered pale whereas in the Large White breed only 25% were considered pale.[23] Care must be taken in using pH_1 measurements to classify PSE carcasses. Recent work has shown that fibre optic probes, particularly those which have spectral scanning, may be useful in predicting PSE meat at 45 min preslaughter.[24–27] Table 1 shows the relationships between pH measurements, drip loss, and appearance for pig carcasses. The classification of

Table 1 *Relationships of meat quality class to* longissimus dorsi *pH values, drip loss, and colour parameters*

Meat quality parameter	Meat quality class			Statistical significance of difference between		
	PSE[a]	Normal	DFD[b]	PSE & normal	DFD & normal	PSE & DFD
(Number of carcasses)	(27)	(103)	(10)			
LD pH_1	5.79	6.08	6.27	***	**	***
LD pH_u	5.59	5.52	6.13	NS	***	***
Drip loss[c] (%)	2.89	1.92	1.84	*	NS	***
Reflex attenuance at 700 nm[d]	0.30	0.37	0.47	*	**	***
Proportion of pigment as:						
Deoxymyoglobin[e]	0.20	0.22	0.31	NS	**	**
Metmyoglobin[e]	0.32	0.29	0.29	NS	NS	NS
Oxymyoglobin[e]	0.48	0.49	0.40	NS	**	**

[a] Carcasses with LD pH_1 of 5.9 or less
[b] Carcasses with LD pH_u of 6.0 or greater.
[c] Expressed as drip lost after 48 h at 4 °C.
[d] Reflex attenuance calculated as $-\log$(Reflectance) (K. Shibata, in 'Methods of Biochemical Analysis IX', ed. D. Glick, Interscience, New York, 1969, p. 217).
[e] Calculated using isosbestic points (K. Krzywicki, *Meat Sci.*, 1979, **3**, 1).

carcasses was undertaken on the basis of pH_1 measurements for PSE and pH_u (ultimate pH) measurements for DFD. There is no difference between pH_u values for PSE and normal pigs, but both were significantly lower than for DFD. Although the pH_1 results were as expected, given that PSE and normal differ only in rate rather than extent of pH decline, the significant differences between pH_1 results for DFD have not previously been reported. These results indicate that high pH_1 values, *e.g.* above pH 6.3, may be predictors of DFD. Depletion of glycogen due to starvation may also influence rate of pH fall, possibly due to a substrate concentration effect on post mortem glycolysis.[18] Drip loss could be used to separate PSE from normal and DFD carcasses, but not to classify DFD separately from normal. In these studies reflectance measurements were made 24 h post mortem; however, they indicate that single wavelength measurements such as reflex attenuance at 700 nm may be suitable for classifying PSE, DFD, and normal carcasses.[18] Multiple wavelength measurements were used to calculate pigment proportions[28] and these results indicate that although this method has only previously been used on beef it may be applicable to pork. Since the difference between PSE and normal pork is due to light scattering changes due to protein denaturation, differences in the proportion of pigments would not be expected between these two classes. In DFD, however, mitochondrial activity is greater at the higher pH, resulting in less oxygen available for oxymoglobin formation, the proportion of deoxymyoglobin increasing relatively. If possible, additional measures should be used along with pH values in order to classify carcasses fully as DFD or PSE.[23,29] Care must be taken to ensure that cut-off points for classification are clearly stated. Measurement of pH by direct probe insertion and by homogenization can give significantly different results.[30–32] Such factors must be taken into account when classifying the meat quality of carcasses.

5 Genetic Potential and Meat Quality

Alteration of genetic potential may be the long-term route to improving meat quality. In pigs it is fairly well established that the incidence of PSE is greater in leaner breeds.[33,34] Table 2 shows the meat quality of a sample of pigs undergoing progeny test. This genetic difference in meat quality has been attributed to the presence of the halothane gene.[35–37] Current testing of pigs in Northern Ireland indicates a low incidence of halothane-positive pigs. In such situations preslaughter handling factors are likely to be relatively more important than genetics. If there is a strong link between factors that control leanness and meat quality, then increased selection for leanness might be expected to result in increased stress sensitivity and higher incidence of PSE. Studies on effect of selection for leanness on meat quality in pigs do not in general show a strong relationship.[23] Quite marked changes in leanness being achieved through selection pressures, with little or no effect on stress sensitivity/meat quality in pigs. In cattle less information is available in relation to genetic selection and meat quality. Any studies on physiological relations between leanness, stress sensitivity, and meat quality in pigs would be expected to bear relationships to those in cattle.

6 Endocrine Function and Meat Quality

Several approaches have been used to study relationships between meat quality and stress sensitivity. The role of thyroid function in PSE condition

Table 2 *Meat quality measurements of progeny test pigs*

	Landrace	Significance	Large White
Drip loss:	(423)[a]		(103)[a]
as percentage of chop wt.	1.28	***	0.94
as percentage of lean wt.	2.03	***	1.65
Percentage of PSE (pH<5.9)	15.9		14.4
Reflectance value (EEL)[b]	45.5	**	43.5

[a] Number of pigs.
[b] Reflectance value measured using EEL (Evans Electroselenium Ltd) smoke stain reflectometer, higher values indicating paler meat.

has been studied by genetic, correlation, pharmacological, clinical, biochemical, and physiological approaches.[38] Such approaches have also been used to study the role of other hormones in relation to PSE.

A number of workers have implicated adrenal function, both medulla and cortex, and thyroid function in relation to meat quality and stress sensitivity.[39–42] Adrenal corticoids are released in response to stressors[3,43] and increase protein breakdown in muscle tissue[44] and promote gluconeogenesis.[45] Thus increased corticosteroid production can result in catabolism of muscle tissue and increased deposition of fat. In humans clinical hyperadrenal states, such as Cushing's syndrome, have typical 'moonface' due to increased fat deposition.[46] Pituitary adrenal function may therefore be a strong contender for the physiological link between leanness, stress sensitivity, and meat quality. Several workers have reported decreased circulating corticosteroid levels in stress-sensitive pigs; some have suggested higher pituitary stimulation as indicated by higher ACTH/corticosteroid ratios.[47,48] There have also been some suggestions as to higher corticosteroid turnover, and the implications of this for gluconeogenesis have been considered in relation to meat quality.[49] Whether breed differences in corticosteroid are significant, and its relationship to meat quality, depend on the time of sampling and stress prior to sampling. Both the circulating level of corticosteroids and their response to stressors show diurnal rhythms.[49,50] The data for Table 3 indicate that in a minimal stress situation there is no significant breed difference in circulating corticosteroid level, but after greater stress (in this case preslaughter handling) a breed difference results. Although both breeds show an increase due to preslaughter handling this increase is greater for the Large White breed than fot the Landrace breed. Other workers were unable to detect any differences in the corticosteroid response to standard hypothalamic, pituitary, or adrenal stimulation between Piétrain (stress sensitive) and Large White pigs.[51] Administration of corticosteroids to stress-sensitive pigs has been shown to reduce the tendency to PSE.[39] β-Endorphins have also been shown to increase in relation to preslaughter handling and stress situations[52,53] and may have a role in the preslaughter physiological responses affecting post mortem glycolysis.

Thyroid function has also been investigated in relation to PSE. Early

Table 3 *Cortisol and thyroxine levels in Landrace and Large White pigs*

Breed	Plasma cortisol (μg dl^{-1})			Thyrosine (μg dl^{-1})		
	In pens[a]		At slaughter[b]	In pens		At slaughter
Landrace	2.3	(***)	4.6	4.0	(NS)	4.2
Large White	2.8	(***)	6.0	4.6	(*)	4.1
Breed differences	NS		*	**		NS

[a] Pigs restrained by noose round upper jaw and blood collected from ear vein.
[b] Blood collected on exsanguination.

studies in this area showed that administration of iodocasein to pigs reduced the incidence of PSE.[39] Typically hyperthyroid humans have an ectomorphic body type and are excitable; thus one might expect that leaner and more stress-sensitive pigs might be hyperthyroid. Several studies indicated that circulating blood thyroxine levels were lower in stress-sensitive pigs than other pigs. Other studies have indicated increased thyroid secretion in stress-sensitive pigs.[54-57] The role of thyroid function in the aetiology of the PSE condition has been reviewed, with the suggestion that thyroid hormone turnover, particularly utilization, might be higher in stress-sensitive pigs.[38] Breed differences in circulating thyroxine (T_4) levels depend on whether the pigs are stressed prior to sampling. In the leaner, more stress-sensitive Landrace breed, preslaughter handling has no significant effect on T_4 levels, whereas in the Large White breed T_4 levels are reduced due to a similar preslaughter stressor (Table 2). When pigs were held in lairage overnight T_4 levels at slaughter were significantly higher than after a short lairage (2 h).[58] The role of thyroxine will be discussed further in relation to sex differences and leanness.

7 Influence of Gender and Sex Status on Meat Quality

In farm animals we not only have gender differences (male/female) but also sex status, *i.e.* entire and castrated males. In general entire males have a greater lean to fat ratio, and better food conversion ratio.[58,59] Their use in animal production has been limited because of other factors, principally their aggressive nature. An obnoxious odour in the meat of entire male pigs, boar taint, has also slowed down their uptake for meat production.[60,61] Other differences in meat texture and fat quality in boars have been reported.[62] Sex differences in thyroid function in pigs have been reported, with lower levels in boars than gilts. If we consider the thesis that leaner animals are more stress sensitive and produce poor meat quality, then the relationship between leanness and stress sensitivity should be independent of the manner by which leanness is achieved. Thus the leaner gender or sex type should have physiological profiles typical of stress-sensitive pigs and greater incidence of PSE. Boars were found to have higher thyroid stimulating hormone (TSH) levels, lower T_4 levels, and higher serum enzyme activities at slaughter than gilts, which is typical of stress-sensitive animals; however, PSE levels were lower in boars.[58] Boars are more aggressive than castrated males or females and it has been suggested that in the case of boars a greater sensitivity to stress may be exhibited as greater excitability and lowered threshold for aggression.[63]

This greater excitability and aggression leads to glycogen depletion preslaughter and thus a greater incidence of DFD than PSE. Bruising on pig carcasses is higher in boars than in gilts and is clear evidence of fighting preslaughter.[64] Increased bruising has been related to both PSE and DFD depending on the time at which fighting occurred prior to slaughter and its duration.[65,66]

A closer examination of the relationship of leanness and stress sensitivity may be undertaken by considering differences due to castration. Table 4 shows thyroid function in relation to both breed and sex status. Within a breed the leanest sex type boars have the lowest T_4 levels, lowest free thyroxine index FTI, highest TSH, and highest TSH/T_4 ratios. The leaner breed Landrace has lower T_4, lower FTI, and higher TSH/T_4 ratio. Ranking the sex types on leanness over both breeds resulted in a similar order to TSH/T_4 ratios, the leanest animals, Landrace boars, having the highest TSH/T_4 ratios and the fattest, Large White gilts, having the lowest TSH/T_4 ratios. Significant differences between litters occurred and the calculated intraclass correlation[67] indicated that thyroid function may have a high heritability coefficient.[38] There is thus strong evidence for the role of thyroid function in the aetiology of the PSE condition; however, the exact physiological mechanism is not clear. In sheep and laboratory species hyperthyroidism has been shown to be related to excitability.[68,69] Thyroid function data on emotional strains of rats indicate that such animals may provide a laboratory species model for studying endocrine function and stress sensitivity.[70-72] Thyroid function has also been shown to alter fibre type distribution.[73] Thus differences in thyroid function may also alter intrinsic biochemical properties of the muscle and how these react in the post mortem period. Histochemical fibre type distribution has been implicated in the aetiology of PSE by several workers.[74-76]

8 Meat Quality Problems in Cattle

In cattle there is little information on genetic differences in stress sensitivity. It is generally accepted that bulls of dairy breeds tend to be more

Table 4 *Sex and breed differences in thyroid function in pigs*

Breed/sex	Thyroxine (T_4) μg dl^{-1}	FTI[a]	TSH[b] μU ml^{-1}	TSH/T_4 ratio
Landrace:				
Boars	3.37	5.32	5.68	1.75
Barrows	3.60	6.58	4.03	1.16
Gilts	3.85	6.47	4.69	1.26
Significance of difference:				
Litters	***	***	NS	NS
Sex	NS	*	NS	NS
Large white:				
Boars	4.12	6.02	6.08	1.50
Gilts	4.59	7.13	5.17	1.14
Significance of difference:				
Litters	NS	NS	NS	NS
Sex	NS	NS	NS	NS

[a] FTI = free thyroxine index: higher values indicate higher proportion of 'free' hormone circulating.
[b] TSH = thyroid stimulating hormone, assayed using human (hTSH) radio-immunoassay kit.

aggressive than those of beef breeds. Since cattle generally have lower rates of post mortem glycolysis than pigs, the conditions of high temperature, low pH do not normally arise. The major meat quality problem is dark cutting, DFD meat, when glycogen is depleted prior to slaughter. The combination of low pH, high temperature in beef carcasses only occurs when carcasses are electrically stimulated post mortem. Under conditions of electrical stimulation, some protein denaturation has been observed but this did not result in increased drip loss.[77] Thus the conditions in electrically stimulated beef were not typical of PSE. Various cut-off points have been used for classifying carcasses as DFD, depending often on the end use of the beef.[78] For instance it is generally recommended that beef with a pH_u of 6.0 or greater should not be vacuum packed, as it will spoil readily. Above a pH_u of 6.2 or 6.3 the meat will be perceived as dark by the consumers and thus rejected or sales reduced.

The high-pH_u meat will not 'bloom up' when cut due to the greater mitochondrial activity at higher pH, utilizing available oxygen (Figure 3). The respiration rate of mitochondria extracted from high-pH_u *longissimus dorsi* was 2.5 times higher than for those extracted from normal-pH_u beef. Even after five days the respiration rate of mitochondria extracted from high-pH_u meat had only declined to that equivalent to normal beef 24 h post mortem.[18]

9 Preslaughter Handling and Dark Cutting Beef

Higher incidences of DFD have been reported after holding in overnight lairage and in cattle sold through auctions marts rather than transported directly from farm to factory.[18] These higher incidences would appear to be largely due to increased aggression and physical activity depleting glycogen rather than lack of feed. Cattle penned in lairage for 27 h in their own producer lots without food but with access to water depleted *longissimus dorsi* glycogen levels by 20%, which resulted in an increase in mean pH_u from 5.6 to 5.7.[18] Mixing of steers even after a short lairage of 3 h resulted in a decrease in glycogen to 70% of controls. Mixing and overnight lairage reduced glycogen reserves to 25% of controls and the levels decreased with further mixings (Table 5). Recovery from such stress, as measured by physiological factors and reduction in DFD, may take several days.[79] The higher incidence of DFD in bulls has been attributed to increased aggression. Recent studies have shown that when steers, bulls, and vasectomized bulls were mixed prior to slaughter, vasectomized bulls showed greater mounting activity than bulls or steers.[80] The number of mounts exhibited was significantly correlated with pH_u values in the *longissimus dorsi*.[81] The activities of creatine phosphokinase (CPK) and lactate dehydrogenase (LDH) were also significantly correlated with pH_u.

Intrinsic factors, particularly muscle histochemical parameters, have been related to DFD in cattle.[82-84] Sex differences in fibre type distribution have been reported.[85] Steers had a significantly higher percentage area of

Table 5 *Influence of the degree and duration of mixing on the meat quality of bovine steers*

	Treatment			
	Control	A	B	C
Number of mixings	0	1	1	3
Lairage duration (h)	3	3	24	24
Number of animals	11	11	11	11
LD pH_u	5.49	5.67	5.95	6.19
Number of carcasses with:				
LD $pH_u > 6.0$	0	1	3	5
LD $pH_u > 6.3$	0	0	0	3
Brightness	64.9	65.4	62.9	56.5
Muscle glycogen[a]	13.8	8.2	3.8	2.3

[a] Glycogen expressed as μmol glucose equivalents per g of muscle tissue.

glycolytic fibre, smaller diameters of oxidative fibre, and consequently higher anaerobic fibre ratio than bulls. In these studies vasectomized bulls had fibre type distributions intermediate between those of bulls and steers.[86]

Although statistically significant correlations between blood parameters such as CPK, LDH, and meat quality have been reported, such correlations are generally only found when the animals have been considerably stressed before slaughter. Little or no correlation is generally observed between serum enzyme activity and meat quality in animals with minimal stress prior to slaughter. Since stress induces an interaction of physiological responses one approach has been to consider multidimensional statistical approaches. Stepwise discriminant analysis has been undertaken and, although it produced selected wavelengths which were suitable for predicting DFD, no combination of blood parameters was found to be suitable.[87] This approach was based on initial classification of carcasses as normal (<5.8), intermediate (5.8–6.0), and dark cutting, (>6.0) and it may be better to consider classification schemes based on how the consumer perceives the beef. Such an approach requires a large data bank, with sensory, metabolic (muscle and blood), and colour parameters included.

10 Conclusion

In conclusion production of leaner animals, particularly by use of intact males but also by changes in genetic potential, may lead to animal welfare problems and poorer meat quality. By being aware of these implications husbandry systems should be able to be developed such that lean meat can be achieved with minimal reduction in meat quality.

11 References

1. K. McCracken, this volume, p.16.
2. F.W.R. Brambell, Technical Committee to Enquire into the Welfare of Animals kept under Intensive Livestock Husbandry systems (Chairman, F.W.R. Brambell) London, HMSO, 1965.
3. H. Selye, 'Stress', Acta Endocrinologica Inc., Montreal, 1951.
4. J.R. Bareham, *Vet. Rec.*, 1973, **92**, 682.
5. A.F. Fraser, *Vet. Rec.*, 1973, **93**, 27.
6. 'Stress in Farm Animals', Proceedings of a Joint symposium with the Royal Society for Prevention of Cruelty to Animals, London, 1973, *Br.Vet. J.*, 1973, **130**, 85.
7. J.L. Barnett and P.H. Hemsworth, *Appl. Anim. Behav. Sci.*, 1990, **25**, 117.
8. R. Ewbank, *Vet. Rec.*, 1973, **26**, 709.
9. A.M. Barrett and M.A. Stockham, *J. Endocrinol.*, 1963, **26**, 97.
10. P.J. Knott, P. Hutson, and G. Curzon, *Pharm. Biochem. Behav.*, 1977, **7**, 245.
11. P.D. Warriss, S.C. Kestin, and J.M. Robinson, *Meat Sci.*, 1983, **9**, 271.
12. T. Grandin, Proceedings of the 35th International Congress of Meat Science and Technology, Copenhagen, 1989, Vol.3. p.971.
13. J. Gray, 'The Psychology of Fear and Stress', World University Library, Weidenfield & Nicolson, London, 1971, p.97.
14. P.H. Hemsworth and J.L. Barnett, *Appl. Anim. Behav. Sci.*, 1991, **30**, 61.
15. R.A. Lawrie, 'Meat Science', 4th Edn., Pergamon Press, Oxford, 1985, p.51.
16. R.A. Lawrie, this volume, p.43.
17. D. Lister, N.G. Gregory, and P.D. Warriss, in 'Developments in Meat Science', ed. R.A. Lawrie, Applied Science Publishers, London, Vol.2, p.61.
18. W.S. Campbell, Ph.D. Thesis, The Queen's University of Belfast, 1984.
19. J.P. Chadwick and A.J. Kempster, *Meat Sci.*, 1983, **9**, 101.
20. B.W. Moss, *J. Sci. Food Agric.*, 1980, **31**, 308.
21. R.K. Scopes, *Biochem. J.*, 1964, **91**, 201.
22. G. Offer, *Meat Sci.*, 1991, **30**, 157.
23. B.W. Moss, in 'The Evaluation of Meat Quality in Pigs', ed. P.V. Tarrant, G. Eikelenboom, and G. Monin, Martinus Nijhoff, The Hague, 1987, p.225.
24. L.E. Anderson, Proceedings of a Scientific Meeting 'Biophysical PSE-Muscle Analysis', ed. H. Pfutyner, Vienna Technical University, 1984, p.173.
25. P. Barton-Gade and E.V. Olsen, Ref.24, p.192.
26. J.R. Andersen, C. Borggard, and P. Barton-Gade, Ref. 12, Vol.1, p.208.
27. D.B. Macdougall, Ref. 24, p.162.
28. K. Krzywicki, *Meat Sci.*, 1979, **3**, 1.
29. P.G. Van der Waal, A.H. Bolink, and G.S.M. Merkus, *Meat Sci.*, 1988, **24**, 79.
30. C.F. Yndgard, *J. Food Technol.*, 1973, **8**, 485.
31. H. Korkeala, O. Maki-Petays, T. Alanko, and O. Sorvettula, *Meat Sci.*, 1986, **18**, 121.
32. C. Sommers, P.V. Tarrant, and J. Sherrington, *Meat Sci.*, 1985, **15**, 63.
33. D.B. Macdougall and J.G. Disney, *J. Food Technol.*, 1967, **2**, 285.
34. P. Jensen, H.B. Craig, and O.W. Robison, *J. Anim. Sci.*, 1967, **26**, 281.
35. A.J. Webb, O.I. Southwood, and S.P. Simpson, Ref.23, p.469.
36. P. Sellier, in Proceedings of a Meeting 'Pig Carcass and Meat Quality', Università di Bologna, Reggio Emilia, 1988, p.145.

37. O. Matassion, Ref.36, p.165.
38. B.W. Moss, Ref.23, p.51.
39. J. Ludvigsen, *Acta Endocrinol.*, 1957, **26**, 406.
40. J. Ludvigsen, in 'Recent Points of View on the Condition and Meat Quality of Pigs for Slaughter', ed. W. Sybesma and P.G. Van der Waal, IVO, Zeist, Netherlands, 1968, p.113.
41. K. Fischer, *Fleischwirtschaft,* 1974, **54**, 1212.
42. M.D. Judge, J.C. Forrest, J.D. Sink, and E.J. Briskey, *J. Anim. Sci.*, 1968, **27**, 1247.
43. R.A. Baldwin and D.B. Stephens, *Physiol. Behav.*, 1973, **10**, 267.
44. D.P. Cuthbertson, in 'Mammalian Protein Metabolism', ed. H.N. Munro and J.B. Allison, Academic Press New York, 1964, Chapter 19.
45. E.L. Smith, R.L. Hill, I.R. Lehman, R.J. Lefkowity, P. Handler, and A. White, 'Principles of Biochemistry. Mammaliain Biochemistry', McGraw-Hill, New York, 1983, p.554.
46. W.F. Ganong, in 'Review of Medical Physiology', 2nd Edn., Blackwell Scientific Publications, Oxford, 1965, p.302.
47. D.N. Marple and R.G. Cassens *J. Anim. Sci.,* 1972, **35**, 1139.
48. M.D. Jude, R.G. Cassens, and E.J. Briskey, *J. Food Sci.*, 1967, **32**, 565.
49. D.N. Marple and R.G. Cassens, *J. Anim. Sci.*, 1973, **35**, 205.
50. S.C. Whipp, R.L. Wood, and N.C. Lyon, *Am. J. Vet. Res.*, 1973, **31**, 2105.
51. D. Lister, J.N. Lucke, and B.N. Perry, *J. Endocrinol.*, 1972, **53**, 505.
52. D.D Fordham, G.A. Lincoln, E. Ssewannyana, and R.G. Rodway, *Anim. Prod.*, 1989, **49**, 103.
53. R.K. Tume and F.D. Shaw, *Meat Sci.*, 1992, **31**, 211.
54. M.D. Judge, E.J. Briskey, R.G. Cassens, J.C. Forrest, and R.K. Meyer, *Am. J. Physiol.*, 1968, **214**, 146.
55. R.K. Meyer, *Am. J. Physiol.*, 1968, **214**, 146.
56. P.H. Sorensen, in 'Use of Radioisotopes in Animal Biology and Medical Sciences', Academic Press, New York, 1961, p.455.
57. F.E. Romack, C.W. Turner, J.F. Laskley, and B.N. Day *J. Anim. Sci.*, 1964, **23**, 1143.
58. B.W. Moss and J.D. Robb, *J. Sci. Food. Agric.*, 1978, **29**, 689.
59. A. Fortin, S. Simpendorfer, J.T. Reid, H.J. Ayala, R. Anrique, and A.F. Kerty, *J. Anim. Sci.*, 1980, **51**, 604.
60. M. Ellis, W.C. Smith, J.B.K. Clark, and N. Innes, *Anim. Prod.*, 1983, **37**, 1.
61. A.J. Kempster, *Pig Vet Soc. Proc.*, 1988, **21**, 140.
62. P.A. Barton-Gade, *Livest. Prod. Sci.*, 1987, **16**, 187.
63. B.W. Moss, *Appl. Anim. Ethol.*, 1978, **29**, 689.
64. B.W. Moss and D. Trimble, *Rec. Agric. Res.*, 1988, **36**, 95.
65. B.W. Moss and D. Trimble, *Rec. Agric. Res.*, 1988, **36**, 101.
66. P.D. Wariss and S.N. Brown, *J. Sci. Food Agric.*, 1985, **36**, 87.
67. G.W. Snedecor, 'Statistical Methods', Iowa State University Press, 1965, p.283.
68. H.S. Liddell, *Sci. Am.*, 1954, January, p. 48.
69. T. Lidz, *Psychosom. Med.*, 1949, **11**, 1.
70. G. Feuer and P.L. Broadhurst, *J. Endocrinol.*, 1962, **24**, 127.
71. G. Feuer and P.L. Broadhurst, *J. Endocrinol.*, 1962, **24**, 253.
72. G. Feuer and P.L. Broadhurst, *J. Endocrinol.*, 1962, **24**, 385.
73. D. Ianuzzo, P. Patel, V. Chen, P. O'Brien, and C. Williams, *Nature (London)*, 1977, **270**, 74.

74. D.D. Didley, E.D. Aberle, J.C. Forrest, and M.D. Judge, *J. Anim. Sci.*, 1970, **31**, 681.
75. C.C. Cooper, R.G. Cassens, and E.J. Briskey, *J. Food Sci.*, 1969, **34**, 299.
76. S. Rahelic and S. Puac, *Meat Sci.*, 1980, **5**, 439.
77. A.R. George, J.R. Bendall, and R.C.D. Jones, *Meat Sci.*, 1980, **4**, 51.
78. P.V. Tarrant, in 'The Problem of Dark Cutting in Beef', Current Topics in Veterinary Medicine and Animal Science, Vol. 10, ed. D.E. Hood and P.V. Tarrant, Martinus Nijhoff, The Hague, 1981, p. 3.
79. P.D. Warriss, S.C. Kestin, S.N. Brown, and L.J. Wilkins, *Meat Sci.*, **10**, 53.
80. A.B. Mohan Raj, B.W. Moss, W.J. McCaughey, W. McLauchlan, D.J. Kilpatrick, and S.J. McCaughey, *Appl. Anim. Behav. Sci.*, 1991, **31**, 157.
81. A.B. Mohan Raj, B.W. Moss, D.A. Rice, D.J. Kilpatrick, W.J. McCaughey, and W. McLauchlan, *Meat Sci.*, 1992, **32**, in press.
82. A. Lacourt and P.V. Tarrant, *Meat Sci.*, 1985, **15**, 85.
83. A. Lacourt and P.V. Tarrant, Ref.78, p. 417.
84. A. Talmant, G. Monin, M. Briand, M. Dadet, and Y. Briand *Meat Sci.*, 1986, **18**, 23.
85. H.W. Ockerman, D. Jaworek, B. Van Stavern, N. Parrett, and C.J. Pierson, *J. Anim. Sci.*, 1984, **59**, 981.
86. A.B. Mohan Raj, B.W. Moss, W.J. McCaughey, W. McLauchlan, S.J. McCaughey, and S. Kennedy, *J. Sci. Food Agric.*, 1991, **54**, 111.
87. A.B. Mohan Raj, Ph.D. Thesis, The Queen's University of Belfast, 1989.

Chemistry of Raw and Cooked Products

Structural Aspects of Raw Meat

N.F.S. Gault

DEPARTMENT OF AGRICULTURE FOR NORTHERN IRELAND AND THE
QUEEN'S UNIVERSITY OF BELFAST, NEWFORGE LANE, BELFAST BT9 5PX, UK

1 Introduction

There are clear differences in the visual appearance and eventual eating quality of meat from domesticated farm animals, poultry, and fish. To a large extent, these differences reflect the physiological diversity of muscle structure which has evolved to help maintain body posture and control body movement in the particular environment in which the animal lives.[1] Despite these differences, the striated musculature of vertebrates shares a common structure in which the contractile cells or muscle fibres, which make up the bulk of the muscle mass, are symmetrically aligned within a complex connective tissue network. It is through this connective tissue network that the contractile forces generated by the muscle fibres are subsequently transmitted to the skeleton to produce movement in the live animal.[2]

In mammals and birds, individual muscles are surrounded by a sheath of connective tissue known as the epimysium which extends into the tendons. From the epimysium, the perimysium, a network of finer connective tissue containing the larger blood vessels and nerves, penetrates throughout each muscle, separating groups of muscle fibres into muscle fibre bundles. Continuous with the perimysium is the finer network of the endomysium (Figure 1). This connective tissue sheath surrounds each individual muscle fibre and contains the blood capillaries and nerve connections vital for muscle function in the live animal. In meat animals, muscle fibres can vary in length from a few mm up to 30 cm or more.[3] In contrast, muscle fibre diameters are comparatively minute, varying between 10 and 100 μm.[4] These muscle fibres may run parallel to, or at an angle to, the length of the muscle depending on its size and anatomical location.

In contrast to meat animals, in which the musculature is organized into several anatomically distinct units, fish muscle is different in that the major proportion of the muscle mass along the length of the fish's body is segmented into several concentric blocks of muscle known as myotomes.[5] These are separated from each other by relatively thick sheets of connective tissue known as myocommata. The individual muscle fibres which

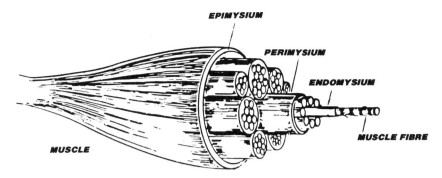

Figure 1 *Diagram of gross muscle structure showing the arrangement of the connective tissue in relation to muscle fibres and muscle fibre bundles*

compose the myotomes are therefore much shorter and less variable in length than the muscle fibres of meat animals although their diameters are similar. The muscle fibres of fish are separated from each other by a fine sheath of endomysial connective tissue and organized into muscle fibre bundles which are surrounded by thicker sheaths of connective tissue functionally equivalent to the perimysium found in the striated musculature of mammals and birds.[5,6]

Having briefly outlined the essential morphological features of vertebrate striated muscles, it is the intention of this review to describe the structural organization and biochemical behaviour of the living muscle which primarily determines its behaviour as raw meat. The extent to which raw meat structure is modified by post mortem biochemical changes that influence the organization and stability of its component parts will also be discussed.

2 The Muscle Cell

Underneath the endomysium, each muscle cell is surrounded by the sarcolemma, a triple-layered cell membrane *ca.* 75 nm thick and composed mainly of phospholipids and glycoproteins. The major function of the sarcolemma is to transmit the nerve impulses for muscle contraction through its associated system of transverse tubules (T-tubules) which are in intimate contact with the individual contractile units of the muscle cell.[7]

Occupying *ca.* 80% of the muscle cell volume is the contractile apparatus, made up of numerous myofibrils which run the length of the cell.[8] Each myofibril is *ca.* 1.0 µm in diameter. It is the ordered alignment of the myofibrils within the muscle cell which gives it its characteristic striated appearance.[8] The myofibrils are surrounded by the sarcoplasm which contains the muscle cell nuclei, mitochondria, lysosomes, glycogen granules, numerous enzymes, and other soluble substances essential for the normal metabolic activities of the muscle cell.[9] In more intimate contact with the myofibrils is the cytoskeletal framework which is responsible for holding the contractile apparatus in register,[10] and the sarcoplasmic reticulum, closely associated with the T-tubules and of functional importance in controlling muscle contraction and relaxation through its ability actively to release and reabsorb calcium ions as a result of muscle stimulation.[11]

The Contractile Apparatus

Under phase contrast microscopy, the contractile apparatus of the muscle fibre appears banded or striated in longitudinal section. This is due to the manner in which the structural components of each myofibril are held in lateral and longitudinal register. In relaxed muscle, the dark bands represent areas of dense protein structure which are anisotropic under polarized light and as such have been termed the A-bands. The I-bands derive their name from the isotropic refractiveness of the lighter bands in which the constituent proteins are less dense. In the central region of each A-band is the less dense H-zone which itself is bisected by the dark M-line. The highly refractive Z-line is also clearly visible as a sharp dark band within the middle of each I-band.[12,13]

Electron microscopy has shown that this banded appearance is due to the extent of overlap of two sets of filaments. Thin F-actin filaments, 8 nm in diameter and 1.0 µm in length, extend from either side of the Z-line and make up the I-band. Thick myosin filaments, 10–12 nm in diameter and 1.6 µm in length, define the extent of each A-band. These thick filaments lie in the centre of the individual contractile units of the myofibril. The contractile units themselves are known as sarcomeres, each sarcomere lying between two Z-lines such that the I-bands are shared by adjacent sarcomeres (Figure 2).

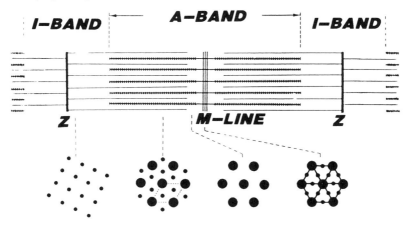

Figure 2 *Diagram of the ultrastructural organization of the muscle sarcomere bounded on each side by a Z-line. Underneath the sarcomere, from left to right, are cross-sectional representations of the organization of (i) the thin filaments near the Z-line, (ii) the overlap regions of the thick and thin filaments, (iii) the thick filaments, and (iv) the M-line in the centre of the sarcomere*

The Thick Filaments. The thick filaments are formed by the spontaneous association of up to 400 individual molecules of the filamentous protein myosin.[14] The myosin molecule is characterized by having two globular head sections attached to a long tail.[15,16] At the quaternary level, the molecule is composed of two large sub-units (the heavy chains) each having a molecular weight of *ca.* 200 000, and four smaller sub-units (the lighter chains) of variable molecular weight in the region of 20 000.[17,18] Both heavy chains are arranged as α-helices along *ca.* 50% of their length and coiled around each other in a rope-like manner to form the tail portion of the myosin molecule. The remainder of each heavy chain folds separately into the globular head regions of the myosin molecule. Each of these head regions binds two of the light chains. Consequently, the myosin molecule is *ca.* 140 μm long with each head region *ca.* 9 nm in diameter at its widest point.

Detailed electron microscopy and low-angle X-ray diffraction studies in the late 1960s gave some insight into the manner in which the myosin molecules assemble to form thick filaments. On each side of the central clear zone of the A-band, all the myosin molecules are symmetrically orientated with their heads pointing away from the centre of the filament. This bipolar structural arrangement is responsible for the smooth central zone (150–200 nm) of the thick filaments which consists of the tail segments of overlapping myosin molecules pointing in opposite directions. The ordered packing arrangement of the myosin molecules also accounts for the manner in which three or four pairs of myosin heads protrude from around the circumference of the thick filaments at 14.3 nm intervals tracing a helical pathway with a repeat distance of 42.9 nm per helical twist.[19]

At least seven other proteins are associated with the thick filaments. Immunofluorescence studies have shown the presence of M-protein, myomesin, and the enzyme creatine kinase in close association with the M-line.[20-22] However, the manner in which these proteins may contribute to the structural integrity of the thick filaments within the M-line region is not known. C-Protein, which appears as several regularly spaced stripes along the length of the thick filaments on either side of the M-line, is closely associated with the tail segments of the myosin molecules and may have a structural role in controlling the formation of the thick filaments.[23] No structural role has yet been established for the minor F- and H-proteins of the A-band,[24,25] while the so called I-protein, localized near each end of the A-band, is known to inhibit myosin ATP-ase activity.[26]

The Thin Filaments. The thin filaments are formed when the spherical monomers of G-actin, 5.5 nm in diameter and with a molecular weight of 42 000, condense to form the double-helical filament known as F-actin. The two actin strands which make up each thin filament twist across each other every 36.5 nm, there being about 13 G-actin monomers within each of these repeating sections of the filament.[27,28] The other major proteins of the thin filaments are the regulatory proteins tropomyosin and the troponin complex.[29,30] Tropomyosin is a thin filamentous protein, 41 nm in length, and composed of two α-helical polypeptide chains coiled around each other to give the molecule a rope-like appearance. It forms tightly bound end-to-end aggregates which run the entire length of the F-actin filament along each of its helical grooves.[31,32] The globular troponin complex, composed of the three sub-units troponin T, I, and C, is found attached to each tropomyosin molecule at 38.5 nm intervals along both sides of the F-actin double helix.[33,34] The minor proteins are β-actinin,[35] which caps the free ends of the thin filaments, γ-actinin,[36] known to inhibit actin polymerization, and the more recently discovered paratropomyosin,[37] located at the A–I band junction.

Table 1 summarizes the relative amounts, molecular weights, and functions of the myofibrillar proteins associated with the contractile apparatus.

Muscle Contraction. The sliding filament theory regards muscle contraction as being due solely to the interdigitation of the thick and thin filaments within each sarcomere.[38,39] When a muscle shortens during contraction, the width of the I-bands decreases as the thin filaments are drawn into the spaces between the thick filaments in the centre of each sarcomere. Throughout contraction, the A-bands maintain their constant length. Sarcomere length thus depends on the extent of overlap between the thick and thin filaments which themselves remain at constant length. Consequently, under physiological conditions in the live animal, a completely extended sarcomere in which there is no overlap of the thick and thin filaments will have a length of 3.6 μm. The sarcomere length at maximum shortening, corresponding to the complete width of the A-band, will therefore be 1.6 μm.[40]

In cross-section, the thick filaments of the A-band are arranged in a regular hexagonal lattice with a centre to centre spacing of *ca.* 40 nm in the resting state.[41] The thin filaments are located symmetrically between the thick filaments at the trigonal points of this lattice structure where both sets of filaments overlap in the outer regions of the A-band. This double hexagonal lattice, in which each thick filament is surrounded by six thin filaments, expands laterally during muscle contraction as the degree of overlap of thick and thin filaments increases with sarcomere shortening.

The Cytoskeletal Framework

There has been considerable interest in recent years in the cytoskeletal proteins which help maintain the structural framework within which the contractile proteins of the muscle cell function.[42,43] Although some of the proteins which constitute the Z-line are associated with actin, they will be considered in this review as cytoskeletal proteins (Table 2).

Connectin (or titin), the third most abundant protein in muscle after

Table 1 Relative amounts, molecular weights, and function of myofibrillar proteins associated with the contractile apparatus

Location	Protein	Amount (%)	Molecular weight (kDa)	Major function
A-band	Myosin	43	520	Muscle contraction
	c-Protein	2	140	Binds myosin filaments
	F-, H-, I-Proteins	<1	121/74/50	Binds myosin filaments
M-line	M-Protein	2	165	Binds myosin filaments
	Myomesin	<1	185	Binds myosin filaments
	Creatine kinase	<1	42 × 2	ATP synthesis
I-band	Actin	22	42	Muscle contraction
	Tropomyosin	5	33 × 2	Regulates muscle contraction
	Troponins T, I, C	5	31/21/18	Regulates muscle contraction
	β-, γ-Actinins	<1	37:34/35	Regulates actin filaments

Compiled from:

T. Obinata, K. Maruyama, H. Sugita, K. Kohama, and S. Ebashi, *Muscle Nerve*, 1981, **4**, 456.
R.M. Robson and T.W. Huiatt, Proceedings of the 36th Annual Reciprocal Meat Conference, Fargo, North Dakota, 1984, p. 116.
K. Maruyama, in 'Developments in Meat Science', ed. R. Lawrie, Elsevier Applied Science, London, 1985, Vol. 3, p. 22.

Table 2 Relative amounts, molecular weights, and function of myofibrillar proteins associated with the cytoskeletal framework

Location	Protein	Amount (%)	Molecular weight (kDa)	Major function
GAP filaments	Connectin (Titin)	10	2800 (2100)	Links myosin filaments to Z-line
N_2-Line	Nebulin	5	500	Unknown
By sarcolemma	Vinculin	<1	130	Links myofibrils to sarcolemma
Z-Line	α-Actinin	2	100×2	Links actin filaments to Z-line
	Eu-actinin, filamin	<1	$42/240 \times 2$	Links actin filaments to Z-line
	Desmin, vimentin	<1	55/57	Peripheral structure of Z-line
	Synemin, Z-protein, Z-nin	<1	230/50/400	Lattice structure of Z-line

Compiled from:

T. Obinata, K. Maruyama, H. Sugita, K. Kohama, and S. Ebashi, *Muscle Nerve*, 1981, **4**, 456.
R.M. Robson and T.W. Huiatt, Proceedings of the 36th Annual Reciprocal Meat Conference, Fargo, North Dakota, 1984, p. 116.
K. Maruyama, in 'Developments in Meat Science', ed. R. Lawrie, Elsevier Applied Science, London, 1985, Vol. 3, p. 22.

myosin and actin, is the dominant cytoskeletal protein. It is characterized by its very high molecular weight and elastic properties.[44,45] In the native state, connectin exists in the α-form. However, only the properties of β-connectin, produced by the proteolysis of α-connectin, have been studied in detail.[46,47] Recent evidence suggests that this thin filamentous protein can be up to 1.1 μm long in the relaxed state and is capable of being stretched to 3.0 μm.[48] Immunofluorescence studies have shown that connectin is found mainly at the A–I band junction.[49] It then extrudes as thin filaments on either side of the centre of the thick myosin filaments through the thin actin filaments to the Z-lines (Figure 3). There now seems little doubt that connectin is the structural component of the so-called 'gap filaments' observed when muscle fibres are stretched beyond the natural overlap length of the thick and thin filaments.[50,51]

Nebulin, which represents as much as 5% of the total myofibrillar proteins, has been identified with the N-lines.[52] These thin structures run transversely across the myofibril parallel to the Z-lines at three different locations: near the Z-line itself, in the middle of the I-band, and at the A–I band junction.[53] However, nebulin has only been positively identified as a structural component of the N_2-lines in the middle of the I-band.[54] These N_2-lines control the geometrical organization of the thin filaments, changing them from a hexagonal lattice at the A–I band junction to a square lattice at the Z-line.[52] As yet, there is little firm evidence to suggest that the transverse N-lines are intimately attached to the connectin filaments to form a three-dimensional cytoskeletal framework within each sarcomere.

α-Actinin is the major protein found within the Z-line and is thought to have a structural role in attaching actin filaments to the Z-line.[55-57] Also found within the Z-line are the minor proteins eu-actinin and filamin,[58,59] while the lattice structure within the interior of the Z-lines is composed of the so-called Z-protein,[60] Z-nin,[61] and the more recently discovered protein zeugmatin.[62] Encircling the Z-lines in the form of a peripheral network are the intermediate filaments, made of the proteins desmin and vimentin.[57,63,64] These intermediate filaments are thought to have a major structural role in holding adjacent myofibrils in lateral register across the

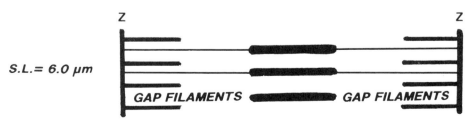

Figure 3 *Diagram of the location of the gap filaments (connectin filaments) in a stretched (6.0 μm) sarcomere, running from near the centre of each side of the thick filaments, between the thin filaments to the Z-lines*

width of a muscle fibre (Figure 4). Also located at the periphery of the Z-lines are minute amounts of the protein synemin.[65]

Another recently discovered cytoskeletal protein, vinculin, has been shown to form lattice structures known as costameres. These run transversely across sarcomeres on either side of the Z-line but only become firmly attached to the myofibrils at the sarcolemma.[66,67]

The Sarcoplasmic Reticulum

The sarcoplasmic reticulum of most vertebrate skeletal muscle cells can be considered as a series of transverse and longitudinal vesicles which form a complex membranous network around each myofibril (Figure 5). Located parallel to the Z-lines and covering much of the I-band are the terminal cisternae, the transverse elements of the sarcoplasmic reticulum from which the longitudinal vesicles emanate on either side. These longitudinal vesicles normally meet at the centre of each sarcomere where they fuse together to form the perforated region of the sarcoplasmic reticulum known as the fenestrated collar. The terminal cisternae are usually found in close association with the T-tubules which run across the muscle cell parallel to

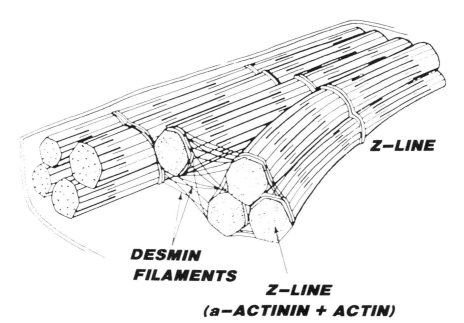

Figure 4 *Diagram showing the location of desmin filaments at the Z-line of adjacent myofibrils*
(Reproduced by permission from R.M. Robson, M. Yamaguchi, T.W. Huiatt, F.L. Richardson, J.M. O'Shea, M.K. Hartzer, W.E. Rathbun, P.J. Shreiner, L.E. Kasang, M.H. Stromer, Y.-Y. S. Pang, R.R. Evans, and J.F. Ridpath, Proceedings of the 34th Annual Reciprocal Meat Conference, Corvallis, Oregon, 1982, p.5.)

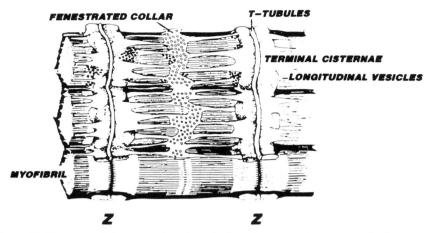

Figure 5 *Diagram of the sarcoplasmic reticulum incorporating the terminal cisternae, longitudinal vesicles, and fenestrated collar. The transverse (T) tubules are seen on top of the terminal cisternae in the region of each Z-line* (Reproduced by permission from L.D. Peachey, *J. Cell Biol.*, 1965, **25**, 209.)

the Z-lines of individual myofibrils. Although the junctions which link the T-tubules to the sarcoplasmic reticulum are arranged differently in other types of muscle fibre, they all share the common function of detecting the nerve impulse transmitted through the T-tubules. This stimulates the entire sarcoplasmic reticulum to release the Ca^{2+} necessary for muscle contraction to proceed. When the nerve impulse is released, Ca^{2+} is actively reabsorbed into the entire sarcoplasmic reticulum network allowing the muscle fibres to relax.[68]

3 Connective Tissue

The structural stability which the connective tissue network imparts to the muscle fibres is primarily determined by the properties of its component collagen fibres and to a lesser extent by those of elastin. Both types of fibre are embedded in an amorphous ground substance rich in proteoglycans and glycoproteins.[69] The ground substance also contains numerous cells including the fibroblasts from which the fibrous proteins collagen and elastin are synthesized. Since collagen fibres are the major constituents of the connective tissue of vertebrate striated muscle, it is hardly surprising that they have been the subject of intensive and fruitful research over many years.

Collagen Structure

In recent years, it has been established that four genetically distinct types of collagen are found in the connective tissue of skeletal muscle (Table 3).

Table 3 Genetic types of collagen found in skeletal muscle

Type	Molecular length (nm)	Molecular composition	Aggregation characteristics	Major location
I	300	$[\alpha 1(I)]_2 \alpha 2(I)$	Striated fibres	Epimysium
III	300	$[\alpha 1(III)]_3$	Striated fibres	Perimysium
IV	420	$[\alpha 1(IV)]_2 \alpha 2(IV)$ and $\alpha 3(IV)$	Non-fibrous	Endomysium (basement membrane)
V	300	$[\alpha 1(V)]_2 \alpha 2(V)$ plus $[\alpha 1(V) \alpha 2(V) \alpha 3(V)]$	Striated fibres	Epimysium

From: ref. 72.

Types I, III, and V collagen, which are composed of different α-chains, share a common molecular length of 300 nm and aggregate into the form of striated fibres. In contrast, Type IV collagen, which has a molecular length of 420 nm and does not aggregate in the fibrous form, is a major constituent of the amorphous basement membrane of the endomysium. Consequently, it is predominantly the fibrous forms of collagen and the chemical and physical interactions of its component molecules which are of interest to the meat scientist concerned with the contribution made by collagen to raw meat structure.

The different fibre-forming types of collagen share a similar basic structure.[70,71] The molecules consist of three α-helical protein chains wound together to form the characteristic triple helix of the tropocollagen molecule. This molecular structure is achieved by the unique amino acid content and sequence of the component α-chains. Here, every third amino acid residue is glycine in a repeating Gly-X-Y sequence where X and Y are frequently proline and hydroxyproline. The regular appearance of these imino acids restricts the rotation of each polypeptide chain causing it to twist into a left-handed helix with three amino acid residues per turn. The three component α-chains are then wound together to form the characteristic right-handed super helix of the tropocollagen molecule. Stability is conferred on this triple helix by internal H-bonds which arise from the regular sequence of glycine residue along each α-chain.

The asymmetric charge distribution along the outer surface of the tropocollagen molecules contributes to the manner in which collagen fibres are formed by the aggregation of these molecules in a quarter-stagger end-overlap manner (see Figure 4 on p. 111). Hydrophobic residues present in the quarter-stagger regions of adjacent tropocollagen molecules, together with similar groups in the amino- and carboxy-terminal end regions of the tropocollagen molecules, help maintain this structural organization. In this form, however, collagen fibres do not possess any mechanical strength. This is only achieved by the polymerization of the molecules within the fibre through the formation of covalent intermolecular cross-links.[70-72]

Intermolecular cross-links are formed primarily by the oxidative deamination of specific lysine or hydroxylysine residues in the short non-helical (teleopeptide) end regions of each tropocollagen molecule by the enzyme lysyloxidase.[73,74] Aldimine cross-links, which are heat- and acid-labile, are formed when the ε-NH_2 group of a lysine residue is converted into an aldehyde which then condenses with a hydroxylysine residue on the triple helix portion of an adjacent tropocollagen collagen. In contrast, keto-imine cross-links, which are heat- and acid-stable, are formed by the condensation of a teleopeptide hydroxylysine aldehyde with a hydroxylysine residue in the same location as above. However, in this case, the initial aldimine bond which is formed spontaneously undergoes an Amadori rearrangement to form the stable keto-imine cross-link. These chemically reducible bifunctional cross-links contribute to the longitudinal stability and inextensibility of collagen fibrils by giving them considerable tensile strength.

As collagen matures with age, the proportion of these aldimine and keto-imine cross-links gradually decreases.[75] This coincides with a proportional increase in the tensile strength of the collagenous tissue. Consequently, it has been suggested that both the lysine-derived aldimine cross-links and the hydroxylysine-derived keto-imine cross-links found in immature collagen interact to form a series of tranverse trifunctional cross-links across the collagen fibre.[76] These mature cross-links are non-reducible and stable. Detailed chemical analysis has revealed the presence of a high molecular weight cross-linked peptide in mature Type I collagen which contains a repeating sequence of similar peptides. This suggests that the mature cross-links occur between tropocollagen molecules which lie in register. Within the quarter-stagger model of collagen fibril formation, such transverse linkages could be achieved between the end-terminal regions of those tropocollagen molecules in closest proximity to each other, thereby giving the fibre three-dimensional stability.[77] More recently an amino acid of molecular weight 460, and known as compound M, has been isolated from such cross-linked peptides which is consistent with the hypothesis that a trifunctional cross-link exists in mature collagen.[78] Amongst other proposals which have been made on the identity of the non-reducible mature collagen cross-link, histidinohydroxylysinonorleucine,[79,80] 3-hydroxylysylpyridinoline,[81] and lysylpyridinoline have also received a lot of attention.[82]

Elastin Structure

Elastin only forms a minor component of the connective tissue of most mammalian muscles, being mainly associated with blood vessels and certain ligaments where elasticity is required.[83] Elastin fibres are composed of an amorphous protein fraction, elastin itself, surrounded by a microfibrillar component rich in glycoproteins.[84]

Elastin has a molecular weight of 70 000 and apart from the absence of hydroxylysine and the virtual absence of histidine has a similar amino acid composition to collagen.[84] However, its molecular structure is completely different in that it forms a random coil. Polymerization of the monomer occurs through the lysyl oxidase mediated oxidation of specific lysine residues to form the reactive aldehyde allysine. This condenses with either another lysine residue to form dehydrolysinonorleucine, or with another allysine to form allysine aldol. In addition to these bifunctional cross-links, the trifunctional cross-link, dehydromerodesmosine, can also occur.[84] The most familiar cross-linking compounds, however, are the tetra-functional desmosine and isodesmosine.[84,85] These are formed either by the repeated addition of single lysine or allysine residues, or from dehydrolysinonorleucine and allysine aldol. Such cross-linking, which occurs spontaneously, gives elastin its characteristic attributes of elasticity and chemical insolubility.

4 Mechanical Properties of Raw Meat

In the living animal, muscles can freely contract and relax under normal physiological conditions. However, when the animal is slaughtered the anaerobic conditions which then exist in the musculature bring about the series of biochemical and biophysical changes that result in a lowering of muscle pH and the onset of *rigor mortis*.[86] It is at this stage, when muscle has entered *rigor* and reached its ultimate pH (pH_u), that it can now be considered as meat. However, meat gradually loses its rigid structure and becomes more flexible with time, a process formerly described as the resolution of *rigor* but now more frequently referred to as ageing or conditioning.

The extent to which muscle contracts prior to *rigor* onset, intermuscular differences in connective tissue content and maturity, and the changes brought about by conditioning all have a marked effect on the physical characteristics of raw meat as described in some detail below.

Shortening and *Rigor Mortis*

In *pre-rigor* muscle, the thick and thin filaments of the sarcomere are capable of sliding past each other while the concentration of ATP in the fibres remains high. Consequently, *pre-rigor* muscle can be stretched by up to 20% of its resting length by loads as small as 5.0 kPa.[87] Once muscle enters *rigor*, however, it effectively becomes inextensible, and stretches by less than 1% under similar loads. This is due to the irreversible formation of actomyosin brought about by the gradual depletion of ATP. Even by increasing the load to 50 kPa, muscle extensibility barely reaches 2% of its rest length, but it will recover its original length when the load is removed. However, attempts to stretch *rigor* muscle any further will result in fibre fracture and loss of recovery.[88]

It is well established that temperature has a major effect on the extent to which muscles can shorten prior to the onset of *rigor mortis*.[89] However, its effect on the *rigor* process in mammalian muscle is anomalous, in that ATP turnover rate is only linear between 38 °C and *ca.* 25 °C, below which its temperature coefficient (Q_{10}) decreases rapidly. In the red muscles of cattle, sheep, pigs, and rabbits, Q_{10} actually becomes negative below 5 °C such that the rate of ATP turnover is as high in these muscles at 2 °C as it is at 15 °C.[88] This is caused by an increase in the rate of actomyosin ATP-ase activity below 25 °C brought about by an increased susceptibility of the sarcoplasmic reticulum to leak Ca^{2+} into the intra-myofibrillar space. However, since the Ca^{2+} concentration only reaches a sufficiently high level to induce muscle contraction below 11 °C,[88,90] this has been used to help explain the phenomenon of cold-shortening first demonstrated with *pre-rigor* beef *sternomandibularis* (StM) muscle.[91] Here, minimum shortening of the excised muscle occurred at 14–19 °C before increasing to a maximum of 50% as the temperature was lowered to 0 °C. In contrast, with increasing temperature, shortening increased only gradually to a maximum of 30% at 40 °C.

Unlike the situation in warm muscle, where shortening only begins as the level of ATP starts to fall quickly during the fast phase of *rigor*, cold shortening is a reversible phenomenon.[88] Cold-shortened muscle, in fact, is capable of relaxing again if the temperature is raised above 11 °C while the muscle pH is still relatively high, usually greater than pH 6.5.[88] Indeed, cold-shortening can be considered as a biphasic process. Provided that the level of ATP and muscle pH are high, cold-shortening will occur very rapidly in excised muscle and maintain this shortened condition until the onset of *rigor*. If the muscle is loaded after this initial cold-shortening phase, however, it will relax for a time, but it then begins to shorten gradually over the next 20 h or so as the second phase of shortening leading to the onset of *rigor* sets in.[88] By increasing the load on such excised muscles, the initial rapid phase of cold-contraction is prevented, although the muscle will eventually shorten to the same extent as unloaded muscle by the time *rigor* sets in. In general, red muscles tend to cold-shorten faster and more extensively than white muscles, while the second phase of shortening does more work than the initial phase.[88]

By applying this rationale to carcass meat, the majority of muscles would not show this initial rapid phase of cold-shortening because of the restraint imposed by their skeletal attachment.[88] Consequently, the second phase of cold-shortening is perhaps of greater significance in determining to what extent carcass meat cold-shortens. Nevertheless, even when muscles are mechanically prevented from cold-shortening, severe crimping of individual myofibres within such muscles can frequently be seen.[92,93] This is indicative of the localized cold-induced contraction which can occur in certain individual muscle fibres but not in others. Thus, the extent of muscle shortening *per se* is not a reliable index of sarcomere length.

Mechanical Characteristics in Rigor: *Mammals.* By tensile testing strips of muscle until they completely rupture it is possible to assess the contribution which the muscle fibres and the connective tissue elements make to the structural stability of raw muscle in *rigor*.[94–96]

It is clear from Table 4 that the connective tissue component of beef muscles is much stronger and more flexible than the muscle fibres. Also, the differences in peak force values between these muscles reflect the differences in their connective tissue content. At one extreme, the StM has 7.8% collagen on a dry matter basis,[83] while the *psoas major* (PM) only has 2.4%.[97] The similar FY:IY ratios for the StM, *triceps brachii* (TB), and *infraspinatus* (IS) muscles probably reflect their contractile states, in which their sarcomere lengths are comparatively short. The connective

Table 4 Tensile behaviour of raw post-rigor beef muscles* subjected to normal chilling regimes on intact carcasses

Muscle	Initial force (IF) (kPa)	Peak force (PF) (kPa)	Initial yield (IY)	Final yield (FY)	FY:IY ratio
StM	143	485	120%	486%	4.1
TB	78	298	100%	416%	4.2
IS	68	273	106%	444%	4.2
PM	58	173	136%	347%	2.5

*Muscles removed from chilled carcasses 2 days post mortem and stored a further 2 days at 4 °C.

Initial force is the first major inflexion on the force–distance or stress–strain curve.
Peak force is the maximum force recorded on the force–distance curve.
Initial yield is the distance from the first registering of force to the IF point.
Final yield is the distance from the first registering of force to the point where the sample had completely broken.
FY:IY ratio is the ratio of the final yield distance to initial yield distance.

StM = *sternomandibularis*   TB = *triceps brachii*
IS = *infraspinatus*   PM = *psoas major*

Compiled from:

M.V. Rao and N.F.S. Gault, *J. Texture Stud.*, 1990, **21**, 455.

tissue network around the muscle fibres of these muscles would be relatively loose, giving them a greater potential to stretch more before taking up tension.[98] In contrast, the PM muscle sets in *rigor* in a very stretched state,[99] giving the PM less scope for further elongation before the tensile forces are taken up by the stretched connective tissue network.

The fact that muscles can enter *rigor* at different degrees of stretching or shortening is very much influenced by carcass posture after slaughter.[99] For example, in beef carcasses which are hung in the usual vertical position from the achilles tendon, the muscles of the outer hindquarter such as the *semimembranosus* (SM), *semitendinosus* (ST), and *biceps femoris* (BF), and also the *longissimus dorsi* (LD), are relatively free to shorten during *rigor* and generally have sarcomere lengths of less than 2.0 μm. In contrast, many of the forequarter muscles become passively stretched in the region of the shoulder and foreamen and set in *rigor* with sarcomere lengths of between 2.3 and 3.1 μm. The PM, which lies convex to the curvature of the spine of the carcass, is completely stretched, having a sarcomere length of *ca.* 3.6 μm. If, however, the carcass is hung in a horizontal position from the obturator foreamen, the converse is true, and muscles such as the LD now go into *rigor* in a more stretched position while others such as the PM, being free to shorten, go into *rigor* in a less stretched state.

The influence which stretching or shortening can have on the tensile characteristics of an individual muscle is more clearly shown in Table 5. These results clearly indicate the increased contribution which the orientation of the connective tissue network makes to both the initial force and initial yield values of muscle which has gone into *rigor* in the stretched state. Similarly, peak force values are also greater in the stretched muscle while final yield values are inversely proportional to the degree of stretch. The FY:IY ratios further highlight the greater extensibility of the connective tissue network in shortened muscle, where its looser orientation allows it to stretch more extensively before the collagen fibres take up the tensile stress.

In a similar manner, it is interesting to note the effect of muscle shortening on the shear force characteristics of different beef muscles (Table 6). Here, shear force is directly proportional to muscle length for the deep pectoral (DP), StM, and SM muscles, while muscle length has no effect on the force required to shear the PM. This is in sharp contrast to the situation with cooked meats where shortening generally increases its resistance to shear.[100] This behaviour has been explained on the basis that the shorter muscles have proportionately less connective tissue per unit cross-sectional area than their longer counterparts and thus their overall resistance to shear will be less.

Mechanical Characteristics in Rigor: *Fish.* In contrast to the situation with mammalian and poultry meat, the onset of *rigor mortis* in certain species of fish can have a more profound effect on muscle structure, leading to the

Table 5 Tensile behaviour of raw beef deep pectoral muscle* as influenced by muscle (sarcomere) length

	Initial force (IF) (kPa)	Peak force (PF) (kPa)	Initial yield (IY)	Final yield (FY)	FY:IY ratio
Shortened (1.28 μm)	80	200	30%	162%	5.4
Control (2.24 μm)	109	339	32%	126%	3.9
Stretched (3.34 μm)	443	630	52%	84%	1.6

*Muscles stored at 1 °C for 3 days post mortem.

For definitions of IF, PF, IY, FY and FY:IY ratio see Table 4.

Compiled from: ref. 95.

Table 6 *Shear force values of raw* post-rigor *beef muscles as influenced by muscle length*

Treatment (*Length relative to* pre-rigor *rest length*)	DP^a (kPa)	StM^b (kPa)	SM^b (kPa)	PM^b (kPa)
Shortened (0.5)	390	400	200	150
Control (1.0)	540	650	410	140
Stretched (1.5)	820	1100	500	150

DP = Deep pectoral muscle   StM = *sternomandibularis*
SM = *semitendinosus*   PM = *psoas major*

Compiled from:

[a] Ref. 95.
[b] D.N. Rhodes and E. Dransfield, *J. Sci. Food Agric.*, 1974, **25**, 1163.

phenomenon known as 'gaping'. This occurs when the forces exerted by muscle contraction during the onset of *rigor* are so strong that the connective tissue ruptures at the myocommatal junctions between the individual myotomes.[101] This has been shown to occur with cod, haddock, and catfish,[102,103] the extent of gaping being proportional to increasing temperature over the range 0–31 °C. With cod muscle, this not only induces a six-fold increase in *rigor* tension (up to 200 kPa), but also results in a five-fold decrease in the breaking stress (down to 50 kPa) of the myocommata/myotome junctions over the same temperature range.[101]

However, unlike the situation with cold water species of fish, placing tropical species such as tilapia in ice immediately after death induces a 'cold shock', in which the flesh stiffens rapidly in a manner similar to the cold-shortening stiffness described earlier for mammalian muscle.[104] The biochemical changes associated with this 'cold shock' concept have been discussed in detail elsewhere.[105,106]

Conditioning

After muscle has gone through *rigor*, it becomes extensible again, even under loads of less than 5.0 kPa.[88] However, unlike the situation with *pre-rigor* muscle, the extensibility of meat *post-rigor* is irreversible.[89,107] Such changes are primarily due to the action of proteolytic enzymes on the myofibrils.

Tensile tests on raw beef muscle have established that conditioning brings about a significant lowering of the initial yield point.[108] For example, with raw beef StM muscle it has been shown to decrease from *ca.* 120–140 kPa at *rigor* to 15–23 kPa after seven days conditioning at 15 °C.[108] Conditioning has a similar effect on the initial yield point of cold-shortened muscle.[109] In general, histological examination has established that the initial yield point coincides with a fracturing of the I-band regions within the myofibrils.[96,109] This promotes a series of random breaks across the muscle fibre, leading to a decrease in initial yield points and the irreversible extensibility of conditioned meat under small loads.

A novel compression testing system has been developed in recent years to study the rheological characteristics of raw meat.[110,111] However, the most interesting contribution to our understanding of the changes which initiate myofibrillar weakening during conditioning has perhaps come from the cyclical testing of strips of muscle under small tensile loads (usually 1.0–11.0 kPa) throughout the *pre-rigor, rigor,* and conditioning periods.[112] The rate and extent to which these changes occur in meat from different species is illustrated in Figure 6. After the onset of *rigor mortis*, which was most rapid for chicken and slowest with beef, the rate and extent of the changes induced by conditioning varied considerably between species, chicken meat changing the most rapidly. Muscle extensibilities brought about by conditioning reached plateaus at between 20% and 40% stretch.

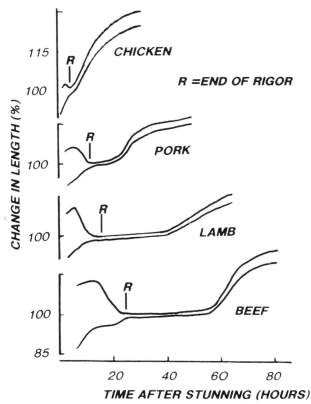

Figure 6 *Changes in length and inextensibility of strips of* pectoralis *muscle of different species relative to length at* rigor *during cycling between 1.0 and 11.0 kPa at 15 °C*
(Reproduced by permission from D.J. Etherington, M.A.J. Taylor, and E. Dransfield, *Meat Sci.*, 1987, **20**, 1.)

It was assumed that further elongation of the meat was limited by the inextensibility of the connective tissue network at these relatively low stress values. Histological examination revealed that conditioning induced some fracturing of muscle fibres in the I-band region. However, after mechanical testing, transverse fracturing across sarcomeres in the I-band region was shown to occur more frequently at a distance of every five sarcomeres or so. Initiation of this facture behaviour has been calculated to occur in conditioned beef muscle at a stress of only 0.7 kPa, indicating the extensive weakening brought about by proteolytic enzyme action on the myofibrillar and cytosketal structural proteins during conditioning.[112]

Over the past ten years or so, considerable research effort has been devoted to the study of the enzymology of meat conditioning.[113-115] It is now generally agreed that two proteolytic systems may be involved. These are the sarcoplasmic calpains, or Ca^{2+}-dependent proteases, and the lysosomal cathepsins, of which cathepsins B, L, and D are particularly relevant to skeletal muscle (see Table 7). The major difference between these two sets of proteases is that the calpains specifically attack certain proteins of the Z-line, more particularly desmin and filamin.[56,116,117] The calpains are also known to attack nebulin,[52] and to a lesser extent connectin.[118-120] They are also thought to release α-actinin.[118,121] In contrast, the cathepsins preferentially attack myosin and actin. In addition, the calpains have an optimum pH range for activity near neutrality, while the cathepsins have pH optima more closely associated with the normal pH_u range found in many skeletal muscles.[114,115]

Under normal refrigerated storage conditions, conditioning only appears to affect the structural components of meat in the region of the Z-line, and in particular their lateral attachment,[122-124] there being no evidence to suggest that proteolysis of actin or myosin occurs under these conditions.[125-127] At temperatures of 15 °C or higher, though, there are several reports which suggest that myosin degradation does indeed occur during high-temperature conditioning.[125,128-130] On the basis of this evidence, it seems reasonable to suppose that the calpains are primarily responsible for the degradative changes which occur during the conditioning at chill temperatures, despite the apparent adverse pH conditions which exist in *post-rigor* meat of normal pH_u. Cathepsins, however, are likely to be involved in degradation of the contractile proteins during prolonged high-temperature *post-rigor* conditioning. However, the development of techniques capable of isolating calpains, cathepsins, and their natural inhibitors from skeletal muscle tissues of various species, and their mode of action *in vitro*, continues to throw further light on the complexities of this interesting area of meat biochemistry.[131-133]

As intimated earlier, conditioning does not appear to affect the overall tensile strength or extensibility of raw meat.[94] This would suggest that the physical characteristics of the connective tissue components are little affected by post mortem proteolytic activity. There is, nevertheless, some evidence to suggest that when mature collagen fibres from beef muscle are

Table 7 Endogenous proteases implicated with meat conditioning

Location	Protease	MW	pH range	Activity
Sarcoplasm	Calpain I[a]	110 000	6.5–7.5	Releases α-actinin, Z-nin
	Calpain II[a]	110 000	6.5–7.5	Degrades desmin, filamin, connectin, nebulin
				Degrades troponins T and I, tropomyosin
				Degrades C- and M-proteins
Lysomal	Cathepsin B[a]	25 000	3.5–6.0	Degrades myosin, actin, troponin T
				Degrades collagen
	Cathepsin L[a]	28 000	3.0–6.0	Degrades myosin, actin, troponins T and I
				Degrades tropomyosin, α-actinin
				Degrades collagen
	Cathepsin D[b]	42 000	3.0–6.0	Degrades myosin, actin, α-actinin
				Degrades troponins T and I, tropomyosin
				Degrades collagen

[a] Thiol protease.
[b] Aspartate protease.

Compiled from:

A. Asghar and A.R. Bhatti, *Adv. Food Res.*, 1987, **31**, 343.

incubated with lysosomal enzymes, an increase in hydrothermal solubility and a reduction in thermal shrinkage temperature is brought about.[134] More recent evidence also suggests that limited proteolysis of endomysial collagen occurs in beef muscle during conditioning,[135] while a reduction in the tensile strength of raw beef perimysium after 14 days conditioning at 1 °C has also been reported.[136]

5 References

1. R.A. Lawrie, 'Meat Science', 4th Edn., Pergamon Press, Oxford, 1985, p.1.
2. A.J. Bailey, in 'Recent Advances in the Chemistry of Meat', ed. A.J. Bailey, Royal Society of Chemistry, London, 1984, p.22.
3. C.L. Davey, Ref.2, p.1.
4. R.P. Gould, in 'The Structure and Function of Muscle', 2nd Edn., ed. G.H. Bourne, Academic Press, New York, San Francisco, London, 1973, Vol.II, Part 2, p.186.
5. T. Suzuki, 'Fish and Krill Protein Processing Technology', Applied Science Publishers, London 1981, p.2.
6. G.W. Rodger and P. Wilding, in 'Food Gels', ed. P. Harris, Elsevier Applied Science, London, 1990, p.361.
7. W. Bloom and D.W. Fawcett, 'A Textbook of Histology', W.B. Saunders, Philadelphia, 1968, p.281.
8. C.R. Bagshaw, 'Muscle Contraction', Chapman and Hall, London, 1982, p.24.
9. A.M. Pearson and R.B. Young, 'Muscle and Meat Biochemistry', Academic Press, San Diego, 1989, p.296.
10. R.M. Robson, J.M. O'Shea, M.K. Hartzer, W.E. Rathbun, F. Lasalle, P.J. Schreiner, L.E. Kasang, M.H. Stromer, M.L. Lusby, J.F. Ridpath, Y.-Y. Pang, R.R. Evans, M.G. Zeece, F.C. Parrish, and T.W. Huiatt, *J. Food Biochem.*, 1984, **8**, 1.
11. L.D. Peachey, *J. Cell. Biol.*, 1965, **25**, 209.
12. C.L. Davey and R.J. Winger, in 'Fibrous Proteins: Scientific, Industrial and Medical Aspects', ed. D.A.D. Parry and L.K. Creamer, Academic Press, London, 1979, Vol.1, p.97.
13. P.J. Bechtel, in 'Muscle as Food', ed. P.J. Bechtel, Academic Press, San Diego, 1986, p.2.
14. W.F. Harrington, 'Muscle Contraction', Carolina Biological Supply Co., Burlington, 1981, p.11.
15. H.S. Slayter and S. Lowey, *Proc. Natl. Acad. Sci. USA*, 1967, **58**, 1611.
16. A. Elliot and G. Offer, *J. Mol. Biol.*, 1978, **123**, 505.
17. S. Lowey and J.C. Holt, *Cold Spring Harbour Symp. Quant. Biol.*, 1972, **37**, 19.
18. A. Weeds, *Eur. J. Biochem.*, 1976, **66**, 157.
19. J.M. Squire, Ref.12, p.27.
20. T. Masaki and O. Takaiti, *J. Biochem.*, 1974, **75**, 367.
21. J. Trinick and S. Lowey, *J. Mol. Biol.*, 1977, **113**, 343.
22. D.C. Turner, T. Walliman, and H.M. Eppenberger, *Proc. Natl. Acad. Sci. USA*, 1973, **70**, 702.
23. G. Offer, C. Moos, and R. Starr, *J. Mol. Biol.*, 1973, **74**, 653.
24. M. Miyahara, K. Kishi, and H. Noda, *J. Biochem.*, 1980, **87**, 1341.
25. K. Yamamoto, *J. Biol. Chem.*, 1984, **259**, 7163.

26. K. Ohashi and K. Maruyama, *J. Biochem.*, 1985, **97**, 1323.
27. J. Hanson and S. Lowey, *J. Mol. Biol.*, 1963, **6**, 46.
28. E.J. O'Brien, J.M. Gillis, and I. Crouch, *J. Mol. Biol.*, 1975, **99**, 461.
29. S. Ebashi and A. Kodama, *J. Biochem.*, 1965, **58**, 107.
30. S. Ebashi, A. Kodama, and F. Ebashi, *J. Biochem.*, 1968, **64**, 465.
31. S. Ebashi, M. Endo, and I. Ohtsuki, *Q. Rev. Biophys.*, 1969, **2**, 351.
32. I. Ohtsuki, *J. Biochem.*, 1974, **75**, 753.
33. D.J. Hartshorne and H. Mueller, *Biochem. Biophys. Res. Commun.*, 1968, **31**, 647.
34. M.L. Greaser and J. Gergely, *J. Biol. Chem.*, 1971, **246**, 4226.
35. K. Maruyama, S. Kimura, T. Ishii, M. Kuroda, K. Ohashi, and S. Muramatsu, *J. Biochem.*, 1977, **81**, 215.
36. M. Kuroda and K. Maruyama, *J. Biochem.*, 1976, **80**, 315.
37. K. Takahasi, F. Nakamura, A. Hattori, and M. Yamanoue, *J. Biochem.*, 1985, **97**, 1043.
38. H.E. Huxley and J. Hanson, *Nature (London)*, 1954, **173**, 973.
39. A. Huxley, 'Reflections on Muscle', Liverpool University Press, 1979, p.37.
40. A.M. Gordon, A.F. Huxley, and F.J. Julian, *J. Physiol.*, 1966, **184**, 170.
41. H.E. Huxley, Ref. 4, Vol.I, p.301.
42. M.L. Greaser, S.-M. Wang, and L.F. Lemanski, Proceedings of the 34th Annual Reciprocal Meat Conference, Corvallis, Oregon, 1982, p.12.
43. I.W. Ohtsuki, K. Muruyama, and S. Ebashi, in 'Advances in Protein Chemistry', ed. C.B. Anfinsen, J.T. Edsall, and F.M. Richards, Academic Press, Orlando, 1986, Vol.38, p.1.
44. K. Wang, J. McClure, and A. Tu, *Proc. Natl. Acad. Sci. USA*, 1979, **76**, 3698.
45. K. Maruyama, S. Kimura, K. Ohashi, and Y. Kuwano, *J. Biochem.*, 1981, **89**, 701.
46. K. Wang, R. Ramirez-Mitchell, and D. Pelter, *Proc. Natl. Acad. Sci. USA*, 1984, **81**, 3685.
47. J. Trinick, P. Knight, and A. Whiting, *J. Mol. Biol.*, 1984, **180**, 331.
48. M.S. Wang and M.L. Greaser, *J. Muscle Res. Cell Motil.*, 1985, **6**, 293.
49. K. Maruyama, T. Yoshioka, H. Higuchi, K. Ohashi, S. Kimura, and R. Natori, *J. Cell. Biol.*, 1985, **101**, 2167.
50. R.H. Locker, Proceedings of the 35th Annual Reciprocal Meat Conference, Blacksburg, Virginia, 1983, p.92.
51. R.H. Locker, *Meat Sci.*, 1987, **20**, 217.
52. K. Wang and C.L. Williamson, *Proc. Natl. Acad. Sci. USA*, 1980, **77**, 3254.
53. R.H. Locker and D.J.C. Wild, *J. Ultrastruct. Res.*, 1984, **88**, 207.
54. K. Wang, *J. Cell Biol.*, 1981, **91**, 355a.
55. S. Ebashi and F. Ebashi, *J. Biochem.*, 1965, **58**, 7.
56. B.L. Granger and E. Lazarides, *Cell*, 1978, **15**, 1253.
57. R.M. Robson, M. Yamaguchi, T.W. Huiatt, F.L. Richardson, J.M. O'Shea, M.K. Hartzer, W.E. Rathbun, P.J. Shreiner, L.E. Kasang, M.H. Stromer, Y.-Y. S. Pang, R.R. Evans, and J.F. Ridpath, Ref.42, p.5.
58. M. Kuroda, T. Tanaki, and T. Mosaki, *J. Biochem.*, 1981, **89**, 297.
59. P. Bechtel, *J. Biol. Chem.*, 1979, **254**, 1755.
60. K. Ohashi, T. Mikawa, and K. Maruyamu, *J. Cell. Biol.*, 1982, **95**, 85.
61. A. Suzuki and Y. Nonami, *Agric. Biol. Chem.*, 1982, **46**, 1103.
62. P.A. Maher, G.F. Cox, and S.J. Singer, *J. Cell Biol.*, 1985, **101**, 1871.

63. E. Lazarides and B.D. Hubbard, *Proc. Natl. Acad. Sci. USA*, 1976, **73**, 4344.
64. B.L. Granger and E. Lazarides, *Cell*, 1979, **18**, 1053.
65. B.L. Granger and E. Lazarides, *Cell*, 1980, **22**, 727.
66. B. Geiger, K.T. Tokuyasu, A.H. Dutton, and S.J. Singer, *Proc. Natl. Acad. Sci. USA*, 1980, **77**, 4127.
67. J.V. Pardo, J.D. Siliciano, and S.W. Craig, *Proc. Natl. Acad. Sci. USA*, 1983, **80**, 1008.
68. C. Franzini-Armstrong, *Fed. Proc. Fed. Am. Soc. Exp. Biol.*, 1980, **39**, 2403.
69. A.M. Pearson and R.B. Young, Ref.9, p.338.
70. A.J. Bailey and D.J. Etherington, in 'Comprehensive Biochemistry', ed. M. Florkin and E.H. Stotz, Elsevier, Amsterdam, 1980, Vol.19B, p.299.
71. K.A. Piez, in 'Extracellular Matrix Biochemistry', ed. K.A. Piez and A.H. Reddi, Elsevier, New York, 1984, p.1.
72. A.J. Bailey, in 'Advances in Meat Research', ed. A.M. Pearson, T.R. Dutson, and A.J. Bailey, Van Nostrand Reinhold, New York, 1987, Vol.4, p.1.
73. D.R. Eyre, M.A. Paz, and P.M. Gallop, *Ann. Rev. Biochem.*, 1984, **53**, 717.
74. A.J. Bailey, Proceedings of the 42nd Annual Reciprocal Meat Conference, Guelph, Canada, 1990, p.127.
75. A.J. Bailey, S.P. Robbins, and G. Balian, *Nature (London)*, 1974, **251**, 105.
76. N.D. Light and A.J. Bailey, *Biochem. J.*, 1980, **185**, 373.
77. T. Sims and A.J. Bailey, this volume, p.106.
78. K. Barnard, N.D. Light, T.J. Sims, and A.J. Bailey, *Biochem. J.*, 1987, **244**, 303.
79. T. Housley, M.L. Tanzer, E. Henson, and P.M. Gallop, *Biochem. Biophys. Res. Commun.*, 1975, **67**, 824.
80. M. Yamaguchi, R.E. London, C. Guenat, F. Hashimoto, and G.L. Mechanic, *J. Biol. Chem.*, 1987, **262**, 11428.
81. D. Fujimoto, K.Y. Akiba, and N. Nakamura, *Biochem. Biophys. Res. Commun.*, 1977, **76**, 1124.
82. D.R. Eyre, Ref.72, Vol.4, p.69.
83. J.R. Bendall, *J. Sci. Food Agric.*, 1967, **18**, 553.
84. J.M. Gosline and J. Rosenbloom, Ref.71, p.191.
85. S.M. Partridge, D.F. Elsden, and J. Thomas, *Nature (London)*, 1963, **197**, 1297.
86. R.A. Lawrie, this volume, p.43.
87. E.C. Bate-Smith, *J. Physiol.*, 1939, **96**, 176.
88. J.R. Bendall, Proceedings of the 19th European Meeting of Meat Research Workers, Paris, 1973, p.1.
89. J.R. Bendall, Ref.4, Vol.II, p.244.
90. R.G. Cassens and R.P. Newbold, *J. Food Sci.*, 1967, **32**, 269.
91. R.H. Locker and C.J. Hagyard, *J. Sci. Food Agric.*, 1963, **14**, 787.
92. B.B Marsh and N.G. Leet, *J. Food Sci.*, 1966, **31**, 450.
93. C.A. Voyle, in 'Food Microscopy', ed. J.G. Vaughan, Academic Press, London, 1979, p.193.
94. D.W. Stanley, G.P. Pearson, and V.E. Coxworth, *J. Food Sci.*, 1971, **36**, 256.
95. P.E. Bouton, P.V. Harris, and W.R. Shorthose, *J. Texture Stud.*, 1975, **6**, 297.
96. R.W Currie and F.H. Wolfe, *Meat Sci.*, 1980, **4**, 123.
97. E. Dransfield, *J. Sci. Food Agric.*, 1977, **28**, 233.

98. R.W.D. Rowe, *J. Food Technol.*, 1974, **9**, 501.
99. H.K. Herring, R.G. Cassens, and E.J. Briskey, *J. Food Sci.*, 1965, **30**, 1049.
100. C.L. Davey, H. Kuttel, and K.V. Gilbert, *J. Food Technol.*, 1967, **2**, 53.
101. J.R. Burt, N.R. Jones, A.S. McGill, and G.D. Stroud, *J. Food Technol.*, 1970, **5**, 339.
102. R.M. Love, J. Lavéty, and N.G. Garcia, *J. Food Technol.*, 1972, **7**, 291.
103. K. Yamaguchi, J. Lavéty, and R.M. Love, *J. Food Technol.*, 1976, **11**, 389.
104. C.A. Curran, R.G. Poulter, A. Brueton, and N.S.D. Jones, *J. Food Technol.*, 1986, **21**, 289.
105. C.A. Curran, R.G. Poulter, A. Brueton, N.R. Jones, and N.S.D. Jones, *J. Food Technol.*, 1986, **21**, 301.
106. P.W.H. Parry, M.V. Alcasid, and E.B. Pauggat, *J. Food Technol.*, 1987, **22**, 637.
107. C.L. Davey and K.V. Gilbert, *J. Food Sci.*, 1969, **34**, 69.
108. R.H. Locker and D.J.C. Wild, *J. Texture Stud.*, 1982, **13**, 71.
109. R.H. Locker and D.J.C. Wild, *Meat Sci.*, 1982, **7**, 93.
110. J. Lepetit, P. Salé, and A. Ouali, *Meat Sci.*, 1986, **16**, 161.
111. J. Lepetit, *Meat Sci.*, 1991, **29**, 271.
112. E. Dransfield, D.K. Lockyer, and P. Prabhakaran, *Meat Sci.*, 1986, **16**, 127.
113. C.L. Davey, Proceedings of the 36th Annual Reciprocal Meat Conference, Fargo, North Dakota, 1984, p.108.
114. D.J. Etherington, *J. Anim. Sci.*, 1984, **56**, 1644.
115. M. Koohmaraie, Proceedings of the 41st Annual Reciprocal Meat Conference, Laramie, Wyoming, 1989, p.89.
116. I.F. Penny, D.J. Etherington, J.L. Reeves, and M.A.J. Taylor, Proceedings of the 30th European Meeting of Meat Research Workers, Bristol, 1984, p.133.
117. P.J.A. Davies, D. Wallach, M.C. Willingham, I. Paston, M. Yamaguchi, and R.M. Robson, *J. Biol. Chem.*, 1978, **253**, 403.
118. A. Suzuki, M. Saito, H. Sato, and Y. Nonami, *Agric. Biol. Chem.*, 1978, **42**, 2111.
119. M.L. Lusby, J.F. Ridpath, F.C. Parrish, Jnr., and R.M. Robson, *J. Food Sci.*, 1983, **48**, 1787.
120. N.L. King, *Meat Sci.*, 1984, **11**, 27.
121. C. Chin-Sheng and F.C. Parrish Jnr., *J. Food Sci.*, 1978, **43**, 46.
122. C.L. Davey and K.V. Gilbert, *J. Food Sci.*, 1969, **34**, 69.
123. C.L. Davey and M.R. Dickson, *J. Food Sci.*, 1970, **35**, 56.
124. A.J. Moller, T. Vestergaard, and J. Wismer-Pedersen, *J. Food Sci.*, 1973, **38**, 824.
125. N. Arakawa, D. Inagaki, T. Kitamura, S. Fujiki, and M. Fujimaki, *Agric. Biol. Chem.*, 1976, **40**, 1445.
126. D.G. Olson and F.C. Parrish, Jnr., *J. Food Sci.*, 1977, **42**, 506.
127. I.F. Penny, in 'Developments in Meat Science', ed. R. Lawrie, Elsevier Applied Science, London, 1980, Vol.1, p.115.
128. P.J. Bechtel and F.C. Parrish, Jnr., *J. Food Sci.*, 1983, **48**, 294.
129. L.D. Yates, T.R. Dutson, J. Caldwell, and Z.L. Carpenter, *Meat Sci.*, 1983, **9**, 157.
130. L.P. Yu and Y.B. Lee, *J. Food Sci.*, 1986, **51**, 774.
131. A. Ouali, N. Garrel, A. Obled, C. Deval, and C. Valin, *Meat Sci.*, 1987, **19**, 83.

132. D.J. Etherington, M.A.J. Taylor, D.K. Wakefield, A. Cousins, and E. Dransfield, *Meat Sci.*, 1990, **28**, 99.
133. A. Ouali and A. Talmant, *Meat Sci.*, 1990, **28**, 331.
134. J. Kopp and C. Valin, *Meat Sci.*, 1981, **5**, 319.
135. C. Stanton and N. Light, *Meat Sci.*, 1990, **27**, 41.
136. G.J. Lewis, P.P. Purslow, and A.E. Rice, *Meat Sci.*, 1991, **30**, 1.

Structural Aspects of Cooked Meat

T.J. Sims and A.J. Bailey

DEPARTMENT OF VETERINARY MEDICINE, UNIVERSITY OF BRISTOL, LANGFORD, BRISTOL BS18 7DY, UK

1 Introduction

The most obvious structural changes that can be observed following the cooking of meat are the shrinkage in volume of the muscle, the consequent loss of fluid, and that the meat is clearly much stiffer than the flaccid feel of the raw state. In other words a change in the texture of the meat, which is the most important attribute of its eating quality, has developed during cooking. The extent of these changes and hence the texture vary considerably with the particular muscle studied, and consequently there is a considerable variation in the texture of meat not only from different parts of the animal but also with the age of the animal.

Although taste panel tests must be the ultimate criteria of eating quality of meat, objective methods can be used and the measurement of the shear force required to break the meat correlates well with the texture as determined by the taste panels. When analysed in this way the shear value can vary over the range 23–46 N, *i.e.* from very tender meat, *e.g. psoas major*, to very tough meat, *e.g. extensor carpi radialis* (Figure 1). A more detailed analysis of the changes taking place during the cooking of the meat reveals two distinct changes in the shear value as the temperature is raised (Figure 2). The first increase occurs at *ca.* 45–50 °C and the second at 65–70 °C. Prolonged heating at these elevated temperatures eventually leads to a decrease in the shear value.

To understand these changes a structural approach is required. In simplistic terms based on light microscope observations the changes at 45–50 °C involve the denaturation of the myofibrillar proteins.[1] At this temperature the collagenous tissues are unaffected. At 65–70 °C the collagen denatures and the second increase in shear value occurs. This increase is greater the older the animal. It would appear, therefore, that both the myofibrillar proteins and the collagen contribute to the texture of meat although the greatest change appears to be due to the collagen, particularly with the older animals. In order to determine the relative contribution of each component it is necessary to study the actual mechanisms involved in the thermal denaturation of meat proteins.

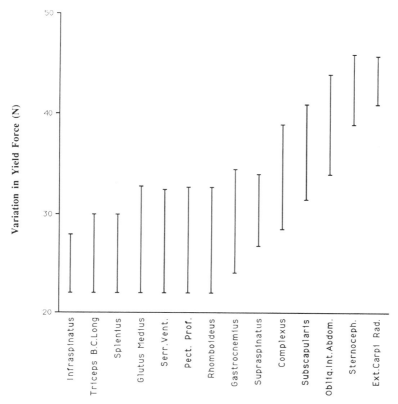

Figure 1 *Graph showing the variation in shear force with of a range of bovine muscles*

Collagen is a minor component of meat, generally *ca.* 2%, but in view of its known high tensile strength, *ca.* 100 times that of the myofibrils, it is not surprising that it has been implicated in the texture of meat. Indeed, it is well known that muscles with a high collagen content are generally tough. Unfortunately such a correlation assumes that the collagen is unaffected by heat. Indeed, when analysed in detail over a wide range of muscles the correlation between collagen content and texture is rather poor[2-5] (Figure 3). We have therefore carried out a complete analysis of the intramuscular collagen in an attempt to identify its role in determining the texture of meat. However, before considering the nature of the intramuscular collagen it is worthwhile at this stage to review briefly the different types of structures comprising the collagen family of proteins.

2 The Structure of the Collagens

Collagen is present in a wide range of connective tissues throughout the body, ranging from the rope-like structures found in tendons and the

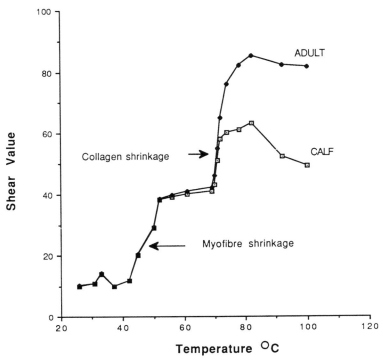

Figure 2 *Change in shear force with increasing temperature during cooking*: ● *adult muscle*, □ *calf muscle*

dermis, through the hard rigid structures such as bone and dentine, to the delicate membranes found in the kidney glomerulus and the lens capsule of the eye. These structures contain genetically distinct types of collagen and often each different tissue contains several collagen types. Over the past few years the number of distinct collagen types identified has increased considerably, primarily through the application of molecular biological techniques. At the present time 15 genetically distinct types of collagen have been characterized,[6-8] but it is certain that a number of others remain to be identified. Some of these molecules aggregate in totally different ways to form a variety of supramolecular structures. The molecules themselves vary in the length of their helices, the charge profile along the helix, and the nature and size of their globular terminal regions. This large variation in the properties of collagen molecules has allowed the collagen chemist, at least partially, to account for the biological diversity of collagenous tissues.

Classification of these collagens can be made on several grounds, but it is probably more helpful to consider the macromolecular structure and possible function rather than the molecular properties. In this way collagens may be classified as (i) fibrous, (ii) non-fibrous, (iii) filamentous, and (iv) fibril-associated (Table 1).

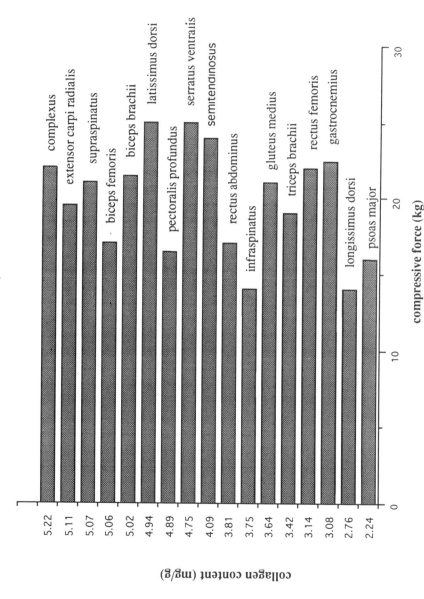

Figure 3 Graph demonstrating the lack of correlation between the collagen content of a range of muscles and compressive force as a measure of textural quality

Table 1 *Tissue distribution of collagen types*

Fibrous	
Type I	Skin, tendon, bone, dentine
Type II	Cartilage
Type III	Skin, aorta, muscle
Type V	Widespread
Non-fibrous	
Type IV	Basement membrane
Type VII	B. M. - skin, amnion
Type X	Cartilage/bone
Filamentous	
Type VI	Widespread
Fibril associated	
Type IX	Cartilage, on Type II
Type XII	Tendon, on Type I
Type XIV	

Fibril-forming Collagens

These collagen molecules have a triple helix 300 nm long and based on their unique charge profile aggregate in a quarter-stagger parallel array with a small end-overlap of 25 nm. This precise alignment of the collagen molecules generates the characteristic 67 nm axial repeat pattern along the fibril[9] (Figure 4). Collagen Types I, II, III, V, and XI all form this type of fibril. Type I collagen is the main collagen of tendon and dermis and of the hard tissues bone and dentine. Type II is the major collagen of articular cartilage. Type III is widely distributed but occurs in the greatest amounts in foetal skin and the vascular system.

Non-fibrous Collagens

Network Structures. The framework of basement membranes is comprised of Type IV collagen. The molecule is 400 nm long and is flexible because of aberrations in the location of the glycine residues,[10-12] and the charge profile does not spontaneously generate a parallel alignment of the molecules. Instead the *N*-terminal regions aggregate in an anti-parallel fashion overlapping by 8 nm to form dimers then tetramers.[10] These tetramers then aggregate to form a network structure by interaction of their *C*-terminal globular domains, thus generating a chicken-wire type of structure[14] (Figure 5a). This thin open structure is ideal for the inclusion of other components of basement membrane, laminin, heparan sulphate, *etc*. Type IV is the major collagenous component of the basement membranes of the kidney glomerulus, lens capsule, and capillaries.

It has been suggested that Type VIII forms a network structure in Descemet's membrane.[14] Preliminary studies on Type X also indicate a network structure.[15]

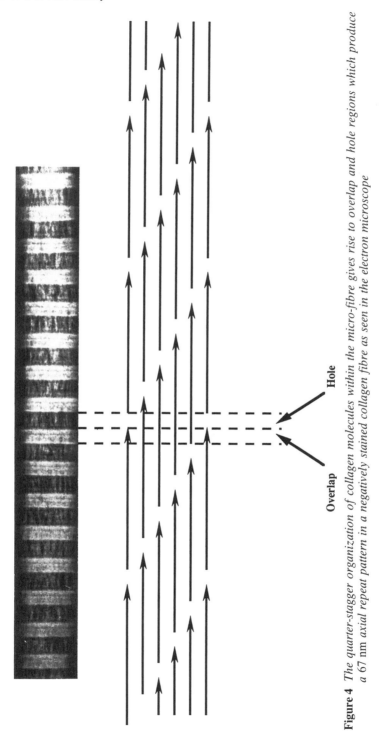

Figure 4 The quarter-stagger organization of collagen molecules within the micro-fibre gives rise to overlap and hole regions which produce a 67 nm axial repeat pattern in a negatively stained collagen fibre as seen in the electron microscope

(a) Type IV Collagen

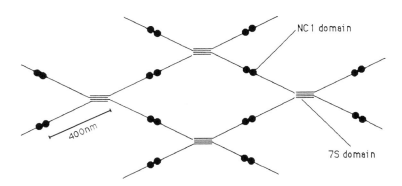

(b) Type VII Collagen

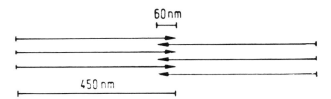

Figure 5 *Diagrammatic representation of* (a) *the proposed Type IV collagen 'chicken wire' network and* (b) *a Type VII collagen micro-fibril showing the 'in register' alignment of anti-parallel dimers*

Dimers. Type VII collagen molecules have a long triple helix, 450 nm, and they interact solely by a single anti-parallel end-overlap of 60 nm to produce dimers with a large globular domain at each end[16-18] (Figure 5b). These dimers tend to align in parallel to form the anchoring fibrils binding the basement membrane to the underlying matrix. These anchoring fibrils are not present on all basement membranes, but are known to be important in the dermal–epidermal membrane and placental membranes.[19]

Filamentous Collagens

Type VI collagen possesses a short triple helix, *ca.* 100 nm, but large globular domains at both the *N*- and *C*-termini.[20] In addition the globular region of one of three peptide chains of the triple helix has a molecular weight nearly double that of the other two chains.[21-25] The chains aggregate by extensive overlapping in an anti-parallel alignment, the dimers so formed then aggregating to form tetramers. These tetramers then polymerize longitudinally to form long filamentous structures[26] with a repeat pattern of 105 nm (Figure 6). Type VI fibres are widely distributed,

Type VI Collagen

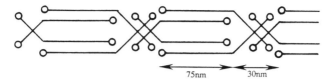

Figure 6 *Molecular organization within a Type VI collagen micro-fibril, showing an anti-parallel dimeric alignment of Type VI molecules which aggregate to form tetramers, which in turn polymerize to form the structure shown above*

albeit as a minor component, throughout the extracellular matrix, but their function has not yet been elucidated.[20-27] It has been suggested that they play a role in the spacing and alignment of the fibres in tissue.[7]

Fibril-associated Collagens

These collagens are found in association with the fibrils formed from other collagen types. This group includes the FACIT collagens - **F**ibril **A**ssociated **C**ollagens with **I**nterrupted **T**riple helices. For example, Type IX collagen first identified in porcine cartilage[28] has been shown to be attached to the surface of the Type II collagen of cartilage[29] (Figure 7), and Type XII is believed to be attached to the surface of some Type I collagen fibres.[30] The function of these collagens is not clear; they may be involved in the regulation of fibre diameter or possibly in the interaction with other components of the extracellular matrix in specific tissues.

Intramuscular Collagens

The major types of collagen present in muscle have been identified biochemically and immunohistochemically[31,32] (Figure 8). The epimysium, or outer muscle sheath, is composed primarily of Type I collagen similar to the tendon. The perimysium fibres surrounding the muscle fibre bundles contain mainly Types I and III, but with smaller amounts, less than 10%, of the minor collagens Types V and VI.[33] The layers of perimysium fibres are arranged at an angle to each other to ensure flexibility on change in

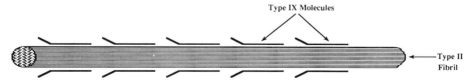

Figure 7 *Attachment of Type IX collagen molecules to the surface of a Type II collagen fibril*

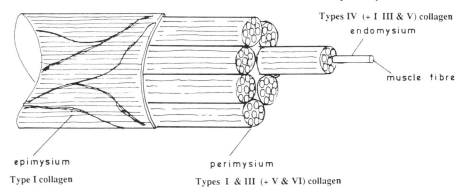

Figure 8 *Distribution of the different molecular types of collagen found in epimysium, perimysium, and endomysium*

muscle shape during muscle contraction. The fibres have a distinct crimp which disappears under tension, and is believed to act as a shock-absorber to take up the strain before the tension is placed on the fibre proper.[34] The endomysium, or sheath surrounding the individual muscle fibres, is made up of a basement membrane of Type IV collagen with an underlying matrix of fine fibres of Type III together with some Types I and V.

3 Non-collagenous Components of Intramuscular Connective Tissue

Elastin

Analysis of the total elastin content of different muscles revealed values less than 5% of the collagen content.[35] From this low content of elastin it was assumed that it is mainly associated with the capillary system of the muscle, and that it was unlikely to play a significant role in determining the texture of meat. In one or two muscles the elastin content is much higher; for example the *latissimus dorsi* and the *semitendinosus* both contain up to 30% of the collagen content, *i.e. ca.* 2% of the dry weight of the muscle. The distribution of this extra elastin in these muscles appears to be associated with the perimysium where thick elastin fibres can be detected.[36] Elastin, unlike collagen, is very stable to heat, its properties being virtually unaffected during cooking. Since it is basically a tough elastic protein it is possible that in these muscles elastin may play a role in determining the texture. However, most of the common muscles contain such a low proportion of elastin that it is unlikely to play a role.

Proteoglycans

The proteoglycan content of muscle, like that of elastin, is very low.[37] Chondroitin sulphate is the predominant proteoglycan, and is generally associated with the collagen fibres when visualized by ruthenium red or cupromeronic complexes. Their regular spacing along the collagen fibres of certain tissues suggests that they may be involved in determining the fibril diameter or the alignment of the fibres in the tissue.[38] However, the large proteoglycan molecules are readily denatured on heating and are unlikely to have a structural role in mechanical terms. On the basis of the present evidence we can discount the role of the proteoglycans in the texture of meat.

4 Collagen Parameters Possibly Affecting the Texture

In an attempt to relate the collagen to meat texture various properties of collagen have been studied to establish a direct correlation.

Total Collagen Content

Early studies were conflicting, some indicating a correlation but others contending the opposite.[39–41] It is possible that at this time cold-shortening could have occurred in some of the muscles and would not have been taken into account thus confusing the picture. More recent studies in which care was taken to avoid cold-shortening failed to demonstrate any correlation with texture[42] (Figure 3).

Collagen Solubility

Other early studies suggested a correlation between solubility of collagen and texture.[3,43–46] In some tissues the solubility of collagen provides a rough estimate of the age of the animal, but unfortunately this does not apply to all tissues. The solubility depends on the nature of the intermolecular cross-links (see later). Tissues cross-linked by the aldimine bonds, *e.g.* young skin, tend to go into solution readily in organic acids or on heating, whereas those stabilized by the keto-imine bond are generally insoluble even in young tissue. Further, both types of tissue decrease in solubility with increasing age owing to the conversion of the above intermediate bonds into more stable bonds in the mature tissue. The intramuscular perimysium contains a high proportion of the keto-imine bonds and is therefore relatively insoluble even in young tissues, and the small proportion extractable cannot provide an accurate correlation with texture.

Fibre Size

The diameter of collagen fibres and the size of the fibre bundles varies from tissue to tissue, and one might expect the thicker fibres to have a greater influence on the toughness of the meat. Recent studies on the fibre diameters in muscles of varying texture showed that tender meat, *e.g.* from *psoas major*, possessed smaller fibres (average diameter 54 nm) than tougher meat *e.g.* from *sternomandibularis* (average diameter 75 nm).[47] However, analysis of intermediate texture muscles failed to confirm a linear relationship with fibre diameter.

Genetic Type of Collagen

Intramuscular collagen contains a number of different types of collagen. The major collagen of the perimysium is Type I, but a high proportion of Type III is present in some muscles. Type III collagen fibres, originally identified as reticulin, are known to be more heat resistant,[48] and it was thought that the level of Type III in the perimysium might correlate with increased toughness. In a study of six muscles ranging from tough to tender a reasonable correlation was obtained for the extremes, but, as in the case of fibre diameter, the correlation breaks down when intermediate texture muscles are included.

Intermolecular Cross-linking

It has been known for a long time that with increasing age the solubility of collagen decreases and the tension generated on heating increases considerably. Our early studies revealed considerable variation in the cross-link profiles of different muscles from a single animal, and between similar muscles from animals of different ages. These differences can now be explained in terms of the nature and extent of the intermolecular cross-links and this has led to a rational explanation of the role of collagen in texture.

5 Formation of the Cross-links

The fibrous structure of collagen is stabilized to withstand mechanical stress by enzyme-induced cross-links. The enzyme, lysyl oxidase,[49] binds to one collagen molecule through the sequence Hyl-Gly-His-Arg- within the helix but close to the N- and C-termini and oxidizes a lysyl or hydroxylysyl residue in the N- or C-non-helical regions of an adjacent molecule aligned in the end-overlap position relative to the first molecule. The aldehyde formed reacts with the ε-amino group of the hydroxylysine residue in the above sequence thus displacing the enzyme and forming an aldimine intermolecular cross-link. If the residue oxidized in the non-helical region is hydroxylysine then the aldimine initially formed undergoes an Amadori rearrangement to form the keto-imine intermolecular cross-link[50] (Figure 9). The aldimine cross-link is heat labile whilst the keto-imine is heat stable.

With increasing age of the animal these intermediate cross-links undergo

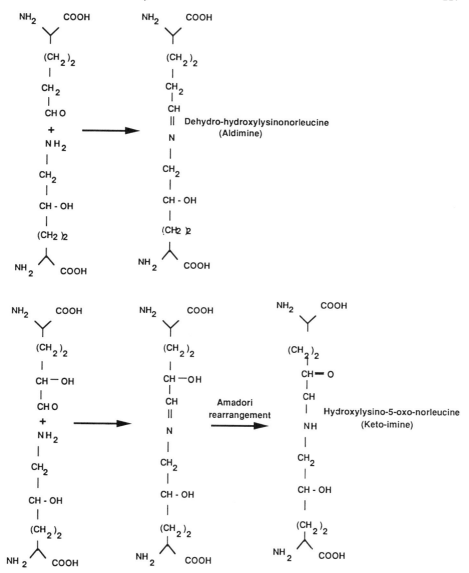

Figure 9 *Formation of the two reducible aldimine and keto-imine cross-links found in collagen*

further reactions to produce trivalent instead of divalent cross-links.[51] These trivalent cross-links, if they link three rather than two collagen molecules, will slowly build by a network of cross-links within the fibre and consequently account for the increase in tensile strength with age. The mechanism of formation of some of these 'mature' cross-links has been elucidated over the past few years. If the aldimine cross-link reacts with histidine from an adjacent molecule the trivalent cross-link formed is histidino-hydroxylysinonorleucine.[52] If the keto-imine cross-link reacts with an hydroxylysine-aldehyde from another molecule then the ring compound hydroxylysyl-pyridinoline[53] is formed (Figure 10). Both these trivalent cross-links are heat stable and can account for the decrease in solubility and the rapid increase in tension generated on heating with increase in age.

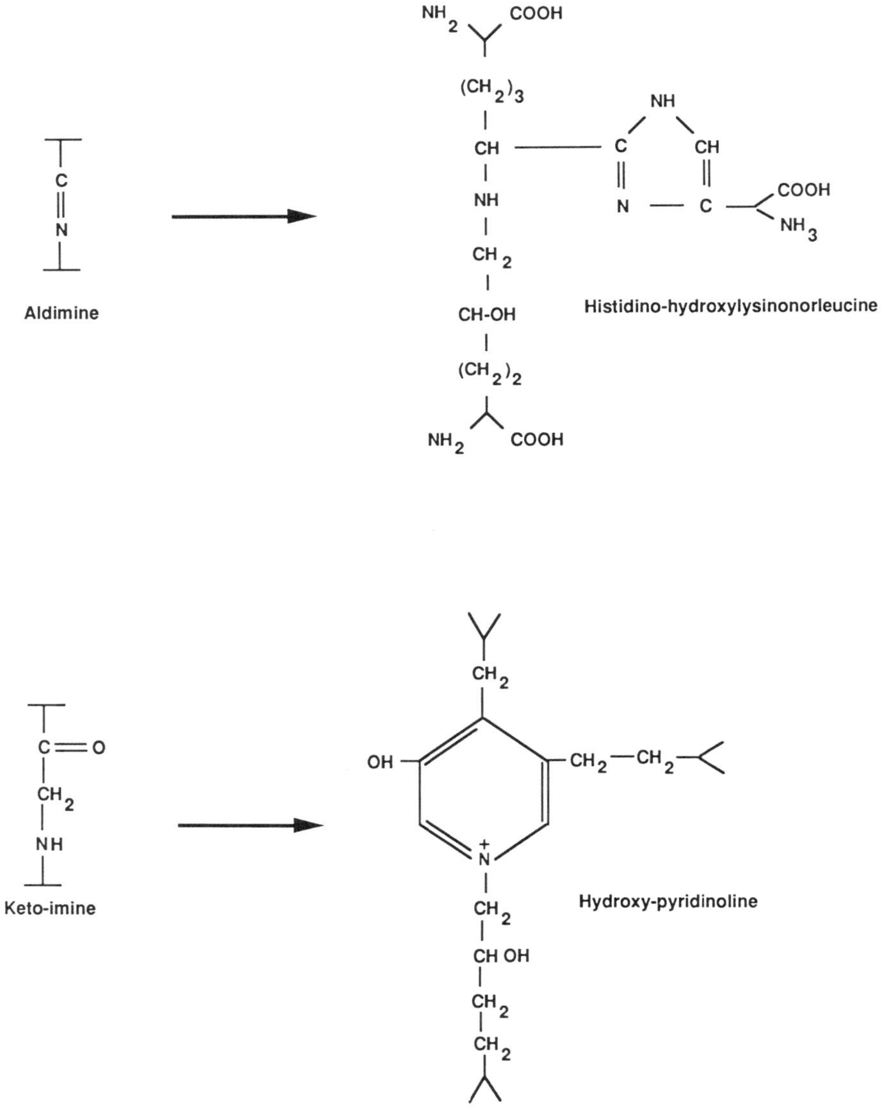

Figure 10 *Formation of two proposed trivalent 'mature' collagen cross-links formed from the divalent aldimine and keto-imine crosslinks*

6 Hydrothermal Shrinkage of Collagen

A collagen fibre is reasonably crystalline and therefore has a fairly sharp melting point at 65–67 °C at which point it shrinks to about one-quarter of its original length as the rigid triple helix collapses. If the fibre is heated under isometric conditions a tension is generated which can be measured. Fibres cross-linked by the aldimine bond do not generate a tension to their full potential owing to the cleavage of this bond following collapse of the helix. In contrast, fibres cross-linked by the keto-imine bond generate a greater tension and do not dissolve on further heating. With increasing age and the formation of the heat-stable trivalent cross-links the tension generated on heating increases dramatically[54] (Figure 11). For example the tension generated by an 18 month old rat tail tendon is 20 times greater than that by a 3 month old tendon. On prolonged heating fibres stabilized by the aldimine bond slowly dissolve to give a solution of gelatin. The fibres stabilized by the keto-imine and the stable 'mature' cross-links do not dissolve and retain a gelatinized 'fibre' with a significant residual strength.

The Role of Collagen in Meat Texture

These fundamental studies have allowed us to propose a rational description of the role of collagen in the texture of meat.[55,56]

As the temperature of the meat is raised to 40–45 °C there is an increase in toughness of the meat as determined by shear force measurements (Figure 2). This is due to the denaturation of the myofibrillar proteins, actin and myosin, which coaggulate to a more rigid gel. Microscopy reveals that the actin and myosin complex shrinks within the collagenous endomysial sheath,[57] which is unaffected at this temperature. This membrane is known to constrain the myofibrils from swelling and is therefore under some tension; hence when the acto-myosin denatures there could be a release of this tension and the endomysial membrane may force fluid out of the space created between the endomysium and the denatured myofibrils, thus accounting for the small loss of fluid at these temperatures.[58]

At 65–70 °C the second increase in shear value occurs owing to the shrinkage of the perimysial collagen. These fibres can be observed to denature to gelatin, losing their characteristic banding pattern and changing from an opaque inelastic fibre to a translucent swollen elastic fibre. The shrinkage of the fibre accompanying denaturation generates a tension against the muscle fibres within the bundles surrounded by the perimysium. The fluid released on denaturation of the acto-myosin will be forced out of the meat by the shrinkage of the perimysium (Figure 12). The extent of this shrinkage and loss of fluid depends on the nature and extent of the intermolecular cross-links stabilizing the perimysial collagen fibres.

The older the animal the greater the tension generated and this is consistent with the higher shear value observed with the older animal

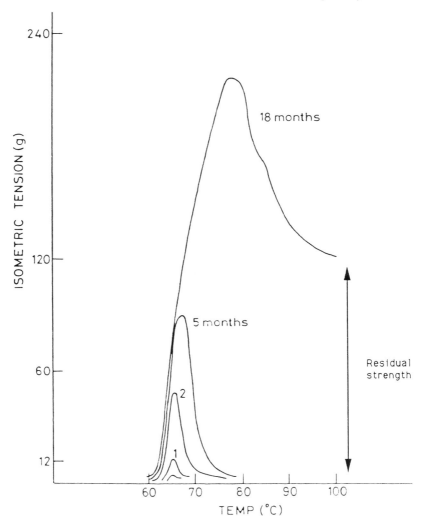

Figure 11 *Isometric tension curves generated by tail tendons from rats of different ages. Note the increased residual strength of the 18 month old tail tendon, due to the presence of heat-stable 'mature' cross-links*

compared with the same muscle from a younger animal. In the case of veal the denaturation of the perimysial collagen results in solubilization of the collagen as gelatin and this is released in the fluids and is seen to set as a gel on cooling. The shrinkage of the meat by the tension generated on thermal shrinkage of the perimysial collagen and the consequent increase in the relative protein concentration of the meat only partly accounts for the change in texture. An equally important factor is the residual strength of the fibre holding the muscle fibres together, which must also contribute

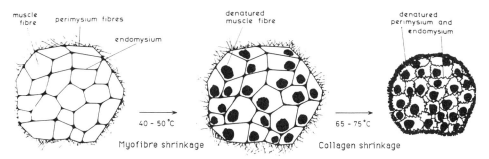

Figure 12 *Diagrammatic representation of the shrinkage of muscle fibres on cooking. Shrinkage of the myofibres occurs at 40–50 °C and this is followed by shrinkage of the connective tissue network at 65–75 °C, thus generating a tension which gives rise to extensive loss of fluid and an increase in toughness*

to the toughness of the meat. Again the greater the proportion of heat-stable cross-links remaining to stabilize the denatured fibre the higher the residual strength.

Prolonged heating of the meat above 70 °C will eventually produce a reduction in the shear value, and this is believed to be due to the cleavage of peptide bonds in the collagen. It has been reported that the shear value of cooked meat measured at 20 °C is greater than that at 70 °C.[59] This is presumably due to the fact that extensively cross-linked collagen readily regenerates the native structure on cooling. This is known as the Ewald reaction. This partial reformation of the inextensible native helical structure increases the residual strength of the fibres binding the muscle fibres together.

It is possible that the myofibrillar proteins, initially denaturing at 40–45 °C, may undergo further denaturation and dehydration as the temperature is increased to 65–70 °C. Although this would increase the shear value it cannot account for the dramatic increase in shear value precisely at the denaturation temperature of collagen. Further, the increased shrinkage in older animals could not be accounted for on the basis of greater denaturation of the myofibrillar proteins related to the age of the animals since the higher metabolic turnover of muscle proteins does not allow age-related changes to occur. It is its very low metabolic rate that permits ageing effects to take place in collagen fibres.

In fresh meat the connective tissue is an order of magnitude stronger than the myofibrillar proteins, but following cooking the muscle fibres denature and become stiffer, whilst the collagen fibres denature and become many times weaker. The question is therefore which is now the weakest component, and consequently the determinant of texture. It has long been known that the tensile properties of cooked meat along the fibre axis are about 10 times greater than the force required to separate the fibres transversely. This clearly indicates that the perimysium binding the muscle fibres laterally is now the weakest component. Quantification of this has been achieved by fracture mechanics. When a transverse slice of meat is pulled apart in the direction perpendicular to the muscle fibre axis, cavities are seen to open up throughout the slice, and these then join to give a line of cleavage between the perimysium and the endomysium (Figure 13). A closer examination of the point of fracture at the scanning electron microscope level indicates that the endomysial–perimysial bonding is the first to fracture, *i.e.* this is the weakest point.[60] When cooked meat is subjected to a tensile force parallel to the muscle fibre direction the breaking strength is substantially greater than in raw meat. The initial event is again debonding of the endomysial–perimysial junction. On increasing the load further the individual muscle fibres carry the load and it is these inextensible units that then fail. The extensible perimysial fibres are then the last to break. In non-aged meat the longitudinal breaking strength is higher than the transverse strength since the muscle fibres make a larger contribution to the breaking strength.

In summary, the mass of meat is provided by the denatured myofibrillar proteins, but the expression of texture is determined by two effects of the collagen perimysial fibres: firstly the squeezing together of the muscle bundles during the shrinkage of the collagen and the subsequent loss of fluid and secondly the tightness of the binding of the muscle fibres due to the residual strength of the denatured collagen fibres. These two effects of the collagen are in turn determined by the nature and extent of the collagen cross-links. We are now attempting to obtain a direct correlation with one or more of the cross-links and texture.

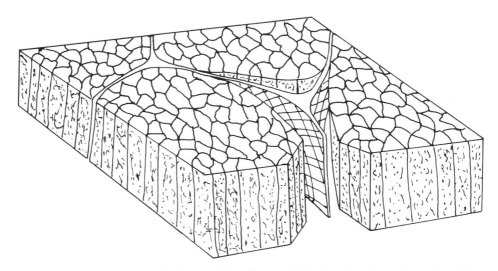

Figure 13 *Propagation of a fracture line, from an initial notch, along the perimysial–endomysial junction in a piece of cooked meat under tension*

8 Pre- and Post-slaughter Treatments

Based on the hypothesis that collagen determines the texture of meat it should be possible to predict the effects of treatment.

Growth

The different rates of growth of animals, e.g. bulls, steers, or double-muscled animals, results in differences both in the amount and cross-linking of the collagen. The rapid 'finishing' of animals increases tenderness, which can be accounted for by the higher proportion of newly synthesized and consequently immature collagen being laid down. Doubled-muscled Charolais steers possess finer and less mature collagen fibres and produce tender meat.[61] The faster rate of growth of bulls than steers would suggest that the collagen should be immature and therefore tender, but the rate of turnover may be slower and allow maturation of the collagen. The conflicting reports on the texture of bulls compared with steers[62,63] indicates that further studies on the collagen should be undertaken. Similarly growth-promoting agents, anabolic steroids, and β-agonists might also be expected to increase tenderness owing to the higher rate. Certainly steroids increase synthesis and produce tender meat, but β-agonists reduce the rate of degradation and the consequent reduction in turnover could allow the collagen to mature, which may account for the increased toughness of meat from animals treated with these drugs.[55]

Conditioning of Meat

The role of collagen in the conditioning (maturation or ageing) of meat can be accounted for on the basis of the above hypothesis. Native collagen fibres are very resistant to proteolytic attack, but could be affected by a limited number of cleavages in the non-helical cross-link regions. Although not resulting in a dramatic increase in solubility this limited cleavage could affect the physical properties of this highly polymerized protein. Indeed, a small but significant decrease in tension generated on shrinkage and in the residual strength has been observed following prolonged conditioning.[64] Evidence for the proteolytic cleavage of peptide bonds has recently been obtained by using two-dimensional gel electrophoresis, when new peptides were observed when compared with controls.[65] Conditioning takes place at pH 5.5 and under these conditions it is the released lysosomal cathepsins that are involved in the cleavage of the non-helical regions[66,67] (Figure 14). The fracture mechanics studies indicated that the weakest component was the endomysium–perimysium junction,[60] and it is possible that the cathepsins could be cleaving these fine Type III fibres. Clearly further studies in this important area are required.

Despite this demonstration that collagen could be affected by conditioning it should be remembered that the enzymes are equally effective in

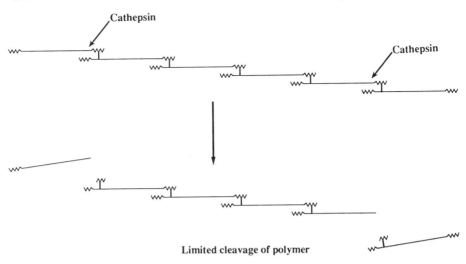

Figure 14 *Release of lysosomal cathepsins during conditioning of meat gives rise to a limited cleavage of the collagenous structures through proteolytic attack on the non-helical regions of the collagen molecules*

degrading the myofibrillar proteins which are probably more susceptible to proteolytic attack. Certainly electron microscopic and biochemical evidence indicates extensive damage to the proteins of the Z-disc,[68] the cleavage of which would dramatically reduce the mechanical properties of the muscle. To elucidate the sequence of structural mechanisms involved in the cleavage of cooked meat researchers have resorted to the simple tensile strength measurements[69] rather than the complex mechanisms of the shear value test. Recent mechanical studies indicate that following conditioning the longitudinal tensile strength of the fibres is markedly reduced, whilst the transverse strength is little affected, suggesting that the predominant effect of conditioning is the degradation of the muscle fibres.[70] Further studies on the relative roles of the muscle and collagen fibres in the texture of meat following conditioning are necessary.

It is clear that collagen, despite being a minor component of meat, does not just play a role in the texture of meat but is the major determinant of texture. On the basis of this rationale it is clear that texture depends on the quality rather than the quantity of collagen in the meat and it should now be possible to correlate the cross-linking of the perimysium with the texture measurements.

9 References

1. J.G. Schmidt and F.C. Parrish, *J. Food Sci.*, 1971, **36**, 110.
2. P.E. McClain, A.M. Mullins, S.L. Hansard, J.D. Fox, and R.F. Boulware, *J.*

Anim. Sci., 1965, **24**, 1107.
3. H.K. Herring, R.G. Cassens, and E.J. Briskey, J. Food Sci., 1967, **32**, 317.
4. R.A. Field, J. Anim. Sci., 1968, **27**, 1149 (abstr.).
5. R.E. Hunsley, R.L. Vetter, E.A. Kline, and W. Burroughs, J. Anim. Sci., 1971, **33**, 933.
6. A.J. Bailey, in 'Collagen as Food', ed. A.M. Pearson, T.R. Dutson, and A.J. Bailey, Van Nostrand Reinhold, New York, 1985, p.1.
7. R.E. Burgeson, Ann. Rev. Cell. Biol., 1988, **4**, 551.
8. R.E. Burgeson and R. Mayne, 'Structure and Function of Collagen Types', Academic Press, Orlando, FL, 1987.
9. D.J.S. Hulmes, A. Miller, D. Parry, K.A. Piez, and J. Woodhead-Galloway, J. Mol. Biol., 1973, **79**, 127.
10. K. Kuhn, H. Wiedemann, R. Timpl, J. Risteli, H. Duringer, T. Voss, and R. Glanville, FEBS Lett., 1981, **125**, 123.
11. W. Babel and R.W. Glanville, Eur. J. Biochem., 1984, **143**, 545.
12. H. Hofmann, T. Voss, K. Kuhn, and J. Engel, J. Mol. Biol., 1984, **172**, 325.
13. R. Timpl, H. Wiedermann, V. van Delden, H. Furthmayer, and K. Kuhn, Eur. J. Biochem., 1981, **120**, 203.
14. V. Labermeier and M.C. Kenny, Biochim. Biophys. Res. Commun., 1985, **116**, 619.
15. A.P.L. Kwan, C.E. Cummings, J.A. Chapman, and M.E. Grant, J. Cell Biol., 1991, **114**, 3.
16. H. Bentz, N.P. Morris, L. Murray, L.Y. Sakai, D.W. Hollister, and R.E. Burgeson, Proc. Natl. Acad. Sci. USA, 1983, **80**, 3168.
17. G.P. Lunstrum, H.J. Kuo, L.M. Rosenbaum, D.R. Keen, and R.W. Glanville, J. Biol. Chem., 1987, **262**, 13706.
18. G.P. Lunstrum, L.Y. Sakai, D.R. Keene, N.P. Morris, and R.E. Burgeson, J. Biol. Chem., 1986, **261**, 9042.
19. R.E. Burgeson, N.P. Morris, K.G. Duncan, D.R. Keene, and L.Y. Sakai, Ann. Rev. Acad. Sci., 1985, **460**, 47.
20. R. Jander and J. Rauterberg, Biochemistry, 1984, **23**, 3675.
21. H. von der Mark, M. Aumailley, G. Wick, R. Fleischmajer, and R. Timpl, Eur. J. Biochem., 1984, **142**, 493.
22. A. Colombatti, P. Bonaldi, K. Ainger, G.M. Bressan, and D. Volpin, J. Biol. Chem., 1987, **262**, 14457.
23. R.A. Heller-Harrison and W.G. Carter, J. Biol. Chem., 1984, **259**, 6858.
24. B. Trueb and P. Bornstein, J. Biol. Chem., 1984, **259**, 8597.
25. E. Engvall, H. Hessle, and G. Klier, J. Cell. Biol., 1986, **102**, 703.
26. R.R. Bruns, W. Press, E. Engvall, R. Timpl, and J. Gross, J. Cell. Biol., 1986, **106**, 393.
27. H. Hessle and E. Engvall, J. Biol. Chem., 1984, **259**, 3955.
28. M. Shimokomaki, V.C. Duance, and A.J. Bailey, FEBS Lett., 1980, **121**, 51.
29. W. Muller-Glauser, B. Humbel, M. Glatt, P. Strauli, and K.H. Winterhalter, J. Cell. Biol., 1986, **102**, 1931.
30. B. Dublet and M. van der Rest, J. Biol. Chem., 1987, **262**, 17724.
31. A.J. Bailey and T.J. Sims, J. Sci. Food Agric., 1977, **28**, 565.
32. A.J. Bailey, D.J. Restall, T.J. Sims, and V.C. Duance, J. Sci. Food Agric., 1979, **30**, 203.
33. T.F. Linsenmayer, A. Mentzer, M.M. Irwin, R. Waldrop, and R. Mayne, Exp. Cell Res., 1986, **165**, 518.

34. L.J. Gathercole and A. Keller, in 'Structure of Fibrous Biopolymers', ed. E.D.T. Atkins and A. Keller, Butterworth, London, 1975, p.153.
35. J.R. Bendall, *J. Sci. Food Agric.*, 1967, **18**, 553.
36. R.W.D. Rowe, *Meat Sci.*, 1986, **17**, 293.
37. T. Nakano and J.R. Thompson, *Can. J. Sci.*, 1980, **60**, 643.
38. E.P. Katz, E.J. Watchel, and A. Maroudas, *Biochim. Biophys. Acta*, 1986, **882**, 136.
39. J.M. Ramsbottom, E.J. Stradine, and G.H. Koonz, *Food Res.*, 1945, **10**, 497.
40. R.G. Kauffman, Z.L. Carpenter, R.W. Bray, and J. Hoekstra, *J. Agric. Food Chem.*, 1964, **12**, 504.
41. A.S. Szczesniak and K.W. Torgenson, *Adv. Food. Res.*, 1965, **14**, 73.
42. A.J. Bailey, T.J. Sims, and N.D. Light, unpublished data.
43. R.A. Field and A.M. Pearson, *J. Anim. Sci.*, 1969, **29**, 121.
44. F. Hill, *J. Food Sci.*, 1966, **31**, 161.
45. J.O. Reagan, Z.L. Carpenter, and G.C. Smith, *J. Anim. Sci.*, 1973, **37**, 270 (abst.).
46. J.R. Williams and D.L. Harrison, *J. Food Sci.*, 1978, **43**, 464.
47. N. Light, A.E. Champion, C. Voyle, and A.J. Bailey, *Meat Sci.*, 1985, **13**, 137.
48. D. Deethardt and H.J. Tuma, *J. Food. Sci.*, 1971, **36**, 563.
49. R.C. Siegel, *Int. Rev. Connect. Tissue*, 1979, **8**, 73.
50. A.J. Bailey, S.P. Robins and G. Balian, *Nature (London)*, 1974, **251**, 105.
51. S.P. Robins, M. Shimokomaki, and A.J. Bailey, *Biochem. J.*, 1973, **131**, 771.
52. M. Yamauchi, R.E. London, C. Guenat, F. Hashimoto, and G.L. Mechanic, *J. Biol. Chem.*, 1987, **262**, 11 428.
53. D. Fujimoto, K.Y. Akiba, and N. Nakamura, *Biochem. Biophys. Res. Commun.*, 1977, **76**, 1124.
54. J.C. Allain, M. LeLous, S. Bazin, A.J. Bailey, and A. Delauney, *Biochim. Biophys. Acta*, 1978, **533**, 147.
55. A.J. Bailey, Proceedings of the 42nd Annual Reciprocal Meat Conference, Guelph, Canada, 1990, p.127.
56. A.J. Bailey and N.D. Light, 'Connective Tissue in Meat and Meat Products', Elsevier, London, 1989.
57. J.R. Bendall and D.J. Restall, *Meat Sci.*, 1983, **8**, 93.
58. A.C. Champion, P.P. Purslow, and V.C. Duance, *Meat Sci.*, 1988, **24**, 261.
59. D.A. Ledward and R.A. Lawrie, *J. Sci. Food Agric.*, 1975, **26**, 691.
60. P.P. Purslow, *Meat Sci.*, 1985, **12**, 39.
61. A.J. Bailey, M.B. Enser, E. Dransfield, D.J. Restall, and N.C. Avery, in 'Muscle Hypertrophy of Genetic Origin and Its Use to Improve Beef Production', ed. J.W.B. King and F. Menissier, Martinus Nijhoff, The Hague, 1982, p.178.
62. S.C. Seideman, H.R. Cross, R.R. Oltjen, and B.D. Schanbacher, *J. Anim. Sci.*, 1982, **55**, 826.
63. D.E. Gerrard, S.J. Jones, A. Aberle, R.F. Lemanager, M.A. Diekman, and M.A. Judge, *J. Anim. Sci.*, 1987, **65**, 1236.
64. A.J. Bailey and T.J. Sims, unpublished data.
65. C. Stanton and N. Light, *Meat Sci.*, 1988, **23**, 179.
66. A.J. Barrett, 'Proteinases in Mammalian Cells and Tissues', North Holland Biomedical Press, 1977.
67. D.J. Etherington and P.J. Evans, *Acta Biol. Med. Ger.*, 1977, **36**, 1555.
68. E. Dransfield and D.J. Etherington, in 'Enzymes and Food Processing', ed.

G.C. Birch, N. Blakebrough, and K.J. Parker, Applied Science Publishers, London, 1981, p.177.
69. P.E. Bouton and P.V. Harris, *J. Food Sci.*, 1972, **37**, 539.
70. G.J. Lewis, P.P. Purslow, and A.E. Rice, *Meat Sci.*, 1991, **30**, 1.

Colour of Raw and Cooked Meat

D.A. Ledward

DEPARTMENT OF AGRICULTURE FOR NORTHERN IRELAND AND THE
QUEEN'S UNIVERSITY OF BELFAST, NEWFORGE LANE, BELFAST BT9 5PX, UK

1 Introduction

The colour of raw (fresh) meat is largely dictated by the concentration and chemical nature of the haemoproteins present and the temperature/pH history of the muscle post-slaughter. The response of the proteins in meat mainly dictates the colour of the cooked product.

2 Raw Meat

Initial (24 hour) Colour

Although haemoproteins constitute only *ca.* 0.5% of the wet weight of red meats such as beef and lamb, and several times less than this of the white meats such as pork, they are of paramount importance in determining meat colour.

In most meats, the haemoprotein in greatest concentration is the muscle pigment myoglobin, although the blood pigment haemoglobin is also present in significant concentrations. Even though the absolute concentration of haemoprotein in meat will obviously affect its colour, the haemoprotein-rich meats such as beef and lamb being much darker than the haemoprotein-poor meats such as pork and chicken, the ratio of haemoglobin to myoglobin has little effect on the colour since their reactivities are similar. The major reason for the wide differences in haemoprotein contents in muscle is variations in myoglobin content.

Myoglobin is apparently distributed uniformly throughout muscle and its role appears to be that of facilitating the diffusion of oxygen from the capillaries to the intracellular structures where the oxygen is used in oxidative processes. In general it would appear that high levels of muscular activity lead to higher concentrations of myoglobin[1] reflecting, in this respect, differences due to species (the muscles of the hare are richer in myoglobin than those of the rabbit), breed, sex (the muscles of the bull contain more myoglobin than those of the cow), age (the muscles of the steer are richer in myoglobin than those of the calf), type of muscle (the leg muscles in the chicken are richer in myoglobin than those of the little used breast), and training (the muscles of stall-fed animals generally contain less myoglobin than those of their free-range counterparts). Diet can also significantly affect the level of myoglobin in meat animals and it has recently been shown that in pigs a restricted feeding regime leads to a significant increase in the haemoprotein content of the *longissimus dorsi*.[2] Although it is possible to rationalize, on the above lines, the wide range of myoglobin concentrations found in different muscles, it is more difficult to explain the variability very occasionally encountered within a specific muscle. For example, it has been claimed that in a given muscle the myoglobin concentrations may vary several hundred-fold over a few centimetres.[1]

Although myoglobin is distributed uniformly throughout muscle and appears to be in true solution, meat is translucent because of the organization of the fibres within the muscle. Thus meat of high pH (>6.0), irrespective of the temperature history during *rigor*, is invariably dark in colour, *e.g.* dark, firm, dry (DFD) beef, since the fibres are able to bind the water in the muscle and, so swollen with little or no free water, the surface reflects little light yielding an almost matt finish. In addition, at these high pH values the oxygen-utilizing enzymes present in the meat are relatively active so that, as discussed later, little surface oxygenation of the myoglobin occurs with the result that the purple, reduced pigment, rather than the bright red oxy-derivative, dominates. At normal pH values (*ca.* 5.5) the fibres hold less water and the oxygen-utilizing enzymes are less active resulting in a brighter, glossier appearance. This situation prevails if the fall in pH, from the *in vivo* value of 7.3 to the ultimate value, occurs at chill temperatures. If, however, the low pH is achieved at higher temperatures then some partial denaturation of myosin may occur, plus perhaps some denaturation of sarcoplasmic proteins, so that the water-holding capacity is decreased still further resulting in pale, soft, exudative (PSE) meat (Figure 1, Hector *et al.*[3]).

Once the inherent colour due to haemoprotein concentration and pH/temperature history is established it cannot be easily modified by further processing. However, colour changes due to the chemical reactivity of the haemoproteins are very dependent on storage and processing conditions.

Effect of Storage on Meat Colour

Structure of Myoglobin All myoglobins possess a single polypeptide chain of *ca.* 155 amino acid residues complexed to a haematin moiety. The protein chain consists of eight α-helical segments, ranging in length from 7 to 24 residues, separated by non-helical regions. The helical regions constitute *ca.* 80% of the molecule.

The haematin moiety in both myoglobin and haemoglobin consists of the protoporphyrin IX ring system (Figure 2) in the centre of which an iron

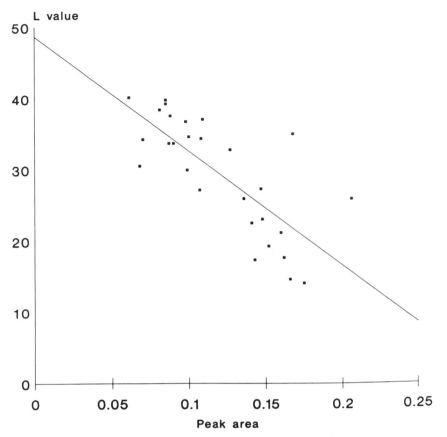

Figure 1 *Relationship between the lightness (L-value) of beef* semimembranosus *muscle and the amount of native myosin (α peak area) in the sample*
(From D.A. Hector, G. Brew-Groves, N. Hassan, and D.A. Ledward, *Meat Sci.*, 1991, **31**, 299.)

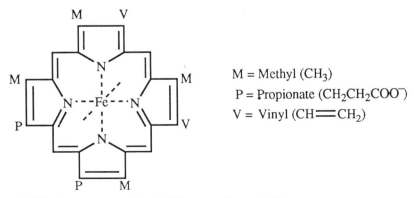

M = Methyl (CH_3)
P = Propionate ($CH_2CH_2COO^-$)
V = Vinyl ($CH\!=\!\!=\!CH_2$)

Figure 2 *The iron protoporphyrin IX group of myoglobin*

atom is attached to the four nitrogen atoms of the pyrroles. However, the iron atom (whether ferrous or ferric) is co-ordinated in an octahedral environment so that it is capable of accepting two further ligands at right angles to the haematin plane. In myoglobin one of these ligands is the nitrogen in the imidazole ring of a globin histidine residue, whereas the other may be any molecule of the correct electronic configuration, small enough to occupy the haematin pocket in the protein. The nature of this sixth ligand and the oxidation state of the iron atom will affect the electronic arrangement of the electrons of the iron which will in turn affect the electronic arrangement of the d-electrons of the iron and consequently the spectral characteristics of the molecule and hence its colour.[4] The haematin moiety itself fits into a space in the hydrophobic interior with both propionic acid side chains being hydrogen bonded to amino acid side chains of the globin. The propionic acid side chains extend through the interior with the carboxyl group on the hydrophilic surface whereas the vinyl groups are buried in the hydrophobic interior. This interaction between the haematin and globin serves to stabilize the whole molecule.

Although the basic structures of all myoglobins are similar the subtle differences that do exist may be important in determining the visual appearance and colour stability of different meats. Thus Satterlee and Zachariah[5] found significant differences in the stability and properties of bovine, ovine, and porcine myoglobins. These workers claimed that whereas ovine and bovine myoglobins were very similar in their stability to both heat and acid porcine myoglobin behaved differently, being more susceptible to acid denaturation, although its stability to heat was similar to that of the other myoglobins. The amino acid composition (and isoelectric point) of porcine myoglobin was also very different to that of both bovine and ovine myoglobin, which although possessing similar compositions had themselves significantly different amino acid sequences. To complicate the picture further it has been reported that myoglobin extracted from PSE pork is different from that obtained from normal pork.[6] This within-species variation may, however, be an artefact created by the unusual pH/temperature environment the molecules experience after the slaughter of stress-susceptible animals.[4]

Formation of Oxymyoglobin. When meat is freshly cut the myoglobin is in the purple reduced form (Mb) in which the iron is in the high-spin ferrous state and the sixth co-ordination site is unoccupied.[7] On exposure to air though myoglobin, because of its great affinity for the oxygen molecule, combines rapidly (and reversibly) with oxygen to form the bright red oxymyoglobin (MbO_2) and it is this pigment that the consumer associates with freshness. Because of its great affinity for oxygen the meat surface 'blooms' to the red colour within minutes of exposure to air and, with time, the small layer of MbO_2 spreads downwards into the meat. The depth to which the oxygen diffuses depends on the activity of the oxygen-utilizing enzymes, *i.e.* the oxygen consumption rate of the meat, the temperature, and external oxygen pressure. After 2 hours exposure to air at 0 °C the MbO_2 layer in different beef muscles was 1–3 mm thick and increased to 7–10 mm after 7 days.[8] The oxygen diffuses through the aqueous environment and enters the hydrophobic haematin cleft to occupy the sixth co-ordination site. Oxygenation-induced conformational shifts are believed to bring a histidine residue, known as the distal histidine to differentiate it from the proximal histidine which occupies the fifth co-ordination site, within interacting range of the liganded oxygen, thereby stabilizing the complex. Recent years have seen some controversy regarding the electronic distribution around the haematin in the oxymyoglobin molecule. The covalent structure ($Fe^{2+}O_2$) is now considered improbable and a great deal of evidence suggests that the complex is best represented as a low-spin superoxide ferric complex, *i.e.* $Fe^{3+}O_2^-$. Giddings[9] reviewing the available evidence suggests that the single best electronic representation of the iron–oxygen complex lies between the completely covalent and ionic extremes and that a dioxygen iron model with substantial, but less than complete, charge transfer to oxygen is most representative.

As might be anticipated 'blooming' is more efficient under conditions which increase oxygen solubility and discourage enzymic activity, *i.e.* at low temperatures and low pHs. Thus meat that has been held (aged) for several weeks *in vacuo* prior to exposure to air blooms more rapidly than fresh meat[8] owing to some loss of activity of the oxygen-utilizing enzymes.

Although beef muscles vary widely in their ability to 'bloom' the variation between species is much greater. For example 'fresh' lamb has a high oxygen consumption rate[10] and blooms far less readily than beef at similar times post-slaughter. There is also some evidence that turkey muscles have very limited ability to bloom since reduced myoglobin is the predominant pigment in several samples analysed by reflectance spectrophotometry (Figure 3). In fact turkey breasts and legs held under high oxygen pressure often appear unnaturally red due to increased formation of oxymyoglobin.

The commercial exploitation of electrical stimulation (ES) in the USA was to some extent helped by the fact that ES-treated beef yielded a better, *i.e.* redder, 'colour' when the *longissimus dorsi* was used to grade the carcass 24 hours after slaughter. This was due to increased formation of MbO_2.[11] Electrical stimulation is a process which effectively accelerates post mortem glycolysis so that the muscle pH is reduced to its ultimate value within a few hours of slaughter, rather than the 20–24 hours necessary at chill temperatures in unstimulated beef carcasses. The lower pH thus permits more oxygenation of the haematin. Although a better colour is seen 24 hours after slaughter there is little difference in the rate or extent of MbO_2 formation in stimulated and unstimulated carcasses 48 hours after slaughter,[11] presumably because the protein environments are now identical. Although electrical stimulation of carcasses generally improves the 24 hour colour of the *longissimus dorsi* (loin) muscle it can have

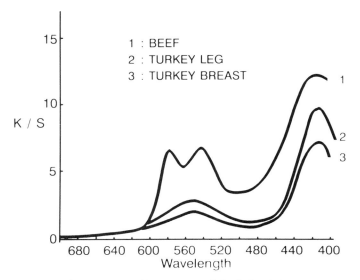

Figure 3 *Absorption/scattering (K/S) spectra of fresh beef (upper curve, typical of oxymyoglobin) and turkey muscles (more typical of reduced myoglobin) after blooming in air for several hours*

adverse effects on the initial colour of the deep seated muscles that cool far more slowly. For example the *semimembranosus* (topside) muscles of low-voltage stimulated beef carcasses can be significantly paler than their unstimulated controls[3,12] owing to partial unfolding of myosin at the high temperatures and low pHs that may co-exist in these muscles (Figure 1). There is no evidence that the haemoproteins themselves are affected by low-voltage electrical stimulations although the work of Bembers and Satterlee[6] suggests that in PSE pork structural modifications of the myoglobin may occur.

Formation of Metmyoglobin. Whatever factors affect the rate and extent of MbO_2 formation the complex readily oxidizes to metmyoglobin (metMb) which is brown and unable to form an oxygen adduct. This brown pigmentation, which downgrades the quality of fresh meat, is formed by the removal of a superoxide anion (or its conjugate acid, HO_2) from the haematin and its replacement by a water molecule to produce a high-spin ferric haematin. The water molecule attached to the ferric iron may ionize so that a hydroxyl group occupies this sixth site to yield a low-spin, red complex. However, this transition has a pK value above 8^{13} and is of little significance in meat and meat products.

In pure solution the autoxidation of MbO_2 is first order with respect to unoxidized myoglobin and depends on several external parameters including ionic strength and oxygen pressure. The rate is maximal at oxygen pressures corresponding to half saturation of the iron,[14] *i.e.* at oxygen pressures of ca. 1–1.4 mmHg. At oxygen pressures above 30 mmHg the rate is independent of the partial pressure of oxygen. In addition the autoxidation is very pH and temperature dependent. At saturated oxygen pressures the half-life of the autoxidation is 2.8 hours at pH 5 and 25 °C whilst at 0 °C and pH 5 the half-life is about 5 days. At pH 9 the values are 7 days and 1 year respectively.[4] It is also well established that metal ions such as Cu^{2+} and Fe^{3+} accelerate the reaction.[15]

The rate of autoxidation does vary to some extent with the type of myoglobin, fish myoglobins being most susceptible and those of mammalian origin least reactive.[16]

However, the results found for myoglobins in solution can not be applied indiscriminately to meat as, in meat, (i) a reducing system is present which is capable of reducing metmyoglobin to the ferrous state, (ii) a catalytic process operates, and (iii) different gradients of oxygen concentration (from 0 to the external pressure) will be present in various muscles. The reduction of metmyoglobin in meat can take place under both anaerobic and aerobic conditions[17] which agrees with the known properties of, for example, a beef heart metmyoglobin reductase.[18] This enzyme has optimum activity at ca. 37 °C and has pH optimum at ca. 6.5, the activity at this pH being about twice that at pH 5.6. Obviously, the reduction of metmyoglobin in meat may be more complex than the studies with the purified enzyme suggest since more than one reaction may be responsible for the reduction observed in meat systems; several possible substrates and intermediates are present.[4]

The ability of different muscles to resist metMb formation during aerobic storage varies greatly and, not unexpectedly, depends on species and, perhaps more surprisingly, on the anatomical location within the carcass.[4] In beef the *longissimus dorsi* and *semitendinosus* muscles are relatively colour stable whilst the *psoas major* discolours very rapidly. O'Keefe and Hood[19,20] extensively studied the factors affecting the formation of metmyoglobin in different beef muscles and claim that the inherent differences are not due to different metmyoglobin reducing activities of the muscles but primarily to variations in the rate of myoglobin oxidation and the ability of the muscle to consume oxygen, *i.e.* muscles with high activities of oxygen-utilizing enzymes, which allow little penetration of oxygen into the tissue, tending to discolour most rapidly.

Recent studies have also shown a significant positive correlation between oxygen consumption rate and colour instability in three 'fresh' (24–48 hour post-slaughter) beef muscles and no significant correlation between reducing activity and colour stability.[21] In fact in beef analysed 1 hour post-slaughter the reducing activities of four muscles were found to be *inversely* proportional to their colour stabilities.[22] However, beef muscles 'aged' *in vacuo* for several weeks brown more rapidly than fresh samples[12] on exposure to air, *i.e.* 'aged' meat blooms and subsequently browns more rapidly than fresh meat. Since 'aged' muscles have decreased oxygen consumption rates compared with their fresh counterparts[19] this observation suggests that, under these circumstances, loss of reducing activity is important in dictating the rate of browning.

Further insight into the relative roles of the reducing and oxidizing systems in the formation of metMb in meat has been afforded by studies carried out on electrically stimulated beef. It is now generally accepted that in muscles allowed to cool relatively rapidly, such as the *longissimus dorsi*, the action of rapidly decreasing the muscle pH by electrical stimulation has little effect on the colour stability,[3,23] presumably because the rapid decrease in temperature protects any temperature-sensitive reactants in the oxidative and/or reducing systems. However, when muscles cool more slowly there is some controversy as to whether rapidly decreasing the pH by stimulation improves or worsens colour stability. Sleper et al.[24] found that, at low oxygen pressures, stimulated *longissimus dorsi* muscles could form metMb more rapidly than non-stimulated muscles whilst other workers have claimed that no significant difference exists between such muscles.[11] In contrast, Claus et al.[25] observed that hot-boned muscles tend to produce more metMb when freshly sliced than do muscles which have undergone electrical stimulation prior to hot-boning. However, Ledward et al.[12] found that low-voltage electrical stimulation decreased the colour stability of beef *semimembranosus* muscles when displayed several days (7 or 28) after slaughter.

The proposed rationale to explain these apparently conflicting results is as follows. In fresh meat exposed to air 24 or 48 hours after slaughter the rate of metMb formation is primarily governed by the oxygen consumption rate of the muscle. Thus, exposing the meat to high temperature and low pH, as occurs in electrical stimulation, may decrease the activity of the oxygen-utilizing enzymes and thus improve the colour stability of the muscle. The oxygen consumption rate of meat though decreases exponentially with time at 2 °C[26] and thus in meat exposed to air several days post-slaughter the rate of enzymic reduction, which decreases only slowly with time,[23] largely dictates the rate of metMb formation. Faustman and Cassens[22] have demonstrated that in beef held in 1% O_2 for 2–6 days the reducing ability, when measured *aerobically*, apparently *increases* quite significantly. This supports the contention that the oxidative ability decreases far more rapidly than the reducing ability over this period. Thus in 'aged' beef any damage to the enzymic reduction system, as may occur by exposure to high temperature and low pH, will diminish the colour stability.

The rapid rate of decline of the catalytic, oxidative process and the slower loss of enzymic reducing ability will explain why beef muscles have maximum colour stability when displayed for sale *ca.* 5–7 days after slaughter.[23]

As both oxidation and reduction of the haematin pigments may occur in meat, and the relative rates are very dependent on the pre- and post-rigor history, it is not surprising that, in intact meat pieces, the kinetics of metmyoglobin formation are complex. However, consideration of the factors known to affect the autoxidation of myoglobin and enzymic reduction of metmyoglobin does enable the effect of the different storage conditions on the rate of formation to be understood. Thus sterile beef or pork stored aerobically at low temperatures (−1 to +10 °C) accumulates increasing concentrations of metmyoglobin at its surface until a 'pseudoequilibrium' level is attained as the rates of the oxidation and reducing reactions equalize (Figure 4). On prolonged storage though, as the enzymic reducing system becomes exhausted, the pseudoequilibrium is lost and further metmyoglobin accumulates at the surface (Figure 4). Also, as the autoxidation is maximal at low partial pressures of oxygen and the enzymic reducing system is believed to be oxygen independent, when meat is stored in air for a day or so at refrigerated temperatures a brown layer of metmyoglobins forms below the surface where the oxygen concentration corresponds to that required for maximum formation. In some meats, of high oxygen consumption rate, such as certain fresh turkey muscles, maximal formation may be at or very near the surface.

Increasing the oxygen pressure in the outside atmosphere causes the metmyoglobin layer to occur at greater distances below the surface and in atmospheres containing 60% oxygen beef pieces 2 cm thick do not normally possess such a layer.[27] Storage at lower oxygen pressure causes the layer to move closer to the surface. If meat can be stored in oxygen-containing atmospheres that prevent the formation of this brown metmyoglobin layer or cause it to occur well below the surface (>1 cm) then browning at the surface is much reduced.[27] However, if the layer is

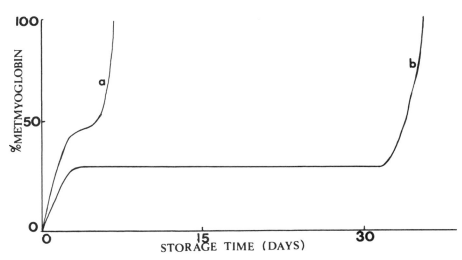

Figure 4 *Schematic representation of typical time courses for the formation of metmyoglobin at the surface of fresh meat during aerobic storage at 0–2 °C. Curve (a) is for a colour-labile muscle and (b) for a colour-stable muscle* (from D.A. Ledward, in 'Developments in Food Protein - 3', ed. B.J.F. Hudson, Elsevier Applied Science, London, 1984, p. 33.)

allowed to form within a few millimetres of the surface the formation of metmyoglobin at the surface appears to be largely independent of external oxygen pressure, *i.e.* from 5 to 20% (Figure 5). This complex dependence on oxygen pressure has led to the development of systems for the transport and storage of fresh meat at high (>60%) oxygen partial pressure. It is usual in these systems to incorporate carbon dioxide (*ca.* 20%) to retard bacterial growth.

As the rate of the autoxidation reaction increases with decreasing pH whilst the enzymic reduction is far less effective at low pH it is not surprising that, in general, muscles of low ultimate pH discolour more rapidly than those of high ultimate pH.[28] However, it must be remembered that, in meat of high pH, owing to the high oxygen consumption rate, the metMb layer will be near to the surface and thus visual discoloration may occur more rapidly than anticipated. Also, as the temperature dependence of the autoxidation is far greater than that of the enzymic reduction and since oxygen penetration will be less at higher temperatures, the observation that a 3–5 °C increase in temperature in a display cabinet may double the rate of discoloration is not unexpected.[4]

Partial dehydration of meat tends to accelerate metmyoglobin formation, although darkening due to haem concentration easily overrides the visual impact of this. High concentrations of oxygen-utilizing bacteria, *e.g.* 10^7 cm^{-2} of *Pseudomonas* sp., can cause increased metmyoglobin formation, presumably due to oxygen depletion at the surface.[4]

As well as thermal oxidation photo-oxidation may also be important at

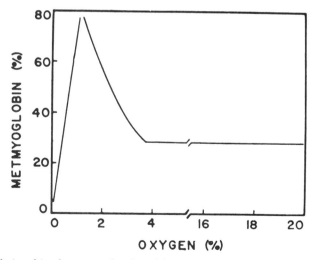

Figure 5 *Relationship between the 'equilibrium' level of metmyoglobin (metMb) formed at the surface of fresh, sterile beef* semitendinosus *muscle stored at 7 °C and the oxygen concentration of the surrounding atomsphere*
(Redrawn from D.A. Ledward, in 'Developments in Food Protein - 3', ed. B.J.F. Hudson, Elsevier Applied Science, London, 1984, p. 33)

low temperatures. For example, during storage at −18 °C photo-oxidation contributes to the formation of metMb when stored under fluorescent light. In chilled meat the effect of visible light is negligible although ultraviolet light is a powerful pro-oxidant.[29] Thus at pH 5.4 and 0 °C the relative rates of photo-oxidation at 546, 366, and 254 nm are 1, 10, and 4700 respectively. The relative rates of metMb formation from oxymyoglobin, due to thermal and photo-oxidation at different wavelengths, are discussed in this volume by Professor Skibsted (p. 266).

Although the formation of metMb in meat can be inhibited by packaging in vacuum or modified (high oxygen) atmospheres such techniques are not ideal and other approaches have been attempted. Formation can be inhibited by feeding α-tocopherol (vitamin E) to the animals or by treatment of the meat with vitamin C.[30]

In view of the relative instability of the red oxymyoglobin pigment attempts have been made to stabilize the red colour associated with fresh meat by the formation of other, more stable, low-spin haematin complexes. Carbon monoxide binds tightly to myoglobin forming the bright red carboxymyoglobin which is more stable than the corresponding oxy derivative.[31] As well as carbon monoxide other strong-field ligands that can be accommodated at the sixth co-ordination site of the haematin group to yield a stable, red, low-spin complex have been suggested, including ammonia and heterocyclic nitrogen-containing compounds such as imidazole.

Other Discolorations in Raw Meat

The formation of metmyoglobin is the major colour problem facing the retailer of fresh (or frozen) meat. However, it is possible for other reactions to occur giving rise to marked colour changes. Of these, reactions leading to saturation of one of the methene bridges and consequent disruption of the conjugation (resonance) around the porphyrin ring are of importance since they usually yield green pigments. For example an H_2O_2-induced oxidation may bring about a hydroxyl substitution of the ring system which may undergo further oxidation to form choleglobin. However, in fresh meat catalase and perhaps peroxidase minimize the H_2O_2 content and thus the formation of choleglobin is not usually a problem in fresh and frozen meats. Of more concern, especially with the increasing use of vacuum packaged meat, is the ability of H_2S to oxidize the porphyrin ring to green sulfmyoglobin.[32] In this compound each molecule of haematin contains one sulfur atom which is bound to the ring system; however, the precise structure of the complex is not clear though it is likely to possess a hydroxythiol or episulfide structure with the globin still in its native form and the iron in the ferrous state.[33]

Sulfmyoglobin (and haemoglobin) formation has been a problem in undrawn (New York dressed) poultry for many years owing to the production of H_2S by the gut microflora which subsequently reacts with the haematin pigments present in the tissue. However, samples of vacuum packed beef of high pH may 'green' with time and this was shown to occur under conditions of high pH (>6) and low oxygen tensions (1–2%), conditions under which the bacteria were able to produce H_2S. The organism primarily responsible is *Altermonas putrefaciens*.[34] Australian workers have also identified a number of lactobacilli-type organisms that may produce H_2S on meat of pH less than 6 and are probably responsible for the greening occasionally observed in meat of normal pH. The mechanism for the formation of sulfmyoglobin in meat has never been elucidated but it may involve a 'ferryl' ion as an intermediate. In a ferryl complex the iron atom has an apparent valency of 4 and has been described as having only four $3d$ valence electrons and an apparent spin of 1.[4] This quadrivalent iron is complexed to an oxygen atom (*i.e.* it may be called ferrimyoglobin peroxide). This type of complex has been implicated in the haem-catalysed oxidation of lipids in meat and as a product of the irradiation of myoglobin. In the laboratory sulfmyoglobin is usually prepared via the ferryl complex[35] and a possible mechanism for its formation in meat is shown in the Scheme.[36]

$$Mb^{2+}\text{—}O_2/Mb^{3+}\text{—}O_2^- \rightleftharpoons Mb^{3+}\text{—}H_2O + O_2^-$$

Oxymyoglobin   Metmyoglobin + Superoxide anion

$$\downarrow H_2O_2$$

$$Mb^{2+}\text{—}SH \xleftarrow{H_2S \text{ from bacteria}} Mb^{4+}\text{—}O_2$$

Sulfmyoglobin   Ferrimyoglobin peroxide

Scheme

H_2O_2 required for production of the ferryl complex has been detected in ground meat[37] and may in fact be a product of the oxidation of oxymyoglobin to the met form or a product of bacterial or enzymic action, peroxidizing lipids, or flavins.

If further oxidation of the porphyrin ring occurs then a green verdohaem complex may form and ultimately the haematin pigments may degrade to yellow or colourless pyrrole fragments (bile pigments) and globin.[1] The formation of bile pigments though is rarely a problem in fresh meat though it may be significant in dried and semi-dried meat products.

3 Cooked Meat

The rate at which the reactions leading to colour changes on heating take place depends on the rate of heating, and Palombo and Wygaards[38,39] have shown that the changes are essentially complete within 5 min at 100 °C and 10 min at 80 °C, but at lower temperatures (50 and 60 °C) the colour is still changing after 3 hours. These authors have also shown that the increase in lightness of the meat on heating is, at least after the first few minutes, mirrored by decreases in the hue and chroma of the sample. However, small changes (decreases) in hue and chroma appear to take place after the lightness has achieved a steady value. General protein precipitation is largely responsible for the lightening in colour of the meat whilst the hue and chroma are primarily due to the nature of the haemoproteins. These results show the relative stability of myoglobin. It has long been established that myoglobin is one of the most heat stable of the proteins present in meat although its actual stability depends on the source, the nature of the ligand at the sixth co-ordination site, and the oxidation state of the iron.[40]

Over recent years there has been some controversy over the actual nature of the ligands attached to the fifth and sixth co-ordination sites in the cooked meat haemoproteins, although it is well established that the haematin iron is usually in the ferric state.[4,41] At high pH though it is possible to generate the red ferrous derivative.

The current situation has recently been reviewed[36] and the conclusion reached that it is not possible, at present, to define unequivocally the nature of the haemoproteins present in cooked meat and that more than one type of complex may be present. However, the major haematin pigment present probably contains two protein-bound imidazole groups.

An unusual feature of myoglobin 'heat denaturation' in meat is that it tends to occur, as evidenced by precipitation, at temperatures of 60 °C and above whilst in pure solution at pH 5.5–6.0 marked denaturation and precipitation only occurs at temperatures of 75 °C and above.[4] It is now thought that myoglobin undergoes a small conformational change at some temperature below its denaturational temperature[42] and it has been argued that this conformational change permits some of the other, less stable proteins present in meat to attack the partially exposed haematin following, or during, their denaturation.[42,43] This attack displaces the globin which spontaneously denatures. The globin may, however, still supply one of the ligands.

On this hypothesis a whole range of haematin-denatured protein complexes are formed in which the bound protein is unlikely to be globin. Studies on muscle extracts have suggested that the concentration and type of bound protein is dependent on the thermal history of the meat. For example, muscle extracts held at 52 °C, a temperature at which several of the sacroplasmic proteins in meat will denature and precipitate but at which myoglobin will be largely unaffected, was found to form less cooked meat haemoprotein on subsequent heating to 60 °C than samples heated directly to 60 °C.[42] The effectivenes of several of the muscle proteins in forming 'cooked meat haemoprotein' was dependent on the conditions employed. Thus at 65 °C most of the formation was due to reaction with the soluble proteins of molecular weight *ca.* 100 000 whilst at 60 °C proteins of molecular weight *ca.* 80 000 were most reactive. The temperature dependence presumably relates to the different thermal stabilities of the proteins present.

It would also appear that on heating meat there is some cleavage of the porphyrin ring and subsequent release of the iron such that the total haemoprotein content decreases.[44] This may cause some small change in the hue and chroma of the meat. The amount of iron released depends on the severity of heating, there being an apparently linear increase with time, and the presence of other constituents: ascorbic acid accelerates the release whereas nitrite inhibits it.[45]

Whatever the reactions responsible for, and the nature of cooked meat haemoproteins, their formation is used to assess the degree of 'doneness' of red meat. For example, beef cooked to an internal temperature of 60 °C is considered rare, that cooked to 71 °C medium-done and that to at least 77 °C as well done.[4] Some white meats such as turkey and pork have a persistent pink coloration even after heat treatment at 95 °C. This may be due to the formation of carbon monoxide or nitric oxide type pigments generated in gas ovens.[4] However, more recent work suggests that it may be associated with the stable reduced cytochrome present at low concentrations in meat.[46]

It appears that the unreacted myoglobin in rare beef, even after prolonged heating at low temperatures, is in the red, oxymyoglobin form,[40] but the reasons for this are not clear as, at the elevated temperatures and low oxygen pressures assumed to be present in the centre of the meat, one might expect the pigment to be present as the brown oxidized metmyoglobin or possibly the purple reduced myoglobin. It is known though that when aqueous muscle extracts are held at temperatures up to 60 °C the undenatured haemoproteins remaining in solution, although initially oxidizing to the met form, are subsequently reduced back to a red pigment with the same spectral characteristics as oxymyoglobin.[4,36] Thus at these temperatures a reductive mechanism may also operate in meat to convert the undenatured pigment into the ferrous state and thus yield the red colour that consumers associate with 'rare' beef. It is also possible that at these temperatures the oxygen-utilizing enzymes are inactivated and thus oxygen solubility increases so that oxygen can diffuse to the centre of the meat.

Although the colour of cooked meat is primarily due to protein denaturation additional browning involving Maillard reactions may occur.[1] These involve reactions between the amino groups of proteins and reducing sugars or other carbonyls.

Canned Meats and Fish

Following thermal processing of both meat and fish such as tuna, where haemoproteins are the major colour pigments, the product is usually brown in colour owing to formation of the cooked meat haemoproteins. However, the interior of such meat and fish pieces may develop a reddish coloration and this may well relate to the high-temperature reduction of metMb seen in the muscle extracts or may be the reduced (ferrous) form of the cooked meat haemoproteins. The reduced form of the cooked meat haemoprotein may be the cause of the pinkness occasionally seen in cooked poultry meat.[47] In white fish the major colour change on heating, apart from protein denaturation leading to loss of translucency, involves non-enzymic (Maillard) browning.[48] It is for this reason that few white fish species are canned.

Greening in canned fish products such as tuna and oysters results from reaction of the porphyrin pigments with sulfur compounds. In tuna the reactant is believed to involve a cysteine on the myoglobin;[49] mammalian myoglobins do not possess a cysteine group and thus do not undergo this reaction.

The above changes relate to the porphyrin pigments in fish and as such have much in common with meat products. However, in crustaceans the coloured pigments are carotenoid in nature and are found in the exoskeletons of such species as crabs and lobsters. The major such pigment in marine life is the red xanthophyl astaxanthin. In nature this pigment is bound to a simple protein (carotenoprotein) to yield components that are usually blue or blue–grey in colour. On heat denaturation (as occurs when crabs or lobsters are immersed in boiling water) the bright red astaxanthin is released.[50]

On canning such crustaceans may undergo one of several undesirable colour changes and Boon[51] has summarized these as (i) a blue discoloration due to reaction of copper ions with crab blood, (ii) Maillard browning, (iii) black iron sulphide production from H_2S and iron, and (iv) discolorations due to diffusion of pigments.

4 References

1. R.A. Lawrie, 'Meat Science', 4th Edn., Pergamon Press, Oxford, 1985.
2. P.D. Warriss, S.N. Brown, J.M. Adams, and D.B. Lowe, *Meat Sci.*, 1990, **28**, 321.
3. D.A. Hector, C. Brew-Graves, N. Hassan, and D.A. Ledward, *Meat Sci.*, 1991 (in press).
4. D.A. Ledward, in 'Developments in Food Protein - 3', ed. B.J.F. Hudson, Elsevier Applied Science, London, 1984, p.33.
5. L.D. Satterlee and N.Y. Zachariah, *J. Food Sci.*, 1972, **37**, 909.
6. M. Bembers and L.D. Satterlee, *J. Food Sci.*, 1975, **40**, 40.
7. A. Trautwein, R. Zimmermann, and F.E. Harris, *Theor. Chim. Acta,* 1974, **36**, 67.
8. M. O'Keefe and D.E. Hood, *Meat Sci.*, 1982, **7**, 209.
9. G.G. Gidding, *CRC Crit. Rev. Food Sci. Nutr.*, 1977, **9**, 81.
10. J.L. Atkinson and M.J. Follet, *J. Food Technol.*, 1973, **8**, 51.

11. M.W. Orcutt, T.R. Dutson, D.P. Cornforth, and C.G. Smith, *J. Anim. Sci.*, 1985, **58**, 1366.
12. D.A. Ledward, R.F. Dickinson, V.H. Powell, and W.R. Shorthose, *Meat Sci.*, 1986, **16**, 245.
13. P. George and G. Hanania, *Biochem. J.*, 1952, **52**, 516.
14. W.D. Brown and L.B. Mebine, *J. Biol. Chem.*, 1969, **244**, 6696.
15. H.E. Synder and H.B. Skrdlant, *J. Food Sci.*, 1966, **31**, 468.
16. D.J. Livingstone and W.D. Brown, *Food Technol.*, 1981, **35**, 244.
17. D.A. Ledward, *J. Food Sci.*, 1972, **37**, 634.
18. L. Hagler, R.I. Coppes, Jr., and R.H. Helman, *J. Biol. Chem.*, 1979, **254**, 6505.
19. M. O'Keefe and D.E. Hood, *Meat Sci.*, 1980–81, **5**, 27.
20. M. O'Keefe and D.E. Hood, *Meat Sci.*, 1980–81, **5**, 267.
21. M. Renerre and R. Labas, *Meat Sci.*, 1987, **19**, 151.
22. C. Faustman and R.G. Cassens, *J. Food Sci.*, 1990, **55**, 1278.
23. D.A. Ledward, *Meat Sci.*, 1985, **15**, 149.
24. P.S. Sleper, M.C. Hunt, D.H. Kropf, C.L. Kastnel, and M.E. Dikeman, *J. Food Sci.*, 1983, **48**, 479.
25. J.R. Claus, D.H. Kropf, M.C. Hunt, C.L. Kastnel, and M.E. Dikeman, *J. Food Sci.*, 1984, **49**, 1021.
26. J.R. Bendall and A.A. Taylor, *J. Sci. Food Agric.*, 1972, **23**, 707.
27. A.A. Taylor and D.B. MacDougall, *J. Food Technol.*, 1973, **8**, 453.
28. J.E. Owen and R.A. Lawrie, *J. Food Technol.*, 1975, **10**, 169.
29. G. Bertelsen and L.H. Skibsted, *Meat Sci.*, 1987, **19**, 243.
30. M. Mitsumato, C. Faustman, R.G. Cassens, R.N. Arnold, D.M. Schaeffer, and K.K. Scheller, *J. Food Sci.*, 1991, **56**, 194.
31. D.L. Gee and W.D. Brown, *J. Agric. Food Chem.*, 1978, **26**, 273.
32. D.J. Nicol, M.K. Shaw, and D.A. Ledward, *Appl. Microbiol.*, 1970, **19**, 937.
33. A.W. Nichol, E. Hendry, D.B. Morrell, and P.S. Clezy, *Biochim. Biophys. Acta*, 1968, **128**, 97.
34. A.F. Egan, B.J. Shay, and G. Stanley, Meat Research in CSIRO Annual Report, 1980, p.30.
35. P. Nicholls, *Biochem. J.*, 1961, **81**, 374.
36. D.A. Ledward, in 'Progress in Food Biochemistry - Proteins', ed. B.J.F. Hudson, Elsevier Applied Science, London, 1991 (in press).
37. S. Harel and J. Kanner, *J. Agric. Food Chem.*, 1985, **33**, 1186.
38. R. Palombo and G. Wygaards, *J. Food Sci.*, 1990, **55**, 601.
39. R. Palombo and G. Wygaards, *J. Food Sci.*, 1990, **55**, 604.
40. H.N. Draudt, Proceedings of the Annual 22nd Reciprocal Meat Conference 1969, American Meat Science Association, Chicago, p.180.
41. A.L. Tappel, *Food Res.*, 1957, **22**, 404.
42. D.A. Ledward, *J. Food Sci.*, 1971, **36**, 138.
43. D.A. Ledward, *J. Food Technol.*, 1974, **9**, 59.
44. B.R. Schricker and D.D. Miller, *J. Food Sci.*, 1983, **48**, 1340.
45. B.R. Schricker, D.D. Miller, and J.R. Stouffer, *J. Food Sci.*, 1982, **47**, 740.
46. B. Girard, J. Vanderstoep, and J.F. Richards, *J. Food Sci.*, 1990, **55**, 1249.
47. D.U. Ahn and A.J. Maurer, *Poult. Sci.*, 1990, **69**, 157.
48. H. Yamanaka, M. Bito, and M. Yokoseki, *Bull. Jpn. Soc. Sci. Fish.*, 1973, **39**, 1293.

49. O. Grosjean, B.F. Cobb, B. Mebine, and W.D. Brown, *J. Food Sci.,* 1969, **34**, 404.
50. M.L. Woolfe, in 'Effects of Heating of Foodstuffs', ed. R.J. Priestley, Applied Science Publishers, London, 1979, p.106.
51. D.D. Boon, *J. Food Sci.,* 1975, **40**, 756.

Oxidative Flavour Changes in Meats: Their Origin and Prevention

J.I. Gray and R.L. Crackel

DEPARTMENT OF FOOD SCIENCE AND HUMAN NUTRITION, MICHIGAN STATE UNIVERSITY, EAST LANSING, MICHIGAN 48824, USA

1 Introduction

Consumer acceptance of meat products depends, to a major extent, on their flavour. Flavour quality of meats is influenced by a number of factors including genetics (*e.g.* mutton flavour and swine sex odour), animal feed, processing and storage procedures (*e.g.* irradiation and high-temperature processing), bacterial growth, and oxidation of meat lipids.[1,2] The latter is the most common cause of undesirable meat flavour and has been the focus of much research activity over the past three decades. In addition, oxidative deterioration in meat has been associated with undesirable colour changes, losses of protein functionality and nutritive value as a result of reaction of oxidation products with proteins, and adverse biological effects such as cardiovascular disease and cancer.[3,4]

Development of oxidative off-flavours has long been recognized as a serious problem during the holding or storage of meat products. These flavour changes can be conveniently divided into two categories: one involving oxidation that occurs in meats following cooking, the other in frozen raw meats.[1] Although the latter category is generally fairly resistant to oxidation, rancidity can develop during freezing and thawing.[5] Wide fluctuations in temperature and inadequate protection from oxygen can accelerate the development of off-flavours. Under proper storage conditions, however, lean raw meat is quite stable for periods of several months to a year, depending on the species from which it originated.[5] The more serious flavour defect attributed to the oxidation of lipids in meats is that which occurs in cooked meats during storage. This is a serious concern particularly in light of the increased demand for portion-controlled, precooked convenience products such as restructured meats.[6]

The role of lipids in the development of off-flavours in meats and factors influencing their susceptibility to oxidation will be addressed in this review. These factors include the composition and location of lipids and the presence of pro-oxidative and antioxidative compounds. Particular attention will be directed toward recent studies on the respective roles of haem and non-haem iron as catalysts of lipid oxidation, and the emergence of lipoxygenase as a possible initiator of the oxidative reactions in muscle systems. The susceptibility of cholesterol to oxidation in meat systems will also be addressed.

2 Composition of Lipids in Meats

Lipids comprising animal fats are commonly classified as depot or adipose tissue and as intramuscular or tissue lipids.[7,8] Depot fats are generally localized in subcutaneous deposits, although significant amounts may be located in the thoracic and abdominal cavities and as intermuscular deposits. These lipids consist mainly of triacylglycerols which may vary in amount and fatty acid composition according to species, diet, sex, age, environment, and depot location within the animal.[9,10] Tissue lipids, on the other hand, vary much less in proportion and fatty acid composition. These lipids consist largely of membrane-bound phospholipids and lipoproteins of the cell wall, mitochondria, and other cellular structures.[8] Although the amount of phospholipid varies inversely with the lipid content of meat, when expressed as a percentage of total tissue weight, levels of phospholipids remain fairly constant and range from 0.5 to 1.0%.[10] There is, however, variation in the phospholipid content between species[11] and between locations within the same species.[10,12] Data in Table 1 also confirm the constancy of the phospholipid fractions in different muscle types, even though the total lipid contents are highly variable.[13]

Table 1 *Influence of muscle and lipid content on quantity of intramuscular neutral lipid and phospholipids*

Species	Muscle or type	Content		
		Lipid (%)	Neutral lipids (%)	Phospholipids (%)
Chicken	White	1.0	52	48
	Dark	2.5	79	21
Turkey	White	1.0	29	71
	Dark	3.5	74	26
Beef	*Longissimus dorsi*	2.6	78	22
		7.7	92	8
		12.7	95	5
Pork	*Longissimus dorsi*	4.6	79	21
	Psoas major	3.1	63	37
Lamb	*Longissimus dorsi*	5.7	83	10
	Semitendinosus	3.8	79	17

From ref.13, with permission.

Fatty acids in meats are typically straight chained with an even number of carbons.[10] Generally, the predominant fatty acids of triacylglycerols are saturated or contain only one or two double bonds.[14] Polyunsaturated fatty acids with three or more double bonds constitute less than two percent of the fatty acids in beef triacylglycerols.[8] The fatty acid composition of depot fat of monogastric animals is influenced by a number of factors including the composition of dietary fats.[8] Early studies by Ellis and Isbell[15] showed that depot fat from pigs fattened on rations containing high levels of unsaturated lipids contained degrees of unsaturation varying with the degree of unsaturation of the feed. Recently, it was demonstrated that elevated levels (10 and 20%) of canola oil (with 64% oleic acid and 28% polyunsaturated fatty acids) in a swine diet increased the percentage of monounsaturated fatty acids in the neutral lipids[16] and in the total lipids.[17] However, the dietary canola oil treatments tended to increase the susceptibility of muscle tissue toward lipid oxidation.[17] Miller et al.[18] also showed alteration in the fatty acid profiles of both subcutaneous and intermuscular adipose tissue of swine fed elevated levels of monounsaturated fats (safflower, sunflower, and canola oils) in the diets.

In contrast to triacylglycerol fractions, phospholipids in muscles are altered only slightly by dietary changes. Additionally, phospholipids contain a much larger proportion of C_{20} and C_{22} polyunsaturated fatty acids. Polyunsaturation of the phospholipid fraction is *ca.* 15 times greater than that of the triacylglycerol fraction.[14,19] The unsaturated fatty acid contents of phospholipids in some muscle foods are summarized in Table 2.[20]

3 Lipid-derived Off-flavours in Meats
Description of Off-flavours

When meat is cooked, lipids undergo thermally induced oxidation to produce a range of volatile products which can contribute to meat aroma, but which may also react with components from the lean tissue to give other flavour compounds.[21] The importance of the interaction of lipids in the Maillard reaction to the formation of meat flavour compounds was demonstrated recently by Farmer et al.[22] who reported that phospholipids had a marked effect on the overall odour of reaction mixtures containing a sugar (ribose) and an amino acid (cysteine) and on the nature of the individual aromas detected.

The development of lipid-derived off-flavours during the holding or storage of meat products has long presented a challenge to the meat technologist. Rancidity in meat begins to develop soon after death and continues to increase in intensity until the meat products become rancid.[6] When uncured meat is stored after cooking, flavour changes occur rapidly. The term 'warmed-over flavour' was initially used by Tims and Watts[23] to describe the rapid development of oxidized flavours in refrigerated cooked meats. The oxidized or stale flavour becomes readily apparent within 48

Table 2 *Unsaturated fatty acid content of phospholipids in some muscle foods*

Fatty acid	Content (%)				
	Lamb	Beef	Pork	Chicken	Fish
C18:1	19.51	33.44	12.78	20.25	19.59
C18:2	18.49	10.52	35.08	14.20	5.88
C18:3	0.43	1.66	0.33	0.90	8.07
C20:2	0.34	0.69	–	–	0.20
C20:3	0.62	2.77	1.31	1.30	0.36
C20:4	13.20	8.51	9.51	11.60	3.75
C20:5	–	0.76	1.31	1.55	7.16
C22:4	–	0.88	0.98	2.10	0.65
C22:5	–	0.92	2.30	5.75	2.39
C22:6	–	–	2.30	5.75	2.39

From ref.20, with permission.

hours in contrast to the more slowly developing rancidity that becomes evident only after freezer storage for a period of months.[6] Although warmed-over flavour was first recognized as occurring in cooked meat, it also develops in raw meat that is ground and exposed to air[24] and in unheated products such as mechanically separated and restructured meats in which the muscle structure is disrupted and air is incorporated.[6]

The term warmed-over flavour has been questioned recently in that it does not adequately describe the flavour changes in cooked meat during storage. Because lipid oxidation has been considered the major contributor to warmed-over flavour,[8,23] the terms rancidity and warmed-over flavour are often used interchangeably. The term, 'meat flavour deterioration' has been developed to describe better the complex series of chemical reactions that contribute to an overall increase in oxidative off-flavour notes, with a concomitant deterioration of the desirable beef flavour itself.[25,26] Many of these reactions involve free radicals.[27] Melton et al.[28] reviewed sensory studies dealing with undesirable flavours in meats, including warmed-over flavour, and stated that descriptive terms for this flavour defect were needed, as were accompanying standards to trained panellists to evaluate warmed-over flavour. Love[29] also indicated that terms such as painty, cardboard, and cooked beef brothy flavour better describe the characteristic flavour changes that stored meat, either raw or cooked, undergoes during storage. Similar terminology was reported previously by Johnsen and Civille,[30] who also described a general pattern involved in warmed-over flavour: the disappearance of the fresh flavours, the appearance then disappearance of the cardboard flavour, and the final dominance of other flavours by the oxidized/rancid note. They believe that a descriptive vocabulary would benefit studies to examine causes and prevention of warmed-over flavour by providing an established protocol that would allow comparison and evaluation of data from different laboratories.

Source of Oxidative Off-flavours in Meats

Many researchers have implicated the phospholipids as the major contributors to oxidative off-flavours in poultry, beef, pork, and lamb. For example, by comparing correlation coefficients between thiobarbituric acid (TBA) numbers (an index of lipid oxidation) and phospholipid levels, Wilson et al.[31] suggested that phospholipids play a major role in the development of warmed-over flavour in turkey, chicken, beef, and lamb, but that total lipids were the major contributor to this flavour defect in pork. Working with model systems of lipid-extracted muscle fibres, Igene and Pearson[19] confirmed total phospholipids as the major contributors to warmed-over flavour. Triacylglycerols enhanced the development of warmed-over flavour only when combined with the phospholipids as total lipids.

The development of oxidative rancidity varies with species, most likely because of differences in phospholipid content and fatty acid composition. Poultry and fish have higher contents of phospholipids than red meats.[32,33] They also contain higher proportions of di- and poly-unsaturated fatty acids.[20] This could explain why rancid off-flavours develop more rapidly in fish and poultry than in beef or pork.[31]

The lability of phospholipids is partially due to their high content of unsaturated fatty acids,[34] particularly linoleic and arachidonic acids. Their pronounced susceptibility to oxidation may also be due to their close association in membranes with tissue catalysts of oxidation. Processes which disrupt the membranes such as grinding, chopping, or emulsifying expose the phospholipids to oxygen, enzymes, haem pigments, and metal ions which can cause the rapid development of rancidity, even in fresh raw meat.[24,35]

Numerous studies have shown that lipid oxidation will occur during freezer storage of meats, oxidative stability being determined by the degree of fatty acid unsaturation. Igene et al.[36] showed the stability of various meats to be: beef > chicken white meat > chicken dark meat. These trends were indicated by increased TBA numbers and decomposition of constituent lipids. There are, however, conflicting reports as to whether the phospholipids or triacylglycerols undergo changes in frozen storage. Igene et al.[36] found that changes in total lipids of raw beef and chicken were mainly due to losses in the neutral lipid fraction. The phospholipid fraction remained relatively constant through thirteen months of frozen storage. Contradictory data from Keller and Kinsella[37] indicated negligible decreases in total lipids, but significant losses of phospholipids during frozen storage of ground beef. These trends may have been caused by the grinding of the meat which promoted oxidation of the membrane-bound phospholipids.

In a second study with intact beef and poultry (light and dark muscles), Igene et al.[14] determined that only minor changes occurred in the fatty acid profiles of the triacylglycerols during either freezer storage of the raw meat or subsequent storage after cooking. However, there was a significant decline in the amount of phosphatidylethanolamine and phosphatidylcholine, and total unsaturation of the raw tissues declined from 67.1 to 58.4% after eight months of frozen storage. Gokalp et al.[38] also reported decreases in certain phospholipid classes in frozen beef patties during storage. Trace amounts of lysodiphosphatidylglycerol also became evident after three months of storage. In a subsequent study, Gokalp et al.[39] demonstrated significant losses of linolenic and arachidonic acids from the phospholipid fraction of cooked beef patties during frozen storage. High correlations between phospholipid total saturated and unsaturated fatty acids, TBA numbers, and sensory panel rancidity scores were reported. Because the oxidation of phospholipids was more advanced at three months compared with the neutral lipid fraction between three months and four and one-half months, it was concluded that tissue lipid oxidation occurs in two stages - initial oxidation of the phospholipids, followed by triacylglycerol oxidation, in agreement with El Gharbawi and Dugan.[40]

4 Lipid Oxidation in Meat

Mechanism of Lipid Oxidation

The reaction of unsaturated lipids (RH) with oxygen to form hydroperoxides is generally considered to be a free-radical process that involves three basic steps:[41]

$$\textit{Initiation}: \text{RH} \xrightarrow{\text{initiator}} \text{R} \cdot$$

$$\textit{Propagation}: \text{R} \cdot + \text{O}_2 \xrightarrow{\text{fast}} \text{ROO} \cdot$$

$$\text{ROO} \cdot + \text{RH} \xrightarrow{\text{slow}} \text{ROOH}$$

$$\textit{Termination}: \text{R} \cdot + \text{R} \cdot \longrightarrow \text{R—R}$$

$$\text{R} \cdot + \text{ROO} \cdot \longrightarrow \text{ROOR}$$

$$\text{ROO} \cdot + \text{ROO} \cdot \longrightarrow \text{ROOR} + \text{O}_2$$

During initiation, a labile hydrogen is abstracted from a reactive methylene group adjacent to a double bond in the presence of an appropriate initiator. The reaction of the resulting lipid free radical (R·) with oxygen in the propagation step leads to peroxy radicals (ROO·) which react with unsaturated fatty acids to form hydroperoxides (ROOH). The formation of non-radical products can occur in the termination stage by the interaction of R· and ROO·.

Products of Lipid Oxidation and Their Effect on Meat Quality

Hydroperoxides, the primary initial products of lipid oxidation, are odourless, but will decompose to a variety of volatile and non-volatile secondary products as shown in Figure 1.[42] Step one involves homolytic cleavage of the oxygen–oxygen bond to yield the alkoxy and hydroxy free radicals. Carbon–carbon scission on either side of the alkoxy radical can result in the formation of an aldehyde and a new free radical (reaction 2). Removal of a hydrogen atom from another fatty acid molecule can produce an alcohol and a new free radical (reaction 3). Free radicals from reactions 2 and 3 may participate in the propagation of the chain reaction, or react with each other to form non-radical products which terminate the chain reaction. Ketones are formed from the reaction been an alkoxy radical and other simple free radicals (reactions 4 and 5).

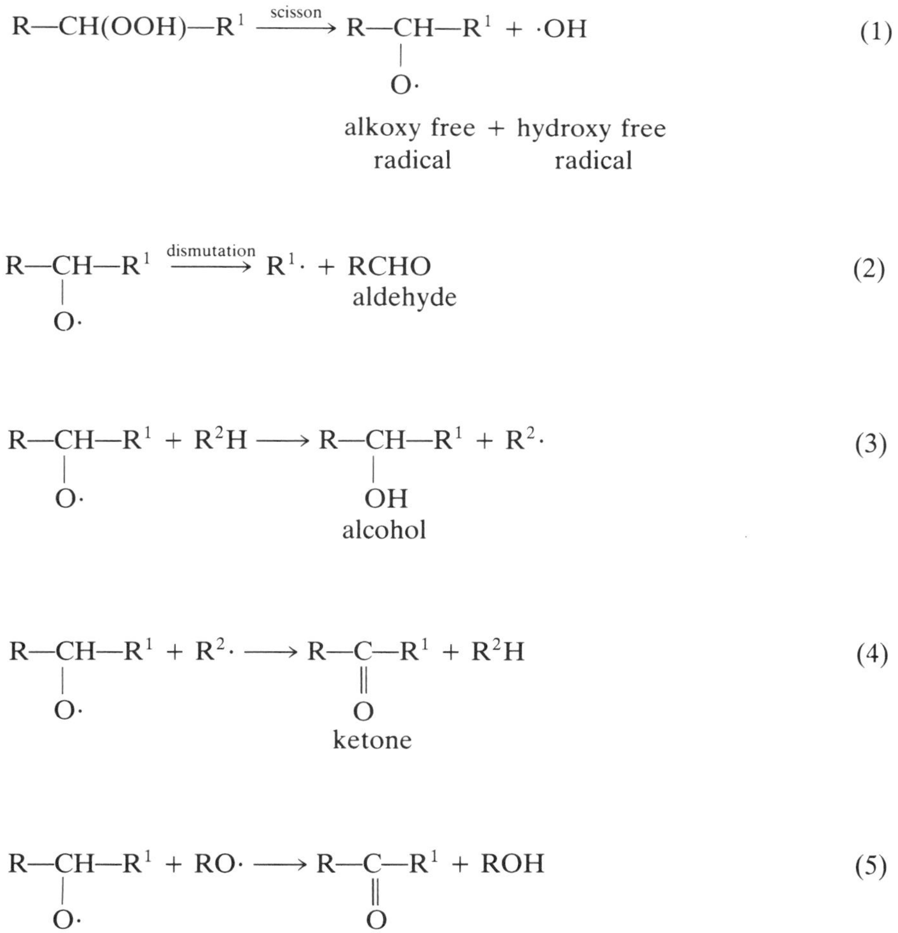

$$R-CH(OOH)-R^1 \xrightarrow{\text{scisson}} R-\underset{\underset{\text{O·}}{|}}{CH}-R^1 + \cdot OH \quad (1)$$

$$\text{alkoxy free radical} + \text{hydroxy free radical}$$

$$R-\underset{\underset{\text{O·}}{|}}{CH}-R^1 \xrightarrow{\text{dismutation}} R^1\cdot + \underset{\text{aldehyde}}{RCHO} \quad (2)$$

$$R-\underset{\underset{\text{O·}}{|}}{CH}-R^1 + R^2H \longrightarrow R-\underset{\underset{\text{OH}}{|}}{CH}-R^1 + R^2\cdot \quad (3)$$

$$\text{alcohol}$$

$$R-\underset{\underset{\text{O·}}{|}}{CH}-R^1 + R^2\cdot \longrightarrow R-\underset{\underset{\text{O}}{\|}}{C}-R^1 + R^2H \quad (4)$$

$$\text{ketone}$$

$$R-\underset{\underset{\text{O·}}{|}}{CH}-R^1 + RO\cdot \longrightarrow R-\underset{\underset{\text{O}}{\|}}{C}-R^1 + ROH \quad (5)$$

Figure 1 *The general decomposition of hydroperoxides into aldehydes and ketones*[42]

Secondary products of lipid oxidation are the major contributors to off-flavours in meats.[43] These compounds do not contribute to desirable meaty flavour but rather impart green, rancid, fatty, pungent, and other off-flavour characteristics to meat.[44] In order to assess the contribution of each of these compounds to the flavour or aroma of meat products, it is necessary to consider the flavour threshold of the compound as well as its concentration and flavour character. Drumm and Spanier[45] provided a literature-derived overview of the relationship between classes of compounds and their flavour thresholds (Table 3), and indicated that, in general, the carbonyl compounds have the greatest impact on flavour owing to their low flavour threshold in comparison with the hydrocarbons, substituted furans, and alcohols. These investigators studied the effects of refrigerated storage on the content and decomposition of the lipid oxidation and sulfur-containing compounds in cooked beef patties and concluded that aldehydes are major contributors to the loss of desirable flavour in meats because of their high rate of formation during lipid oxidation and low flavour threshold. The rate of formation of the saturated aldehydes (*e.g.* hexanal, pentanal, nonanal) was greater than that of the unsaturated aldehydes (*e.g.* nona-2,4-dienal, deca-2,4-dienal) during refrigerated storage of the patties.

The contribution of alcohols to the undesirable flavour quality of the beef patties was measurably smaller than that of the aldehydes, owing to the relatively higher flavour thresholds of the alcohols (Table 3). For example, pentanol and oct-1-en-3-ol, derived from linoleate oxidation, did not exceed their threshold until approximately day 3 of the storage period.[45] Several ketones, including heptan-2-one and octane-2,3-dione, were also produced during the four day storage period, the latter being

Table 3 *Flavour thresholds of classes of lipid oxidation compounds*

Class	Threshold (p.p.m.)
Hydrocarbons	90–2150
Alk-1-enes	0.02–9
Substituted furans	2–27
2-Phenylfuran	1–10
Saturated alcohols	0.3–2.5
Vinyl alcohols	0.05–3
Oct-1-en-3-ol	0.001
Aldehydes	0.014–0.03
Alk-2-enals	0.04–2.5
Alkanals	0.04–1
t,t-Alka-2,4-dienals	0.005–0.5
t,c-Alka-2,4-dienals	0.002–0.6
Methyl ketones	0.16–5.5
Vinyl ketones	0.0002–0.007

From ref.45, with permission.

unique in that it was not detected immediately after cooking but rapidly developed during storage. Drumm and Spanier[45] also determined that the content of heterocyclic sulfur compounds did not change significantly with storage. They concluded that the decrease in desirable flavour observed during the refrigerated storage of cooked meat may be attributed to the masking of desirable flavour notes by the increased content of undesirable flavour compounds rather than to the degradation of desirable flavour compounds.

Hexanal, one of the major secondary products formed during the oxidation of linoleic acid,[46] and other aldehydes have been used successfully to follow lipid oxidation in meat products and are related to off-flavour development. Shahidi et al.[47] reported a linear relationship between hexanal content, sensory scores, and TBA numbers of cooked ground pork, while St.Angelo et al.[48] established a similar correlation for cooked beef. The latter investigators also demonstrated that octane-2,3-dione and total volatiles, as well as hexanal, were significantly correlated to sensory scores and TBA numbers of roast beef (Table 4). They concluded that many compounds usually associated with lipid oxidation reactions could be used as marker compounds to follow the development of warmed-over flavour in cooked beef, although hexanal was the compound which increased most rapidly during the storage period. Earlier, Ruenger et al.[49] had reported that heptanal and n-nona-3,6-dienal are major flavour compounds related to warmed-over flavour in turkey.

5 Catalysts of Oxidative Changes in Meats
Haem and Non-haem Iron

Much effort has been devoted to understanding the catalytic roles of haem and non-haem iron in the oxidative deterioration of meat lipids. Traditionally, lipid oxidation has been attributed to haem catalysts such as haemoglobin, myoglobin, and cytochromes. Tappel[50] proposed that haematin compounds catalysed the decomposition of lipid hydroperoxides into free radicals which could then propagate the free-radical chain mechanism. This mechanism depends on the presence of preformed hydroperoxides and does not explain the initiation process, *i.e.* the abstraction of a hydrogen atom from a methylene group in a polyunsaturated fatty acid molecule.

Non-haem iron has also been implicated for its ability to decompose lipid hydroperoxides. Ferrous iron is known to decompose lipid hydroperoxides, forming very reactive alkoxy radicals for the propagation reactions, whereas ferric ion produces relatively less reactive peroxy (ROO·) radicals from fatty acid hydroperoxides:[51]

$$ROOH + Fe^{2+} \rightarrow RO\cdot + OH^- + Fe^{3+}$$
$$ROOH + Fe^{3+} \rightarrow ROO\cdot + H^+ + Fe^{2+}$$

Table 4 *Mean sensory scores, chemical and instrumental values of cooked beef*

	Sensory score[b]	TBA[c]	Total[c] volatiles	Hexanal[c]	Octane-2,3-dione[c]
Overall mean[a]					
Fresh	2.1	3.75	5.35	2.55	0.19
WOF	5.0	14.71	19.09	11.16	1.60
Correlation coefficients[d]					
r		0.80	0.84	0.80	0.81
r[1]		–	0.84	0.92	0.88

[a] $n = 14$.
[b] Intensity of warmed-over flavour: 0 = no WOF; 10 = extreme WOF.
[c] Parts per million, based on hexane equivalent for GC peaks.
[d] r = correlation coefficient between sensory scores and chemical or instrumental values; r^1 = correlation coefficient between TBA and instrumental values.
From ref.48, with permission.

Sato and Hegarty[24] studied the development of warmed-over flavour in water-extracted muscle tissue and concluded that non-haem iron rather than haem iron was the active catalyst responsible for the rapid oxidation of cooked meat. Haem compounds were reported to have little influence on the development of off-flavours or TBA-reactive substances in meat. Love and Pearson[52] and Igene et al.,[53] working with similar systems, also concluded that myoglobin was not the principal pro-oxidant in cooked meat. The latter study also revealed that the level of free iron greatly increased during cooking and accelerated lipid oxidation in cooked meat. It was concluded that haem pigments serve as a source of free iron, being readily broken down during the heating process and catalysing autoxidation. The increase in the amount of non-haem iron as a result of heating was verified by Schricker et al.[54] and Schricker and Miller.[55] However, as pointed out by Rhee,[56] the enhanced susceptibility of cooked meat to oxidation may also be due to the disruption of meat tissues by cooking, thus bringing the lipids and catalysts into closer contact.

Recently, Kanner and his associates[57-59] have demonstrated a direct involvement of ferric haem pigments in lipid oxidation in muscle tissues. Their results suggest that the haem pigments may only be effective catalysts in the presence of hydrogen peroxide. However, Rhee et al.[60] provided some evidence that the catalytic effect of the metmyoglobin–hydrogen peroxide system may be due, at least in part, to the release of iron from the haem pigment by hydrogen peroxide. They concluded that the activated metmyoglobin was the primary catalyst of oxidation in the raw water-extracted beef muscle system, although the non-haem iron released from the haem pigment was also contributory. When the muscle system was heated, lipid oxidation was slow in the system initially treated with metmyoglobin alone, but was rapid in those systems containing metmyoglobin and hydrogen peroxide. These investigators suggested that haem iron (activated myoglobin) may initiate lipid oxidation in cooked meat as it may do so in raw meat, but non-haem iron plays a greater role in accelerating lipid oxidation in cooked meat than in raw meat. Kanner et al.[61,62] also reported that lipid oxidation in ground turkey muscle is effected by free iron and other transition metals.

Results of recent studies, however, have cast doubt on the apparently superior catalytic properties of non-haem iron.[8,63] Johns et al.[64] demonstrated that haemoglobin was more effective than inorganic iron compounds in stimulating lipid oxidation in emulsions containing refined lard, egg white, and corn starch. They further demonstrated that haemoglobin, when added to and evenly distributed in washed muscle fibres at concentrations comparable to those naturally present in meat, was again a strong catalyst, whereas all forms of inorganic iron appeared to have little pro-oxidant activity. They concluded that the disparity between their results and those of previous researchers[24,52,63] may be due, in part, to the difficulty of evenly dispersing the catalysts in the washed fibres, and that hydrogen peroxide, formed by the oxidation of oxyhaemoglobin, may be necessary for methaem pigments to be effective catalysts. Monahan (unpublished data) confirmed the results of Johns et al.,[64] but also determined that the mode of addition of the catalysts to the washed fibres did not affect their catalytic activity. In addition, similar results were obtained for haemoglobin and myoglobin, thus casting doubt on the suggestion that hydrogen peroxide, produced by the oxidation of haemoglobin, activated the methaemoglobin into an effective catalyst.[64] Monahan further demonstrated that the catalytic effects of haem proteins and ferrous iron increased with increasing catalyst concentration and were similar when present in equimolar iron concentrations. Thus, further studies are necessary to clarify the respective roles of haem and non-haem iron as catalysts in meat model systems.

Initiators of Oxidation

Direct interaction of molecular oxygen with an unsaturated fatty acid is extremely slow because the initiation of the oxidative process requires an active form of oxygen.[65] Ground-state oxygen is paramagnetic (i.e. contains two unpaired electrons that have the same spin but are in different orbitals), while the ground state of unsaturated fatty acids corresponds to the singlet state which is diamagnetic (i.e. contains no unpaired electrons). Hence, there is a spin barrier which prevents the direct addition of triplet state oxygen to singlet state unsaturated fatty acid molecules. This spin restriction can be overcome by several mechanisms including photooxidation, partially reduced or activated oxygen species such as hydrogen peroxide, superoxide anion, and hydroxyl radicals, active oxygen–iron complexes (ferryl iron), and iron-mediated homolytic cleavage of the hydroperoxides which generate organic free radicals.[65]

A number of different initiators of lipid oxidation in biological systems have been suggested,[35,66] but at present the evidence in support of these initiators seems more suggestive than conclusive. Those mechanisms receiving the most attention in recent years include the hydroxyl radical, perferryl or ferryl radical, dioxygen-bridged radical, and the porphyrin cation radical. The hydroxyl radical ($\cdot$OH) is a very potent reactive oxygen species and it has been suggested that it can abstract hydrogen from unsaturated lipids and thus initiate lipid oxygen. Most of the $\cdot$OH generated *in vivo* comes from the metal-dependent decomposition of hydrogen peroxide. The Fe^{2+}-dependent breakdown of hydrogen peroxide (the so-called Fenton reaction) is usually written as

$$Fe^{2+} + H_2O_2 \rightarrow Fe^{3+} + \cdot OH + OH^-$$

Many of the proponents of the hydroxyl radical theory hold the Fenton-type reactions to be responsible for hydroxyl radical formation in biological systems.[67]

Tien and Aust[68] described lipid oxidation initiated by NADPH-cytochrome P-450 reductase and by xanthine oxidase. In either case, $ADP-Fe^{3+}$ is reduced by either the reductase or by superoxide radicals (O_2^-) to form $ADP-Fe^{2+}-O_2$. This complex will either react with polyunsaturated fatty acids to initiate lipid peroxidation, or it may be reduced to the ferryl ion ($ADP-[FeO]^{2+}$) which could initiate oxidation. Minotti and Aust[69] also suggested that a ferrous–dioxygen–ferric complex is the actual initiating species. This complex originates either from ferric ions in the presence of a reducing agent or from ferous ion in the presence of an oxidizing agent. Complete reduction of the ferric ion, however, would inhibit oxidation, while complete oxidation of the ferrous ion would have the same effect.

Recently, Kanner and his associates[57-59] extensively studied muscle membranal lipid oxidation initiated by hydrogen peroxide-activated metmyoglobin. The interaction of hydrogen peroxide and metmyoglobin resulted in an activated species capable of initiating lipid oxidation. Haem pigments or hydrogen peroxide alone did not promote oxidation. They proposed that the oxidation of oxymyoglobin and oxyhaemoglobin leads to the formation of methaem proteins and the superoxide radical which dismutates to hydrogen peroxide. Metmyoglobin is activated by hydrogen peroxide to form a ferryl species called the porphyrin cation radical, $P^+-Fe^{4+}=O$, in which iron has an oxidation number of four. Initiation of lipid oxidation by the porphyrin cation radical proceeds via two-electron reduction of the catalyst.

Enzyme-catalysed Lipid Oxidation

The enzyme-catalysed lipid oxidation system in the microsomal fraction of muscle tissues has been extensively studied[70-73] and reviewed by Hsieh and

Kinsella.[65] Recently, lipoxygenase has been identified as a possible initiator of lipid oxidation in fish tissues. Lipoxygenases in trout gill and skin tissues can oxidize polyunsaturated fatty acids into position-specific hydroperoxides.[74,75] These hydroperoxides may undergo homolytic cleavage and β-scission to produce various volatile compounds such as aldehydes, alcohols, ketones, and hydrocarbons. These compounds influence the flavour quality and palatability of fish products.[76]

Two lipoxygenases, the 12- and 15-lipoxygenases, have been identified in fish gill tissue and provide the potential for development of many different volatile compounds from the same fatty acid precursor.[74,77] The major volatile compounds generated from the oxidation of arachidonic and eicosapentaenoic acids by gill 12-lipoxgenase were oct-1-en-3-ol, oct-2-enal, non-2-enal, nona-2,6-dienal, octa-1,5-dien-3-ol, and octa-2,5-dien-1-ol.[76] The compounds non-2-enal and nona-2,6-dienal are present in freshwater fish,[78] while oct-1-en-3-ol, octa-1,5-dien-3-ol, and octa-2,5-dien-1-ol occur in some species of freshwater and saltwater fish.[78,79] Hsieh and Kinsella[76] concluded that lipoxygenase-initiated oxidation of polyunsaturated fatty acids results in the generation of specific oxidative flavour compounds in fish tissues which may be important in imparting fresh aroma to fish and also in causing off-flavours. They also indicated that proper post-harvest methods such as low-temperature storage on ice, avoidance of bruising or injury, mild heat treatments such as blanching, and application of specific antioxidants may be utilized to control or retard lipoxygenase-initiated lipid oxidation and retard quality deterioration of fish post mortem.

Sodium Chloride

Sodium chloride (salt) is added to processed meats for its sensory, functional, and preservation properties.[6] In restructured meats, salt is critical for the extraction of salt-soluble proteins which provide the cohesiveness of the meat particles in the finished product. It is widely recognized, however, that salt may initiate undesirable colour and flavour changes in meat, although the mechanism remains unclear. Early work suggested that salt catalysed oxidation by lipoxidase[80] or by myoglobin.[81,82] Ellis et al.[83] reported that increasing levels of salt accelerated lipid oxidation in pork but did not alter the decomposition of hydroperoxides to monocarbonyls. However, in samples containing high proportions of lean, lower conversion of peroxides into monocarbonyls was observed. They postulated that salt may activate a component in the lean meat which results in a change in the oxidation characteristics of adipose tissue. Other researchers have proposed that salt-catalysed rancidity may be related to traces of metal impurities in the salt. Olson and Rust[84] reported, though, that using a purified low-metal salt to cure hams did not improve taste panel scores over those for hams cured with conventional salt. They did find that hams cured with salt containing antioxidants were preferred over the control samples and hams cured with low-metal salt.

A recent study has indicated that the catalytic effect of sodium chloride is derived from the enhancement of the pro-oxidant effect of chelatable iron ions. Kanner et al.[85] demonstrated the pro-oxidant effect of salt on the oxidation of minced turkey muscle lipids, an effect which was inhibited by the addition of ceruloplasmin and ethylenediaminetetra-acetic acid. These investigators also demonstrated that in a model system of washed muscle fibres containing ascorbic acid and iron, salt decreased lipid oxidation. However, if the muscle cytosol fraction (the supernatant obtained after homogenizing ground muscle in a sodium acetate solution followed by centrifugation) was added to the model system, salt enhanced lipid oxidation. It was concluded that the effects of salt on both model systems and ground turkey muscle seem to derive from the capability of salt to displace iron ions from binding macromolecules. More studies of this nature are required to elucidate fully the mechanism of salt-catalysed lipid oxidation in processed meat systems.

6 Prevention of Off-flavour Development in Meat Products During Storage

Lipid oxidation leading to rancid off-flavour development in raw and cooked meat and meat products can be reduced or prevented by the appropriate use of antioxidants or curing with nitrite.[86] Gray and Pearson[6] reviewed procedures which prevented the development of warmed-over flavour including the use of nitrite, phosphates or other chelating agents, Maillard reaction products, synthetic and naturally occurring phenolic antioxidants such as those which occur in spices, and ascorbate.

Warmed-over flavour is not a problem in cured meat. Fooladi et al.[87] demonstrated that a nitrite concentration of 156 mg kg^{-1} in meat inhibited oxidative changes in cooked meat, with a two-fold reduction of TBA numbers for beef and chicken, and a five-fold reduction for pork (Table 5). Many investigators have confirmed the effectiveness of nitrite in eliminating lipid oxidation in fresh ham stored under refrigeration by showing that the pattern of the volatiles in the headspace differed from that of nitrite-free controls.[2,88-90] Ramarathnam et al.[91] also reported that the concentration of hexanal was ca. 400 times greater in uncured pork than in cured pork.

The mechanism by which nitrite functions as an antioxidant has also been the focus of much attention.[92-94] Evidence has been presented to indicate that several mechanisms may be operative in cured meats. Nitrite may form a complex with haem pigments to prevent release of iron during cooking, stabilize unsaturated lipids in membranes, or interact with metal ions to prevent their catalysis of oxidation. Freybler et al.[95] confirmed the nitrite stabilization of haem pigments and membrane lipids in cured meats and in addition demonstrated the antioxidative properties of nitric oxide myoglobin. Bailey[96] called nitrite the ultimate antioxidant for preventing

Table 5 *Effect of nitrite on oxidative stability of meat*

Treatment	TBA numbers		
	Without nitrite	With nitrite	Mean difference
Chicken samples			
Raw, 0 days	2.52	1.36	1.16
Cooked, 0 days	3.58	1.06	2.52[a]
Raw, 48 h at 4 °C	5.52	1.47	4.05[a]
Cooked, 48 h at 4 °C	6.98	3.05	3.93[a]
Pork samples			
Raw, 0 days	1.52	0.85	0.67
Cooked, 0 days	1.83	0.72	1.11
Raw, 48 h at 4 °C	2.48	1.42	1.06
Cooked, 48 h at 4 °C	7.85	1.64	6.21[a]
Beef samples			
Raw, 0 days	0.92	0.66	0.26[b]
Cooked, 0 days	1.07	0.75	0.32[a]
Raw, 48 h at 4 °C	1.84	1.17	0.67[a]
Cooked, 48 h at 4 °C	4.12	2.06	2.06[a]

[a] $p < 0.01$.
[b] $p < 0.05$.
Adapted from ref.87, with permission.

warmed-over flavour and indicated that most of the table-ready, precooked meats available to the consumer have been cured with nitrite.

Phosphates are usually added to processed meats because they increase the water-holding capacity and yield of the finished product. Phosphates also delay or prevent oxidation,[23,24] presumably by chelating heavy metals such as iron and copper which are known pro-oxidants. Shahidi et al.[97,98] confirmed that sodium pyrophosphate and sodium tripolyphosphate were effective inhibitors of warmed-over flavour development in cooked meats, and noted a strong synergistic effect when ascorbates were used in combination with polyphosphates. Sodium tripolyphosphate, in combination with the fat soluble ascorbic acid derivatives, was perhaps even more effective.[98] These antioxidant mixtures were at least as effective as sodium nitrite (150 mg kg^{-1}) in the presence of sodium ascorbate.

Generally, in food systems, the most effective antioxidants function by interrupting the free-radical chain mechanism of lipid oxidation.[99] Butylated hydroxytoluene (BHT), butylated hydroxyanisole (BHA), and tertiary butyl hydroquinone (TBHQ) have been used with mixed success in meat systems, and are usually more effective in comminuted meats or emulsions than in intact meat cuts.[96] For example, Greene[100] reported that BHA and propyl gallate offered substantial protection against oxidation of fresh meat pigments and effectively inhibited lipid oxidation in raw ground beef. Phenolic antioxidants have been proven to be effective in restructured meat systems, systems that are very susceptible to oxidation because of the methods used during processing. Chastain et al.[101] showed that the addition of antioxidants (BHA and TBHQ), alone or in combination, increased sensory scores for flavour and acceptability, and decreased discoloration of the raw restructured beef/pork steaks. Crackel et al.[102] also reported that TBHQ was effective in retarding off-flavour development in restructured beef steaks.

Current research has focused heavily on the use of natural ingredients as antioxidants owing to growing concerns about the safety of synthetic antioxidants and a general consumer perception that natural is better. These substances include various edible products from vegetables, fruits, oilseeds, and grains, as well as spices, herbs, and protein hydrolysates. The antioxidant nature of these compounds has been extensively reviewed.[86,103] The most attractive source of natural antioxidants, at present, appears to be spices and herbs. Rosemary, for example, contains a number of compounds possessing antioxidant activity including carnosol, rosmanol, rosmariquinone, and rosmaridiphenol. The latter two compounds, when tested in lard at a level of 0.02%, have been found to be better antioxidants than BHA.[104,105] Rosemary oleoresins have been evaluated in a number of meat products including beef patties,[27] restructured beef steaks,[106] and restructured chicken nuggets.[107] St.Angelo et al.[27] reported that oil-soluble rosemary oleoresin was an effective inhibitor of warmed-over flavour in cooked beef patties stored at 4 °C. The sensory panel rated the cooked beef brothy flavour intensity as very high, whereas the painty and cardboard intensities received low scores. Stoick et al.[106] and Lai et al.[107] also reported that rosemary oleoresin, when used in combination with sodium tripolyphosphate, produced an additive protective effect during frozen storage for restructured beef steaks and chicken nuggets, respectively.

Vitamin E is a powerful antioxidant and functions as a lipid-soluble antioxidant in cell membranes. It is capable of quenching free-radicals and thus protects phospholipids and cholesterol against oxidation. Dietary supplementation of vitamin E for the subsequent benefit of increased lipid stability in animal food products has been extensively reported for poultry,[108,109] pigs,[110,112] and veal calves.[113,114] Faustman et al.[115] also reported improvement of pigment and lipid stability in Holstein steer beef by dietary supplementation with vitamin E (370 I.U./head/day) for ca. 43 weeks. During six days of storage at 4 °C, metmyoglobin accumulation and lipid oxidation were greater in beef from the control animals compared with supplemented animals. These results indicated a very real advantage for supplementing Holstein steers with vitamin E to obtain ground beef with greater oxidative stability. On the other hand, consistent results for supporting the exogenous addition of vitamin E to meat products are lacking.[115]

There is also some interest in Maillard reaction products as inhibitors of lipid oxidation in meat systems. Although several attempts have been made to identify Maillard reaction products that act as antioxidants,[116,117] there have been no definitive conclusions regarding the most important intermediates in the Maillard reaction systems responsible for antioxidant activities.[118] Maltol (3-hydroxy-2-methyl-4H-pyran-4-one) has been identified as one of the compounds produced in a browning reaction that possesses antioxidant potential in cooked meat.[119] Bailey et al.[118] indicated that there is strong support for the concept that the antioxidative effect of Maillard reaction products is due to the formation of free radicals formed during heating of sugars and amines. Other possible mechanisms include the reduction of hydroperoxides formed during lipid oxidation and the chelation of heavy metals including iron. Continued research is needed to determine the most useful Maillard reaction products for preventing off-flavours in cooked meat during storage.

Off-flavour development in meats may be controlled by various packaging techniques such as vacuum packaging, gas flushing, shrink and skin packaging, and use of barriers to oxygen and light. Nolan et al.[120] demonstrated the beneficial effects of vacuum packaging in retarding lipid oxidation in cooked pork and turkey as measured by TBA and sensory analyses. Storage of the meat samples in 100% CO_2 and 100% NO_2 produced less oxidation than that occurring in samples stored in air, yet the samples were not as protected against oxidation as those in vacuum packages. Hwang et al.[121] reported that cooked beef samples held in a freezer in modified atmosphere packages (80% N_2, 20% CO_2) had similar TBA numbers, hexanal contents, and sensory scores to frozen, vacuum-packaged samples.

7 Cholesterol Oxidation and Its Relationship to Meet Quality

Cholesterol oxidation has been the subject of numerous studies in recent years.[122,123] Many of the oxidation products have been reported to produce a variety of adverse biological effects, such as inhibition of cholesterol biosynthesis,[124] atherogenesis,[125-127] cytotoxicity,[128,129] mutagenesis,[130] and carcinogenesis.[131]

Until recently, knowledge of the occurrence of cholesterol oxidation products in foods was somewhat scarce because of limitations in methodology, the trace levels at which they occur in foods, and changes in their concentrations upon isolation and quantitation.[3,132,133] As improvements in the methodology for the isolation and quantitation of cholesterol oxides were made, the significance of cholesterol oxidation in foods was realized. Cholesterol oxides have been reported in a number of foods including dairy products,[134,135] eggs and egg powders,[136,137] heated fats,[138-140] and meat products.[133,141,142]

The mechanism of cholesterol oxidation is similar to that of unsaturated fatty acid oxidation in that hydrogen is initially abstracted, resulting in the formation of a free radical, which then reacts with oxygen to form a peroxy radical.[143] This peroxy radical can abstract hydrogen from another unsaturated molecule to propagate the reaction. Cholesterol will undergo oxidation at C-7 (*i.e.* the carbon adjacent to the double bond), resulting in the formation of 7α- and 7β-hydroperoxides (major primary products) which then form 7α- and 7β-hydroxycholesterol and 7-ketocholesterol (the most predominant products of cholesterol oxidation). Side-chain oxidation also occurs at the tertiary C-25, resulting in 25-hydroxycholesterol, and occasionally A ring oxidation may occur at C-3.[143]

The relationship between off-flavour development (as measured by TBA numbers) and cholesterol oxidation in meats has been studied by several researchers. Park and Addis[133] detected some oxidation products at the low $mg\,kg^{-1}$ level in broiled beef steaks, but not in precooked beef products. As rancidity developed in comminuted and cooked meats during storage, oxidation of cholesterol became more apparent (Table 6). DeVore[144] reported similar trends for raw and cooked ground beef and established correlation coefficients of 0.82 and 0.98 between TBA numbers and 7-ketocholesterol concentrations for raw and cooked ground beef patties, respectively. It was concluded that because polyunsaturated fatty acids and cholesterol are integral components of membrane structure, free radicals from phospholipid oxidation may initiate cholesterol oxidation in the tissue membranes of ground beef.

The role of the membrane lipid environment in cholesterol oxidation in meats was investigated further by Engeseth.[114] She demonstrated that the cholesterol moiety of cholesteryl linoleate was oxidized to a greater extent than that of cholesteryl stearate and free cholesterol. Also, cholesterol was oxidized when dispersed in a model system with phosphatidylcholine or with beef adipose issue. These studies confirmed that the environment surrounding cholesterol may play an important role in determining the susceptibility of cholesterol to oxidation and that the oxidation of nearby

Table 6 *TBA-reactive substances (TBARS) and cholesterol oxidation products (COPS) in comminuted, cooked and stored beef, and turkey[a]*

Days at 4 °C	Beef		Turkey	
	TBARS	Total of COPs, % of cholesterol	TBARS	Total of COPs, % of cholesterol
0	0.52[b]	ND[e]	1.66[b]	ND[e]
3	1.88[c]	0.45[f]	3.53[c]	1.20[f]
8	2.22[d]	1.74[g]	4.68[d]	2.90[g]

[a] Mean values of three determinations. Means in a column with different superscript letter are significantly different at the 1% level.
ND = not detected.
From ref.133, with permission.

unsaturated lipids may result in the formation of free radicals which can attack cholesterol. Engeseth[114] further demonstrated that stabilizing the membrane lipids of veal through dietary supplementation of the calves' diets with vitamin E controlled the rate of cholesterol oxidation. As with other indices of oxidation (*e.g.* TBA numbers), oxide development was greater in the cooked samples. Similarly, the greater concentration of cholesterol oxidation products in turkey meat compared with beef, when expressed as a percentage of the total cholesterol content,[133] may be due to the greater susceptibility to oxidation of the component fatty acids of turkey phosphoplipids compared with beef phospholipids.

8 Conclusions

This overview of flavour deterioration in meat and meat products through oxidative changes has revealed that there is a need for further research on certain aspects of lipid oxidation. As indicated by Drumm and Spanier,[45] meat flavour deterioration involves not only the development of oxidative off-flavours, but also the loss of desirable flavour components. Thus, further studies on warmed-over flavour must include detailed sensory analyses using panellists trained to recognize specific attributes such as the flavour of freshy cooked lean beef and cardboard and painty characteristics. Replacing terms such as warmed-over flavour and stale or off-flavour with these more specific terms should enable researchers in the meat flavour area to obtain more useful information from sensory studies.[29]

It is also very apparent that there still remains some confusion regarding the respective roles of haem and non-haem iron in the lipid oxidation process. Model system studies have not completely resolved this issue, and this stresses once again the difficulties involved in relating data from such studies to the more complex meat systems. The emergence of lipoxygenase as a possible initiator of oxidative reactions in muscle foods also must be more fully explored, particularly with respect to red meats. Evidence to date suggests that the role of lipoxygenases may be greater than was previously believed.

9 References

1. G.A. Reineccius, *J. Food Sci.*, 1979, **44**, 12.
2. M.E. Bailey, H.P. Dupuy, and M.G. Legendre, in 'The Analysis and Control of Less Desirable Flavors in Foods and Beverages', ed. G. Charalambous, Academic Press, Inc., New York, 1980, p.31.
3. A.M. Pearson, J.I. Gray, A. M. Wolzak, and N.A. Horenstein, *Food Technol.*, 1983, **37** (7), 121.
4. P.B. Addis, *Food Chem. Toxicol.*, 1986, **24**, 1021.
5. A.M. Pearson and J.D. Love, *J. Am. Oil Chem. Soc.*, 1971, **48**, 547.
6. J.I. Gray and A.M. Pearson, in 'Advances in Meat Research, Vol.3 - Restructured Meat and Poultry Products', ed. A.M. Pearson and T.R. Dutson, Van Nostrand Reinhold Co., New York, 1987, p.221.

7. B.M. Watts, in 'Symposium on Foods: Lipids and Their Oxidation', ed. H.W. Schultz, E.A. Day, and R.O. Sinnhuber, AVI Publishing Co., Westport, CT, 1962, p.202.
8. A.M. Pearson, J.D. Love, and F.B. Shorland, *Adv. Food Res.,* 1977, **23**, 1.
9. J.M. Eichhorn, C.M. Bailey, and G.J. Blomquist, *J. Anim. Sci.,* 1985, **61**, 892.
10. L.R. Dugan, Jr., in 'The Science of Meat and Meat Products', ed. J.F. Price and B.S. Schweigert, Food and Nutrition Press, Inc., Westport, CT, 1987, p.507.
11. M. Kaucher, H. Galbraith, V. Button, and H.H. Williams, *Arch. Biochem.,* 1944, **3**, 203.
12. G.M. Gray and M.G. MacFarlane, *Biochem. J.,* 1961, **81**, 480.
13. C.E. Allen and E.A. Foegeding, *Food Technol.,* 1981, **35** (5), 253.
14. J.O. Igene, A.M. Pearson, and J.I. Gray, *Food Chem.,* 1981, **7**, 289.
15. N.R. Ellis and H.S. Isbell, *J. Biol. Chem.,* 1926, **69**, 219.
16. L.C. St.John, C.R. Young, D.A. Knobe, L.D. Thompson, G.T. Schelling, S.M. Grundy, and S.B. Smith, *J. Anim. Sci.,* 1987, **64**, 1441.
17. K.S. Rhee, Y.A. Ziprin, G. Ordonez, and C.E. Bohac, *Meat Sci.,* 1988, **23**, 201.
18. M.F. Miller, S.D. Shackelford, K.D. Hayden, and J.O. Reagan, *J. Anim. Sci.,* 1990, **68**, 1624.
19. J.O. Igene and A.M. Pearson, *J. Food Sci.,* 1979, **44**, 1285.
20. S.L. Melton, *Food Technol.,* 1983, **37** (7), 105.
21. D.S. Mottram and R.A. Edwards, *J. Sci. Food Agric.,* 1983, **34**, 517.
22. L.J. Farmer, D.S. Mottram, and F.B. Whitfield, *J. Sci. Food Agric.,* 1989, **49**, 347.
23. M.J. Timms and B.M. Watts, *Food Technol.,* 1958, **12**, 240.
24. K. Sato and G.R. Hegarty, *J. Food Sci.,* 1971, **36**, 1098.
25. A.M. Spanier, J.V. Edwards, and H.P. Dupuy, *Food Technol.,* 1988, **42** (6), 110.
26. A.J. St.Angelo, J.F. Vercellotti, H.P. Dupuy, and A.M. Spanier, *Food Technol.,* 1988, **42** (6), 133.
27. A.J. St.Angelo, K.L. Crippen, H.P. Dupuy, and C. James, Jr. *J. Food Sci.,* 1990, **55**, 1501.
28. S.L. Melton, P.M. Davidson, and J.R. Mount, in 'Warmed Over Flavor of Meat', ed. A.J. St. Angelo and M.E. Bailey, Academic Press, Inc., Orlando, FL, 1987, p.141.
29. J.D. Love, *Food Technol.,* 1988, **42** (6), 140.
30. P.B. Johnsen and G.V. Civille, *J. Sensory Studies,* 1986, **1**, 99.
31. B.R. Wilson, A.M. Pearson, and F.B. Shorland, *J. Agric. Food Chem.,* 1976, **24**, 7.
32. M.T. Younathan and B.M. Watts, *Food Res.,* 1960, **25**, 538.
33. J.O. Igene, A.M. Pearson, L.R. Dugan, Jr., and J.F. Price, *Food Chem.,* 1980, **5**, 263.
34. C.H. Lea, *J. Sci. Food Agric.,* 1957, **8**, 1.
35. A. Asghar, J.I. Gray, D.J. Buckley, A.M. Pearson, and A.M. Booren, *Food Technol.,* 1988, **42** (6), 102.
36. J.O. Igene, A.M. Pearson, R.A. Merkel, and T.H. Coleman, *J. Anim. Sci.,* 1979, **49**, 701.
37. J.D. Keller and J.E. Kinsella, *J. Food Sci.,* 1973, **38**, 1200.

38. H.Y. Gokalp, H.W. Ockerman, R.F. Plimpton, and A.C. Peng, *J. Food Sci.*, 1981, **46**, 19.
39. H.Y. Gokalp, H.W. Ockerman, R.F. Plimpton, and W.J. Harper, *J. Food Sci.*, 1983, **48**, 829.
40. M.I. El Gharbawi and L.R. Dugan, Jr., *J. Food Sci.*, 1965, **30**, 817.
41. E.N. Frankel, *Progr. Lipid Res.*, 1980, **19**, 1.
42. G. Paquette, D.B. Kupranycz, and F.R. van de Voort, *Can. Inst. Food Sci. Technol. J.*, 1985, **18**, 112.
43. D.A. Lillard, Ref.28, p.41.
44. S.S. Chang and R.J. Peterson, *J. Food Sci.*, 1977, **42**, 298.
45. T.D. Drumm and A.M. Spanier, *J. Agric. Food Chem.*, 1991, **39**, 336.
46. E.N. Frankel, W.E. Neff, and E. Selke, *Lipids*, 1981, **16**, 279.
47. F. Shahidi, J. Yun, L.J. Rubin, and D.F. Wood, *Can. Inst. Food Sci. Technol. J.*, 1987, **20**, 104.
48. A.J. St.Angelo, J.R. Vercellotti, M.G. Legendre, C.H. Vinnett, J.W. Kuan, C. James, Jr., and H.P. Dupuy, *J. Food Sci.*, 1987, **52**, 1163.
49. E.L. Ruenger, G.A. Reineccius, and D.R. Thompson, *J. Food. Sci.*, 1978, **43**, 1198.
50. A.L. Tappel, in 'Symposium on Foods: Lipids and Their Oxidation', ed. H.W. Schultz, E.A. Day, and R.O. Sinnhuber, AVI Publishing Co., Westport, CT, 1962, p.122.
51. K.U. Ingold, Ref.50, p.93.
52. J.D. Love and A.M. Pearson, *J. Agric. Food Chem.*, 1974, **22**, 1032.
53. J.O. Igene, J.A. King, A.M. Pearson, and J.I. Gray, *J. Agric. Food Chem.*, 1979, **27**, 838.
54. B.R. Schricker, D.D. Miller, and J.R. Stouffer, *J. Food Sci.*, 1982, **47**, 740.
55. B.R. Schricker and D.D. Miller, *J. Food Sci.*, 1983, **48**, 1340.
56. K.S. Rhee, *Food Technol.*, 1988, **42** (6), 127.
57. S. Harel and J. Kanner, *J. Agric. Food Chem.*, 1985, **33**, 1186.
58. S. Harel and J. Kanner, *J. Agric. Food Chem.*, 1985, **33**, 1188.
59. J. Kanner and S. Harel, *Arch. Biochem. Biophys.*, 1985, **237**, 314.
60. K.S. Rhee, Y.A. Ziprin, and G. Ordonez, *J. Agric. Food Chem.*, 1987, **35**, 1013.
61. J. Kanner, I. Shegalovich, S. Harel, and B. Hazan, *J. Agric. Food Chem.*, 1988, **36**, 409.
62. J. Kanner, B. Hazan, and L. Doll, *J. Agric. Food Chem.*, 1988, **36**, 412.
63. J.Z. Tichivangana and P.A. Morrissey, *Meat Sci.*, 1985, **15**, 107.
64. A.M. Johns, L.H. Birkinshaw, and D.A. Ledward, *Meat Sci.*, 1989, **25**, 209.
65. R.J. Hsieh and J.E. Kinsella, *Adv. Food Nutr. Res.*, 1989, **33**, 233.
66. B. Halliwell and J.M.C. Gutteridge, *Methods Enzymol.*, 1990, **186**, 1.
67. J.M.C. Gutteridge, *FEBS Lett.*, 1984, **172**, 245.
68. M. Tien and S.D. Aust, in 'Lipid Peroxides in Biology and Medicine', ed. K. Yagi, Academic Press, Inc., London, 1982, p.23.
69. G. Minotti and S.D. Aust, *J. Biol. Chem.*, 1987, **263**, 1098.
70. T.S. Lin and H.O. Hultin, *J. Food Sci.*, 1976, **41**, 1488.
71. T.S. Lin and H.O. Hultin, *J. Food Sci.*, 1977, **42**, 136.
72. K.S. Rhee, T.R. Dutson, and G.C. Smith, *J. Food Sci.*, 1984, **49**, 675.
73. K.S. Rhee and Y.A. Ziprin, *J. Food Biochem.*, 1987, **11**, 1.
74. J.B. German and J.E. Kinsella, *J. Agric. Food Chem.*, 1985, **33**, 680.
75. J.B. German and J.E. Kinsella, *Biochim. Biophys. Acta*, 1986, **875**, 12.

76. R.J. Hsieh and J.E. Kinsella, *J. Agric. Food Chem.*, 1989, **37**, 279.
77. J.B. German and R.K. Creveling, *J. Agric. Food Chem.*, 1990, **38**, 2144.
78. D.B. Josephson, R.C. Lindsay, and D.A. Stuiber, *J. Agric. Food Chem.*, 1984, **32**, 1344.
79. D.B. Josephson, R.C. Lindsay, and D.A. Stuiber, *J. Agric. Food Chem.*, 1983, **31**, 326.
80. C.H. Lea, *J. Soc. Chem. Ind. (London)*, 1937, **56**, 376.
81. A.L. Tappel, *Food Res.*, 1952, **17**, 550.
82. A. Banks, *Chem. Ind. (London)*, 1961, **2**, 40.
83. R. Ellis, G.T. Currie, F.E. Thornton, N.C. Bollinger, and A.M. Gaddis, *J. Food Sci.*, 1968, **33**, 555.
84. D.G. Olson and R.E. Rust, *J. Food Sci.*, 1973, **38**, 251.
85. J. Kanner, S. Harel, and R. Jaffe, *J. Agric. Food Chem.*, 1991, **39**, 1017.
86. K.S. Rhee, Ref.28, p.267.
87. M.H. Fooladi, A.M. Pearson, T.H. Coleman, and R.A. Merkel, *Food Chem.*, 1979, **4**, 283.
88. C.K. Cross and P. Ziegler, *J. Food Sci.*, 1965, **30**, 610.
89. M.E. Bailey and J.W. Swain, in 'Proceedings of the Meat Industry Research Conference', American Meat Science Association, Chicago, 1973, p.29.
90. B. MacDonald, J.I. Gray, Y. Kakuda, and M.L. Less, *J. Food Sci.*, 1980, **45**, 889.
91. N. Ramarathnam, L.J. Rubin, and L.L. Diosady, *J. Agric. Food Chem.*, 1991, **39**, 344.
92. M.W. Zipser, T.W. Kwon, and B.M. Watts, *J. Agric. Food Chem.*, 1964, **12**, 105.
93. J.O. Igene, K. Yamauchi, A.M. Pearson, and J.I. Gray, *Food Chem.*, 1985, **18**, 1.
94. P.A. Morrissey and J.Z. Tichivangana, *Meat Sci.*, 1985, **14**, 175.
95. L.A. Freybler, J.I. Gray, A. Asghar, A.M. Booren, A.M. Pearson, and D.J. Buckley, in 'Proceedings of the 35th International Congress of Meat Science and Technology', Vol.3, 1989, p.903.
96. M.E. Bailey, *Food Technol.*, 1988, **42** (6), 123.
97. F. Shahidi, L.J. Rubin, L.L. Diosady, N. Kassam, J.C. Li Sui Fong, and D.F. Wood, *Food Chem.*, 1986, **21**, 145.
98. F. Shahidi, L.J. Rubin, and D.F. Wood, *Food Chem.*, 1987, **23**, 151.
99. L.R. Dugan, Jr., *Food Technol.*, 1961, **15**, 10.
100. B.E. Greene, *J. Food Sci.*, 1969, **34**, 110.
101. M.F. Chastain, D.L. Huffman, W.F. Hsieh, and J.C. Condray, *J. Food Sci.*, 1982, **47**, 1779.
102. R.L. Crackel, J.I. Gray, A.M. Booren, A.M. Pearson, and D.J. Buckley, *J. Food Sci.*, 1988, **53**, 656.
103. C.M. Houlihan and C.T. Ho, in 'Flavor Chemistry of Fats and Oils', ed. D.B. Min and T.H. Smouse, American Oil Chemists' Society, 1985, p.139.
104. C.M. Houlihan, C.T. Ho, and S.S. Chang, *J. Am. Oil Chem. Soc.*, 1984, **61**, 1036.
105. C.M. Houlihan, C.T. Ho, and S.S. Chang, *J. Am. Oil Chem. Soc.*, 1985, **62**, 98.
106. S. Stoick, J.I. Gray, A.M. Booren, and D.J. Buckley, *J. Food Sci.*, 1991, **56**, 597.
107. S.M. Lai, J.I. Gray, D.M. Smith, A.M. Booren, R.L. Crackel, and D.J.

Buckley, *J. Food Sci.*, 1991, **56**, 616.
108. W.L. Marusich, E. DeRitter, E.F. Ogrinz, J. Keating, M. Mitrovic, and R.H. Bunnell, *Poultry Sci.*, 1975, **54**, 831.
109. C.F. Lin, J.I. Gray, A. Asghar, D.J. Buckley, A.M. Booren, and C.F. Flegal, *J. Food Sci.*, 1989, **54**, 1457.
110. J. Buckley and J.F. Connolly, *J. Food Protect.*, 1980, **43**, 265.
111. D.J. Buckley, J.I. Gray, A. Asghar, J.F. Price, R.L. Crackel, A.M. Booren, A.M. Pearson, and E.R. Miller, *J. Food Sci.*, 1989, **54**, 1193.
112. F. Monahan, D.J. Buckley, J.I. Gray, P.M. Morrissey, A. Asghar, T.J. Hanrahan, and P.B. Lynch, *Meat Sci.*, 1990, **27**, 99.
113. F.B. Shorland, J.O. Igene, A.M. Pearson, J.W. Thomas, R.K. McGuffey, and A.E. Aldrige, *J. Agric. Food Chem.*, 1981, **29**, 863.
114. N.J. Engeseth, PhD Dissertation, Michigan State University, East Lansing, MI, 1990.
115. C. Faustman, R.G. Cassens, D.M. Schaefer, D.R. Buege, S.N. Williams, and K.K. Scheller, *J. Food Sci.*, 1989, **54**, 858.
116. H. Lingnert and C. E. Eriksson, *Progr. Food Nutr.*, 1981, **5**, 453.
117. H. Lingnert, C.E. Eriksson, and G.R. Waller, in 'The Maillard Reaction in Foods and Nutrition', ed. G.R. Waller and M.S. Feather, ACS Symposium Series No. 215, Washington, DC, 1983, p.335.
118. M.E. Bailey, S.Y. Shin-Lee, H.P. Dupuy, A.J. St.Angelo, and J.R. Vercellotti, Ref.28, p.237.
119. K. Sato, G.R. Hegarty, and H.K. Herring, *J. Food Sci.*, 1973, **38**, 398.
120. N.L. Nolan, J.A. Bowers, and D.H. Kropf, *J. Food Sci.*, 1989, **54**, 846.
121. S.Y. Hwang, J.A. Bowers, and D.H. Kropf, *J. Food Sci.*, 1990, **55**, 26.
122. L.L. Smith, 'Cholesterol Autoxidation', Plenum Press, New York, 1981.
123. P.B. Addis and S.W. Park, in 'Food Toxicology: A Perspective on the Relative Risks', ed. S.L. Taylor and R.A. Scanlan, Marcel Dekker, Inc., New York, 1989, p.297.
124. A.A. Kandutsch and H.W. Chen, *J. Biol. Chem.*, 1973, **248**, 8408.
125. C.B. Taylor, S.K. Peng, N.T. Warthessen, P. Tham, and K.T. Lee, *Am. J. Clin. Nutr.*, 1979, **32**, 40.
126. H. Imai, N.T. Werthessen, V. Subramanyam, P.W. LeQuesne, A.H. Soloway, and M. Kanisawa, *Science*, 1980, **297**, 651.
127. D.C. Cox, K. Comai, and A.L. Goldstein, *Lipids*, 1988, **23**, 85.
128. S.K. Peng, P. Tham, C.B. Taylor, and B. Mikkelson, *Am. J. Clin. Nutr.*, 1979, **32**, 1033.
129. A. Sevanian and A.R. Peterson, *Food Chem. Toxicol.*, 1986, **24**, 1103.
130. G.A.S. Ansari, R.D. Walker, V.B. Smart, and L.L. Smith, *Food Chem. Toxicol.*, 1982, **20**, 35.
131. H.S. Black and W.B. Lo, *Nature (London)*, 1971, **234**, 306.
132. E.T. Finoccchiaro and T. Richardson, *J. Food Protect.*, 1983, **46**, 917.
133. S.W. Park and P.B. Addis, *J. Food Sci.*, 1987, **52**, 1500.
134. J. Nourooz-Zadeh and L.A. Appelquist, *J. Food Sci.*, 1987, **52**, 57.
135. B.D. Sander, D.E. Smith, P.B. Addis, and S.W. Park, *J. Food Sci.*, 1989, **54**, 874.
136. L.S. Tsai and C.A. Hudson, *J. Food Sci.*, 1984, **49**, 1245.
137. S.R. Missler, B.A. Wasilchuk, and C. Merritt, Jr., *J. Food Sci.*, 1985, **55**, 595.
138. T.C. Ryan, J.I. Gray, and I.D. Morton, *J. Sci. Food Agric.*, 1981, **32**, 305.

139. S.W. Park and P.B. Addis, *J. Food Sci.*, 1986, **51**, 1380.
140. S.W. Park and P.B. Addis, *J. Agric. Food Chem.*, 1986, **34**, 653.
141. M.A. Higley, S.L. Taylor, A.M. Herian, and K. Lee, *Meat Sci.*, 1986, **16**, 175.
142. J.E. Pie, K. Spahis, and C. Seillan, *J. Agric. Food Chem.*, 1991, **39**, 250.
143. G. Maerker, *J. Am. Oil Chem. Soc.,* 1987, **64**, 388.
144. V.R. DeVore, *J. Food Sci.*, 1988, **53**, 1058.

Meat Flavour

L.J. Farmer

DEPARTMENT OF AGRICULTURE FOR NORTHERN IRELAND AND THE QUEEN'S UNIVERSITY OF BELFAST, NEWFORGE LANE, BELFAST BT9 5PX, UK

1 Introduction

The subject of meat flavour is immense, encompassing the effects of a range of factors on the sensory attributes of cooked meat and the incidence of taints and off-flavours, as well as the chemical basis of desirable flavour formation. Many of these aspects of meat flavour have been thoroughly reviewed over the past few years and in the space available it is not feasible to cover the whole subject. Instead, a brief definition of flavour will be followed by a discussion of three aspects of meat flavour:

(1) Chemistry of meat flavour, *i.e.* how flavour compounds are formed.
(2) Effect of breed, feed, pre- and post-slaughter treatment, *etc.* on flavour, as detected by sensory evaluation.
(3) Recent work on the identification of compounds important for meat flavour.

Definition of Flavour

Flavour, whether of meat or any other food, comprises mainly the two sensations of taste and smell, although other sensations such as mouthfeel and juiciness may also affect the perception of flavour. Flavour is an important contributor to the overall acceptability of foods, and the volatile components of aroma often affect our judgement of a food even before consumption.

While receptors in the mouth can recognize four main taste sensations (sweet, salt, sour, and bitter) many hundreds or thousands of odours can be distinguished by the human nose. Thus, much of the flavour of foods is, in fact, detected in the nose as aroma and odour plays a major part in defining the characteristic flavour of foods. The sensation of odour is produced by volatile chemical substances which stimulate the receptors in the nasal epithelium. Odour compounds may reach these receptors either through the nostrils (the anterior nares) or through channels at the back of the nose (the posterior nares) while food is in the mouth.

Raw meat has little flavour and only a blood-like taste; the characteristic flavour of cooked meat only develops after heating.[1] During cooking, numerous chemical reactions occur between the many non-volatile compounds which are present in meat to give many hundreds of volatile compounds.

The total number of volatile compounds identified in cooked red meats and poultry exceeds one thousand.[2] However, it is believed that a relatively small number of compounds actually play an important part in the overall aroma and flavour of the cooked meat.[3] Whether a compound is a key odour impact compound depends on both its concentration and its odour threshold, *i.e.* how sensitive is the human nose to that particular compound.

2 Chemistry of Meat Flavour

The precursors for odour-forming reactions are various components of the raw muscle. Muscle comprises *ca.* 75% water, 19% protein, 2.5% lipid, 1.2% carbohydrates, and 2.3% other water-soluble non-protein substances;[4] these last three categories are the source of meat flavour. The water-soluble fraction includes amino acids, peptides, reducing sugars, vitamins, and nucleotides. The lipid in muscle is made up of the neutral triacylglycerol, in the intramuscular fat, and the polar phospholipids, which are important structural components of the cell membranes. Phospholipids differ from triacylglycerols in that they possess a polar group (*e.g.* choline or ethanolamine), attached to the glycerol backbone via a phosphate linkage, and a higher proportion of polyunsaturated fatty acids.

Adipose tissue is predominantly triacylglycerol but also contains connective tissue and traces of water-soluble compounds. Phospholipids are also present in the cell membranes but constitute a much smaller proportion of the total lipid than in muscle.

Heat-induced Reactions Important for Flavour Formation

Reactions involved in flavour formation include the thermal breakdown of individual muscle components, reactions between amino acids and carbohydrates, the thermal oxidation of lipids, and interactions between these different reactions. The reactions involved in the formation of the volatile components of meat aroma have been recently and extensively reviewed.[5-10]

Breakdown of Individual Substances. The degradation of sugars and amino acids alone to give volatile compounds requires much higher temperatures than the reaction between these two compound classes. While some breakdown of sugars to give furanones and furfurals may occur at the temperatures involved during cooking, amino acids are relatively stable and are unlikely to undergo pyrolysis except at the surface of grilled or roasted meat, where dehydration permits localized high temperatures.[5]

Thiamine (vitamin B_1) is a sulfur-containing compound with thiazole and pyrimidine ring systems. The thermal degradation of thiamine gives a variety of odorous compounds, some of them described as 'meaty'; these include thiophenes, thiazoles, furans, and bicyclic heterocyclic compounds.[5,7,11,12]

Reactions between Amino Acids and Sugars. The reaction between amino acids (or peptides) and reducing sugars was first described by Maillard[13] in 1912, and the Maillard or 'non-enzymatic browning' reaction is a complex network of reactions which yields both high molecular weight brown-coloured products and volatile aroma compounds. The reaction between one amino acid and one sugar will yield well in excess of one hundred volatile compounds.[14,15] The odorous products of these reactions include aldehydes, ketones, furans, pyrroles, pyrazines, pyridines, and, where the amino acid contains sulfur, thiophenes, thiazoles, and thiols.

About 30 years ago it was shown that the reaction between the sulphur-containing amino acid cysteine and reducing sugars gave meat-like aromas.[16,17] Investigations of the products of such reactions have proved of great interest for the formulation of meat flavourings and have resulted in a number of patent applications, especially in the late 1960s and 1970s.[7] These studies have shown that many compounds possess meaty flavours; seventy-eight compounds reported by the literature to possess meat-like aromas have been compiled by MacLeod.[18] However, which, if any, of these compounds were responsible for the flavour of meat itself remained unknown until quite recently (see Section 3).

Oxidation of Lipids. Lipids can contribute to meat flavour in several ways. They can act as solvents for extraneous taint compounds derived from the feed, from metabolism, or from chemical contamination.[19] They can also break down via oxidative reactions of the fatty acids; such reactions give volatile odour compounds which contribute to both undesirable and desirable flavours.[5]

At room temperature and even, to a lesser extent, at refrigeration and freezer temperatures, autoxidation can occur to give rancidity in raw meat or 'warmed-over flavour' in stored cooked meat.[20,21] Rancidity develops slowly, during prolonged storage, while warmed-over flavour can develop quite rapidly and is particularly obvious in reheated cooked meats.

The heating of lipids, however, gives rise to thermal oxidation reactions[22] which, although they follow very similar pathways to autoxidation, contribute to the desirable flavour formed during cooking. The balance between the many pathways which make up lipid oxidation reactions is affected by the application of heat.

One of the factors affecting the rate of oxidation is the degree of unsaturation of the fatty acids; polyunsaturated fatty acids are much more susceptible to oxidation than monounsaturated or saturated fatty acids. The oxidation of phospholipids, which contain high levels of polyunsaturated fatty acids, is thought to be responsible for warmed-over flavour[20] and seems also to contribute to desirable meat flavour.[23]

The oxidation of lipids gives a wide range of aliphatic compounds, including both saturated and unsaturated hydrocarbons, alcohols, aldehydes, ketones, acids, and esters as well as cyclic compounds such as furans, lactones, and cyclic ketones. Many of these possess intense odours and contribute to the overall aromas of many foods.[21]

Interactions between Lipid Oxidation and the Maillard Reaction. Much less work has been done on the interaction between lipid oxidation and the Maillard reaction than on these reactions individually, particularly with reference to the formation of desirable food flavours. However, all foods consist of a complex mixture of components and it is rare, in normal foods, for either of these reactions to occur in isolation from the other. Thus, it may be expected that the course of each reaction would be modified by the reactants, intermediates, and products of the other. Such interactions have been recently reviewed.[24,25]

Interest in the role of reactions between lipid and Maillard constituents in meat was initiated by work investigating the role of triacylglycerols and structural phospholipids in the formation of meat aroma; either the triacylglycerols only, or the total lipids (including the phospholipids), were extracted from meat and the aroma volatiles studied.[23] Removal of the triacylglycerols had little effect on the aroma of the cooked meat or the pattern of volatile compounds observed using gas chromatography–mass spectrometry. However, the additional removal of phospholipids removed the 'meaty' character of the odour and also caused a marked alteration in the volatile products. Thus, it appeared that phospholipids may be important for meat flavour formation. In addition, the nature of the changes in volatile compounds suggested that the route by which phospholipids promote meat flavour may involve their participation in the Maillard reaction. MacLeod and Ames[26] have also observed that the extraction of all lipid removed the meaty odour of cooked meat and increased the quantities of some Maillard products.

The beneficial effect of phospholipids on meat flavour formation has been corroborated by studies using model systems, in which various lipids were reacted with the amino acid, cysteine, and the sugar ribose.[27] Phospholipids conferred a more intense meaty aroma on the model systems than triacylglycerol, and also showed more interaction in the Maillard reaction. Phosphatidylethanolamine gave a more intense meaty aroma than phosphatidylcholine or a beef phospholipid fraction; this phospholipid had the highest content of highly unsaturated fatty acids and of the ethanolamine amino group and it is believed that these factors may dictate the way in which phospholipids contribute to meat flavour formation.

The amounts and proportions of the various precursors of flavour in muscle will dictate the progress of flavour-forming reactions and hence the flavour of the cooked meat. The effect of these precursors will be considered in the following discussion of some of the production factors affecting meat flavour.

3 Aspects of Meat Production Affecting Flavour

The flavour of meat may be affected by a great many factors, including the species, sex, and age of the animal, various aspects of production, slaughter, and the period of ageing of the carcass. Subsequent processing and cooking techniques will also have an effect. In this discussion, a few of these factors have been selected to illustrate the interrelationships between meat production and its eventual cooked flavour.

Species

Early workers fractionated the precursors of flavour and concluded that 'meaty' flavour, considered to be similar in all species, originated in the water-soluble fraction from the lean muscle while the species-specific flavour was derived from the fat.[28-30] However, as lean muscle always contains some lipid, and adipose tissue some water-soluble components, it seems unlikely that either meaty or species-specific aromas will be formed exclusively in one or other component.

One of the few species-specific flavours whose origin has been established is that of lamb and mutton, described as 'sweaty, sour'.[31] This flavour is caused by branched-chain fatty acids, which arise from metabolic processes which occur in the rumen of sheep. The most important of these fatty acids are 4-methyloctanoic and 4-methylnonanoic acids.[32] The same compounds are considered to contribute to the odour of goat meat; fatty acid analyses demonstrated the presence of these acids in adipose tissue from mutton and goat, but little or none in beef.[33]

Pork fat is identifiable on heating due to 'piggy', 'sour', and 'goaty' aromas, but beef species odour is weaker and difficult to identify.[30] The compounds responsible for these odours are unknown. The characteristic species odour of pork is quite distinct from the odour of boar taint, which occurs in pork from entire male pigs.[34]

The flavour of chicken is thought to be characterized by the high levels of unsaturated aldehydes present. The removal of carbonyls from chicken volatiles has been shown to result in a decrease in 'chicken odour' and an increase in 'meaty odour'.[35] Of these carbonyls deca-*trans*-2,*trans*-4-dienal seems to be of particular importance;[36,37] this compound has been shown to have the most potent aroma of the compounds extracted from chicken broth.[37] Many of the unsaturated aldehydes characteristic of chicken aroma are oxidation products of linoleic or arachidonic acids. While the levels of these fatty acids in the phospholipids of chickens are not grossly different from those of other species, triacylglycerols from poultry contain *ca.* 25% linoleic acid,[38] compared with only *ca.* 2% in beef triacylglycerol;[39] this is a probable source of the unsaturated aldehydes responsible for the species-flavour of chicken.

Sink[40] suggests that species effect on flavour is probably expressed through the genetic control of lipid metabolism and composition, and this certainly seems to be the case in lamb and chicken.

Diet

Two recent reviews have assessed the impact of diet on the flavour of red meat[41] and poultry.[42] It is evident that it is much easier to produce off-flavours in meat by feeding an inappropriate diet than it is to increase the intensity of desirable flavour. Past literature has cited numerous examples of dietary components which cause off-flavours in meat;[41] however, these will not be discussed in this chapter.

A considerable number of studies have shown that beef animals reared on high-energy concentrates tend to give meat with a more desirable or intense beef flavour than those reared on pasture.[41,43] The flavour of grass-produced beef is described as 'milky-oily', 'grassy', 'sour', or 'sweet, gamey' compared with that of grain-fed beef.[44-46] Differences have been detected in the volatile aroma compounds from these cooked meats;[45,46] grass-fed beef has higher levels of low molecular weight aldehydes[46] and also certain terpenoid-type compounds derived from the fermentation of chlorophyll in the rumen.[45]

It has also been shown that concentrate-fed lambs have more desirable flavour than those fed on pasture.[47,48] Different types of pasture may also affect flavour; compared with a ryegrass-based pasture, vetch can confer a 'stronger, sweetish' flavour, white clover or lucerne a more intense flavour, while alfalfa can give a flavour which is described by some as 'dirty, sickly' whereas others find it more acceptable.[41,49]

These changes in flavour quality have been linked to changes in the fatty acid composition of the lipids. Compared with grass-fed beef, the more desirably flavoured meat from grain-fed cattle contains more of the $(\omega - 6)$ fatty acids and less $(\omega - 3)$ fatty acids;[41,50-52] this is thought to be due to the high content of $(\omega - 3)$ fatty acids in forage grasses.[50] These classes of fatty acids differ in the positions of their double bonds and give different oxidation products; as they are polyunsaturated they are very reactive in flavour-forming reactions. Thus, the balance of $(\omega - 6)$ and $(\omega - 3)$ acids, especially in the phospholipids, may be important for flavour formation.

Some workers have examined the effect on meat flavour of feeding unsaturated fats; the direct feeding of these substances to ruminants requires protection from hydrogenation by rumen micro-organisms. Australian work involving feeding a protected lipid supplement high in linoleic acid gave meat with a 'sweet-oily' aroma; this was attributed to a particular lactone (4-hydroxydodec-cis-6-enoic acid lactone).[53] It has also been shown that by feeding protected lipids to give specified levels of linoleic acid in the adipose tissue the lamb could be given either a 'pork-like' or a 'nutty' flavour.[44,54] High levels of linoleic acid (9.7%) in the adipose tissue were said to give a 'bland' flavour, while lower levels (4.0%) gave a strong 'lamb' flavour. It is unlikely that the adipose linoleic acid itself is responsible for these changes, rather that the modified dietary lipids have altered the overall balance of fatty acids in the various muscle and adipose lipids and that this alters the availability of fatty acids for flavour-forming reactions.

In one study,[55] diet was reported to enhance the flavour of chickens. In this study, a diet containing an increased proportion of whole wheat and access to green vegetables (brussels sprouts) was reported to give meat with higher scores for flavour; words used to describe this flavour included: 'richer', 'meatier', 'sweeter', 'more roast', 'gamey', and 'off'.

Fatness

While the generation of rancid off-flavours by lipid oxidation is well known, the role of lipid in the formation of desirable meat flavour is rather more contentious.

There is still a widely held opinion that "you've got to have a bit o' fat for flavour" and for those who enjoy the flavour of fat, this will be true. However, current advice, detailed in the NACNE[56] and COMA[57] reports, that the intake of saturated fats in the human diet should be reduced, is encouraging the production of leaner meat; this has raised the question of the importance of lipids in flavour formation. Scientific evidence has shown that lipid may indeed be needed for the flavour of lean meat, but not the visible, white, saturated, subcutaneous, or intramuscular fat; there is sufficient lipid within the muscle for the requirements of flavour formation.[23]

Evidence from sensory work has shown that the flavour of the lean meat from a fat animal is indistinguishable from that from a lean animal.[58,59] No correlation was found between backfat thickness and taste panel scores for flavour. These studies support findings, described earlier,[23,27] which indicated that, while triacylglycerols have little effect on meat aroma formation, phospholipids play an important role.

Stress and pH

It is well known that the handling of meat animals prior to slaughter is not only important from a welfare point of view but also affects the quality of the resulting meat. Preslaughter stress causes depletion of the glycogen reserves, which, in turn gives a high ultimate pH in the meat, and this affects flavour. The higher the ultimate pH, the lower the perceived flavour intensity.[4] In cattle, preslaughter stress causes a condition known as 'dark cutting' or 'dark, firm, dry' (DFD) beef. 'Normal' meat (pH ≤ 5.8) consistently has a significantly stronger meat flavour than DFD beef (pH ⩾ 6.0), and off-flavour notes have been detected in the DFD meat.[60] The effects have been observed in steers as well as bulls.[61]

There are two possible mechanisms to explain the effect of stress on flavour. The less intense flavour of DFD beef may be because activity and stress prior to slaughter have depleted the levels of free sugars available for flavour-forming reactions. Alternatively, pH may have a direct effect on the formation of meat flavour compounds. The latter possibility is discussed later (p.179).

Pork can also show adverse effects due to abnormal pH; as well as high pH, DFD meat, PSE (pale, soft, exudative) meat may have an unusually low pH, possibly due to a build up of lactic acid following rapid post mortem glycolysis.[4,62] However, there is evidence that, for pork, pH may have the opposite effect on flavour than for beef. DFD pork showed a predominance of 'porky, sweet, fatty' notes which enhanced flavour, while the PSE meat possessed a 'sour' note, possibly due to high levels of lactic acid, which detracted from the flavour.[62] Thus, the link between pH and flavour is by no means straightforward. Relatively little is known about the important flavour compounds in pork; further studies in this area may offer an explanation for the effect of pH on their formation.

Conditioning

Various studies have demonstrated an increase in flavour intensity during ageing of beef for a period of up to four weeks.[63-65] It seems clear that conditioning increases the intensity of 'beefy', and sometimes 'fatty', notes, but whether this is deemed desirable depends on the personal tastes of the assessor. There is a risk of oxidation of the lipids to give rancid off-flavours, and this is the likely origin of any undesirable flavours occurring after prolonged ageing.

The concentrations of free sugars and amino acids are increased during post mortem conditioning by the breakdown of glycogen and the action of proteolytic enzymes.[8] Hydrolysis of the phospholipids also takes place to release free fatty acids.[66] As these sugars, amino acids, and fatty acids are all important precursors of flavour-forming reactions, it seems reasonable that there should also be an increased yield of flavour compounds; in fact, this has been demonstrated. The quantity of aroma volatiles collected from boiled beef has been shown to increase with ageing period, and the rate of increase accelerates as ageing progresses; this increase in volatile compounds is concomitant with a change in odour descriptions from 'bland, weak, unappetizing' (no ageing) to 'strong, savoury, appetizing, and roasted' after four weeks ageing.[67]

An increase in flavour has also been found on storage of chickens.[55] Chickens eviscerated after eight days chilled storage, then frozen, were given higher scores for flavour than chickens eviscerated immediately after slaughter and then frozen. Terms used to describe meat from stored, uneviscerated birds included: 'more flavour', 'gamey', and 'livery' while the freshly eviscerated birds gave meat described as 'bland'. In this case it was not clear whether the stronger flavour was caused by the activity of the microbial population of the gut or by autolytic processes occurring in the muscle during storage, as suggested for beef.

4 Key Compounds Responsible for Meat Flavour

The search for the compound or compounds responsible for the characteristic flavour of meat has made considerable progress in the past few years. The quest has focused on a class of heterocyclic thiols and disulfides. These have long been known to possess meat-like odour properties and low odour thresholds and their use in synthetic meat flavourings was patented in the early 1970s.[68,69] The meaty character of these compounds depends on the position of the thiol group and the degree of unsaturation; furans and thiophenes with a thiol group in the 3-position tend to possess a 'meaty' aroma while those with the thiol in the 2-position tend to be 'burnt' and 'sulfurous'.[70] The best meat-like aroma is given when there is a methyl group adjacent to the thiol group and the ring contains at least one double bond.[71]

(1) 2-Methyl-3-(methylthio)furan

Recently, members of this class of compounds have also been detected in meat. The first of these, identified in cooked beef by MacLeod and Ames,[72,73] was 2-methyl-3-(methylthio)furan (1). Gasser and Grosch[3] used dilution studies to locate the key odour impact compounds in cooked beef; seventeen compounds were detected at high dilution and it may be deduced that these are the key odour impact compounds. These included methional (like cooked potato), 2-acetyl-1-pyrolline and 2-acetylthiazole (roasty), four unsaturated aldehydes (fatty, like fried potato), oct-1-en-3-one (mushroom-like), and also 2-methylfuran-3-thiol (2) and bis(2-methyl-3-furyl) disulfide (3) (both meaty). It was concluded that these last two compounds are important contributors to meat flavour. Bis(2-methyl-3-furyl) disulfide (3) has one of the lowest odour thresholds known and can be detected[74] at 2 parts in 10^{14} and it is interesting that the 2-methylfuran-3-thiol (2) could not be detected by mass spectrometry, although it was easily detected by smell.[3]

(2) 2-Methylfuran-3-thiol   (3) Bis(2-methyl-3-furyl) disulfide

A study on chicken flavour by the same authors[37] identified sixteen important odour compounds, including some of those detected in beef; however, the relative importance of the various compounds was altered. While 2-methylfuran-3-thiol (2) was still an important meaty compound, its disulfide (3) was not present at sufficiently high levels to make a significant contribution to the aroma. Instead, deca-*trans*-2,*trans*-4-dienal and other unsaturated aldehydes were of increased importance and the contribution of these compounds to the characteristic species odour of chicken has already been discussed.

Additional heterocyclic disulfides have been identified recently in cooked beef.[75] A dilution technique modified from that used by Grosch's group[3,37] was used to identify the locations of key odour impact compounds in three cooked bovine muscles; heart, *semimembranosus*, and *psoas major*. Heart is reputed to have a particularly strong flavour while *psoas major*, although tender, is thought to be one of the less flavoursome cuts.

Five structurally related, late-eluting disulfides were detected among the volatiles from cooked heart, while lesser quantities of these compounds were detected in one or both of the other muscles.[75] These compounds were bis(2-methyl-3-furyl) disulfide (3), 2-furfuryl 2-methyl-3-furyl disulfide (4), bis(2-furfuryl) disulfide (5), and two compounds tentatively identified as dimethylfuryl 2-methyl-3-furyl disulfide (6) and 2-methyl-3-furyl 2-methyl-3-thienyl disulfide (7). Only the first and last of these had previously

(4) 2-Furfuryl 2-methyl-3-furyl disulfide

(5) Bis(2-furfuryl) disulfide

(6) Dimethylfuryl 2-methyl-3-furyl disulfide

(7) 2-Methyl-3-furyl 2-methyl-3-thienyl disulfide

been reported in the volatiles from meat. Two of these disulfides were detected as 'meaty' at high dilution. Bis(2-methyl-3-furyl) disulfide (3) was reported by Gasser and Grosch;[3] however, it appears that 2-furfuryl 2-methyl-3-furyl disulfide (4) is also a key odour impact compound in beef. The higher levels of these compounds in heart compared with the other muscles[75] offers an explanation for the reputedly strong flavour of cooked heart.

Many heterocyclic disulfides have been detected in model systems containing cysteine (or cystine) and a reducing sugar;[76-80] such compounds frequently possess strong meaty aromas. The monomeric forms of these compounds may be formed either by the reaction of furfurals or hydroxyfuranones with H_2S[81] or from the breakdown of thiamine.[79] The disulfides can then be formed by oxidation of these compounds.[75]

One aspect of the model system studies which proved of particular interest was the effect of pH on the generation of these compounds. The formation of some of these compounds was examined at pH values over the pH range 4.5–6.5.[80] It was apparent that the generation of this class of compound is extremely pH dependent, with maximum formation at low pH. The levels of the monomeric forms, 2-methylfuran-3-thiol (2) and 2-methylthiophene-3-thiol, also decrease with increasing pH, although small quantities were still detected at pH 6.5. As discussed earlier, the flavour intensity of beef is known to increase with decreasing pH. The marked pH dependency of the generation of these compounds may offer some explanation for the relationship between pH and flavour in meat.

5 Summary

Much progress has been made in the past few years toward the identification of the compounds which give cooked meat its characteristic meaty flavour. However, this work has not answered all the questions: for instance, although a great deal is now known about the mechanism by which phospholipids can interact in flavour-forming reactions, it still remains unclear exactly how they increase the meaty character of the aroma. This question is the subject of ongoing research.

These recent developments in the chemistry of meat flavour formation open up the possibility of improving our understanding of the effect of production factors on flavour. Can the effects of diet, age, husbandry, conditioning, *etc.* be rationalized in terms of the known mechanisms of flavour formation? Much of the meat available in the shops is accused of being bland or inconsistent in flavour quality. If fundamental studies of the chemistry of meat flavour and its precursors can be correlated with the factors involved in the production of meat animals, it should be possible to improve the flavour of the meat produced.

6 References

1. E.C. Crocker, *Food Res.*, 1948, **13**, 179.
2. H. Maarse, 'Volatile Compounds in Foods - Qualitative and Quantitative Data', 6th Edn., TNO-CIVO Food Analysis Institute, Zeist, the Netherlands, 1989.
3. U. Gasser and W. Grosch, *Z. Lebensm,-Unters. Forsch.*, 1988, **186**, 489.
4. R.A. Lawrie, 'Meat Science', Pergamon Press, Oxford, 1985.
5. D.S. Mottram, in 'Volatile Compounds in Foods and Beverages', ed. H. Maarse, Marcel Dekker, New York, 1990, p.107.
6. D.S. Mottram, in 'Bioflavour '90', ed. J.R. Piggott and A. Patterson, Elsevier, 1992, in press.
7. G. MacLeod and M. Seyyedain-Ardebili, *CRC Crit. Rev. Food Sci. Nutr.*, 1981, 309.

8. F. Shahidi, in 'Flavor Chemistry - Trends and Developments', ed. R. Teranishi, R.G. Buttery, and F. Shahidi, American Chemical Society, Washington, DC, 1989, p.188.
9. F. Shahidi, L.J. Rubin, and L.A. D'Souza, *CRC Crit. Rev. Food Sci. Nutr.*, 1986, **24**, 141.
10. D.A. Baines and J.A. Mlotkiewicz, in 'Recent Advances in the Chemistry of Meat', ed. A.J. Bailey, Royal Society of Chemistry, London, 1984, p.119.
11. L.M. van der Linde, J.M. van Dort, P. de Valois, H. Boelens, and D. de Rijke, in 'Progress in Flavour Research', ed. D.G. Land and H.E. Nursten, Applied Science, London, 1979, p.219.
12. G.J. Hartman, J.T. Carlin, J.D. Scheide, and C.T. Ho, *J. Agric. Food Chem.*, 1984, **32**, 1015.
13. L.C. Maillard, *C. R. Seances Acad. Sci.*, 1912, **154**, 66.
14. L.J. Salter, D.S. Mottram, and F.B. Whitfield, *J. Sci. Food Agric.*, 1988, **46**, 227.
15. L.J. Farmer, D.S. Mottram, and F.B. Whitfield, *J. Sci. Food Agric.*, 1989, **49**, 347.
16. P.J. Kiely, A.C. Nowlin, and J.H. Moriarty, *Cereal Sci. Today*, 1960, **5**, 273.
17. I.D. Morton, P. Akroyd, and C.G. May, UK Patent 836 694, 1960.
18. G. MacLeod, in 'Developments in Food Flavours', ed. G.G. Birch and M.G. Lindley, Elsevier, London, 1986, p.191.
19. R.L.S. Patterson, in 'Meat', ed. D.J.A. Cole and R.A. Lawrie, Butterworth, London, 1974, p.359.
20. M.J. Tims and B.M. Watts, *Food Technol.*, 1958, **12**, 240.
21. W. Grosch, in 'Food Flavours. Part A', ed. I.D. Morton and A.J. MacLeod, Elsevier, Amsterdam, 1982, p.325.
22. W.W. Nawar, in 'Flavor Chemistry of Fats and Oils', ed. D.B. Min and T.H. Smouse, American Chemical Society, 1985, p.39.
23. D.S. Mottram and R.A. Edwards, *J. Sci. Food Agric.*, 1983, **34**, 517.
24. L.J. Farmer, PhD Thesis, University of Bristol, 1990.
25. F.B. Whitfield, *Crit. Rev. Food Sci. Nutr.*, 1992, **31**, 1.
26. G MacLeod and J.M. Ames, *J. Food Sci.*, 1987, **52**, 42.
27. L.J. Farmer and D.S. Mottram, *J. Sci. Food Agric.*, 1990, **53**, 505.
28. I. Hornstein and P.F. Crowe, *J. Agric. Food Chem.*, 1960, **8**, 494.
29. I. Hornstein and P.F. Crowe, *J. Agric. Food Chem.*, 1963, **11**, 147.
30. A.E. Wasserman and A.M. Spinelli, *J. Agric. Food Chem.*, 1972, **20**, 171.
31. E. Wong, C.B. Johnson, and L.N. Nixon, *Chem. Ind. (London)* 1975, 40.
32. E. Wong, L.N. Nixon, and C.B. Johnson, *J. Agric. Food Chem.*, 1975, **23**, 495.
33. E. Wong, C.B. Johnson, and L.N. Nixon, *NZ J. Agric. Res.*, 1975, **18**, 261.
34. R.L.S. Patterson, *J. Sci. Food Agric.*, 1968, **19**, 31.
35. L.J. Minor, A.M. Pearson, L.E. Dawson, and B.S. Schweigert, *J. Food Sci.*, 1965, **30**, 686.
36. E.L. Pippen and M. Nonaka, *Food Res.*, 1960, **25**, 764.
37. U. Gasser and W. Grosch, *Z. Lebensm.-Unters. Forsch.*, 1990, **190**, 3.
38. M.A. Katz, L.R. Dugan, Jr., and L.E. Dawson, *J. Food Sci.*, 1966, **31**, 717.
39. J.M. Eichorn, C.M. Bailey, and G.J. Blomquist, *J. Anim. Sci.*, 1985, **61**, 892.
40. J.D. Sink, *J. Food Sci.*, 1979, **44**, 1.
41. S.L. Melton, *J. Anim. Sci.*, 1990, **68**, 4421.
42. L.M. Poste, *J. Anim. Sci.*, 1990, **68**, 4414.

43. S.L. Melton, *Food Technol.*, 1983, 239.
44. S.L. Melton, J.M. Black, G.W. Davis, and W.R. Backus, *J. Food Sci.*, 1982, **47**, 699.
45. D.K. Larick, H.B. Hedrick, M.E. Bailey, J.E. Williams, D.L. Hancock, G.B. Garner, and R.E. Morrow, *J. Food Sci.*, 1987, **52**, 245.
46. D.K. Larick and B.E. Turner, *J. Food Sci.*, 1990, **54**, 649.
47. R.L. Summers, J.D. Kemp, D.G. Ely, and J.D. Fox, *J. Anim. Sci.*, 1978, **47**, 622.
48. J.D Kemp, M. Mahyuddin, D.G. Ely, J.D. Fox, and W.G. Moody, *J. Anim. Sci.*, 1980, **51**, 321.
49. A.M. Nicol and K.T. Jagusch, *J. Sci. Food Agric.*, 1971, **22**, 464.
50. W.N. Marmer, R.J. Maxwell, and J.E. Williams, *J. Anim. Sci.*, 1984, **59**, 109.
51. G.J. Miller, R.A. Field, L. Medieros, and G.E. Nelms, *J. Food Sci.*, 1987, **52**, 526.
52. D.K. Larick and B.E. Turner, *J. Food Sci.*, 1990, **55**, 312.
53. R.J. Park, K.E. Murray, and G. Stanley, *Chem. Ind. (London)* 1974, 380.
54. W. Ralph, *The Shepherd*, 1989, Feb., p.9.
55. G.C. Mead, N.M. Griffiths, C.S. Impey, and J.C. Coplestone, *Br. Poult. Sci.*, 1983, **24**, 261.
56. W.P.T. James (chairman), 'Proposals for Nutritional Guidelines for Health Education in Britain', National Advisory Committee on Nutritional Education, 1983.
57. V. Wheelock, 'COMA report: Diet and Vascular Disease', University of Bradford, 1984.
58. J.D. Wood, E. Dransfield, and D.N. Rhodes, *J. Sci. Food Agric.*, 1979, **30**, 493.
59. J.D. Wood, D.S. Mottram, and A.J. Brown, *Anim. Prod.*, 1981, **32**, 117.
60. S. Fjelkner-Modig and H. Ruderus, *Meat Sci.*, 1983, **8**, 203.
61. L.E. Jeremiah, J.A. Newman, A.K.W. Tong, and L.L. Gibson, *Meat Sci.*, 1988, **22**, 103.
62. L.E. Jeremiah, A.C. Murray, and L.L. Gibson, *Meat Sci.*, 1990, **27**, 305.
63. M. Seydi and C. Touraille, *Rev. Tech. Vet. Aliment.*, 1986, **218**, 18.
64. I. Seuss, M. Martin, and K.O. Honikel, *Fleischwirtschaft*, 1990, **70**, 1083.
65. L.E. Jeremiah, G.G. Greer, and L.L. Gibson, *J. Muscle Foods*, 1991, **2**, 119.
66. V. Kesava Rao and B.N. Kowale, *Meat Sci.*, 1991, **30**, 115.
67. B.M. Coppock and G. MacLeod, *J. Sci. Food Agric.*, 1977, **28**, 206.
68. W.J. Evers, US Patent 1256462, 1971.
69. G.A.M. van den Ouweland and H.G. Peer, UK Patent 1283912, 1972.
70. W.J. Evers, H.H. Heinsohn, Jr., B.J. Mayers, and A. Sanderson, in 'Phenolic, Sulfur and Nitrogen Compounds in Food Flavours', ed. G. Charalambous and I. Katz, American Chemical Society, Washington, DC, 1976, p.184.
71. G.A.M. van den Ouweland, E.P. Demole, and P. Enggist, in 'Thermal Generation of Aromas', ed. Y.H. Parliment, R.J. McGorrin, and C.-T. Ho, American Chemical Society, Washington, DC, 1989, 433.
72. G. MacLeod, and J.M. Ames, *Chem. Ind. (London)*, 1986, 175.
73. G. MacLeod and J.M. Ames, *J. Food Sci.*, 1986, **51**, 1427.
74. R.G. Buttery, L.C. Ling, R. Teranishi, and T.R. Mon, *J. Agric. Food Chem.*, 1984, **32**, 674.
75. L.J. Farmer and R.L.S. Patterson, *Meat Sci.*, **40**, 201.
76. T.A. Misharina, S.V. Vitt, and R.V. Golovnya, *Biotekhnologiya*, 1987, **3**, 210.

77. T.A. Misharina and R.V. Golovnya, in 'Characterization, Production and Application of Food Flavours', ed. M. Rothe, Akademie-Verlag, Berlin, 1988, p.115.
78. P. Werkhoff, R. Emberger, M. Guntert, and M. Kopsel, Ref.71, p.460.
79. P. Werkhoff, J. Bruning, R. Emberger, M. Guntert, M. Kopsel, W. Kuhn, and H. Surburg, *J. Agric. Food Chem.*, 1990, **38**, 777.
80. L.J. Farmer and D.S. Mottram, in 'Flavour Science and Technology', ed. Y. Bessiere and A. Thomas, John Wiley and Sons, Chichester, 1990, p.113.
81. G.A.M. van den Ouweland and H.G. Peer, *J. Agric. Food Chem.*, 1975, **23**, 501.

Meat Iron Bioavailability: Chemical and Nutritional Considerations

R.J. Neale

DEPARTMENT OF APPLIED BIOCHEMISTRY & FOOD SCIENCE, UNIVERSITY OF NOTTINGHAM, SUTTON BONINGTON, LOUGHBOROUGH, LEICESTERSHIRE LE12 5RD, UK

1 Introduction

The term iron (Fe) bioavailability is widely recognized as meaning the proportion of the Fe contained in a food or ingested as a supplement which is absorbed and utilized by the human body. With Fe deficiency being the most prevalent nutritional deficiency in the world today a greater understanding of the role of food Fe in its prevention is undoubtedly needed.

Considerable amounts of work have been done on the clinical and medical aspects of the problem of Fe deficiency, but until now very little work has been done to understand more fully the chemical nature of food Fe and the complex interactions which it undergoes in association with other foods (and components) in its passage through the gastrointestinal tract.

In developing countries the much higher incidence of Fe deficiency is due primarily to insufficient intake of dietary iron or consumption of diets in which the Fe is poorly bio-available. Both these factors are linked to dietary composition in a number of interesting ways which will be described further in this review.

2 Food Sources of Fe and Its Dietary Supply

It is recognized by most nutritionists that human Fe requirements are the most difficult to meet from foods available in the UK and western countries even when energy intake is adequate for energy requirements. Recent work commissioned by MAFF and DHSS[1] on the diets of British adults shows that the average daily intake of Fe was 12.3 mg for women and 14.0 mg for men which compares favourably with the recommended daily amount (RDA) of 12 mg for women and 10 mg for men under 55 years of age. The contribution to the total Fe intake provided by the Fe from meat and fish products was 27% for men and 23% for women. These dietary intakes at first sight appear highly satisfactory but further analysis of blood samples showed that 1% of men and 4% of women had low haemoglobin levels ($< 11.0\ \mathrm{g\,dl^{-1}}$) and were considered anaemic and that much larger numbers of women (one in three) had serum ferritin levels below the level considered to indicate low Fe stores. It is therefore clear from this national survey that the iron nutrition of many adults (particularly women) is poor.

In nutritional and chemical terms the total dietary Fe can be divided into two separate pools - the haem and non-haem Fe pools. Earlier work had shown that the Fe available for absorption from these pools was very different in quantitative terms and one of the major prerequisites for the prediction of the bioavailable Fe from whole diets is the measurement of the total dietary haem and non-haem Fe levels.[2]

Meat (including fish and poultry) is the only dietary source of haem Fe and the amounts of haem Fe vary considerably in a range of raw (uncooked) meats (Table 1). The red meats in general have haem Fe contents 2–3 times those of white meats (poultry *etc.*) and on processing (heating) a proportion of the haem Fe in meat is degraded to non-haem Fe. This varies considerably with time, temperature, and processing conditions (Table 2) so that the dietary haem to non-haem Fe ratio would not be known accurately without a full analysis of a duplicate meal prior to consumption.

3 The Nutritional Significance of Haem and Non-haem Fe in Meat and the 'Meat-Effect'

Early work in human subjects by Moore and Duback[9] and by Martinez-Torres and Layrisse,[10] in which foods of both plant and animal origin were

Table 1 *Haem Fe as a proportion of the total Fe in meat (raw); values are percentages*

Beef	Pork	Lamb	Chicken	Reference
71	36	ND	ND	3
62	49	ND	ND	4
74	47	59	29 (thigh)	5

ND = Not determined.

Table 2 *Effect on cooking of haem Fe content of beef*

Conditions	% Reduction in haem Fe	Reference
Cooked to 70 °C in water	18	6
Braised and microwaved (68 °C)	10	7
Boiled for 90 min	25	8
Autoclaved at 121 °C for 90 min	62	8

biosynthetically labelled with radio iron and fed alone to humans, showed that foods with high levels of haem Fe were much more highly bioavailable (*i.e.* were absorbed to a greater degree as a percentage of dose Fe given) than foods with lower or no haem Fe present at all (vegetable foods in Figure 1).

The haem Fe which was available for absorption was said to arise from digestion of haemoproteins (haemoglobin and myoglobin) in both stomach and small intestine such that the haem Fe is released intact in the intestinal lumen (*i.e.* not degraded[11,12]) and is then transferred preferentially by a haem Fe-specific transport pathway with the Fe still locked in the porphyrin ring (Figure 2).

The iron from the porphyrin ring is released within the intestinal mucosal cells by the action the enzyme haem oxygenase, with carbon monoxide and bilirubin as the end products as they are in other tissues. The haem oxygenase is present in higher concentrations in the duodenum than in the remainder of the small intestine. The rate of haem catabolism is more rapid in Fe deficiency and haem oxygenase activity increases.

The other absorption pathway (Figure 2) specific for non-haem Fe is probably available to a wide range of low molecular weight Fe compounds some of which are produced in the intestinal tract by the process of digestion of high molecular weight Fe-containing proteins in meat (ferritin and haemosiderin) and others which arise through the complexation, chelation, and reduction of non-haem Fe species with other food components (vitamin C, organic acids, meat proteins). To be available for uptake

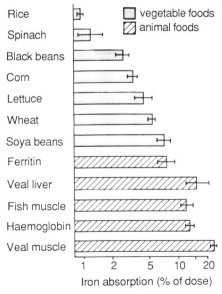

Figure 1 *The absorption of iron from vegetable and animal foods by Venezuelan peasants. Results are expressed as the geometrical mean ($\pm$ s.e.)* (After Martinez-Torres and Layrisse[10])

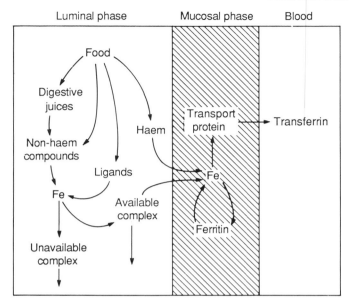

Figure 2 *The effects of various intraluminal factors on the absorption of haem and non-haem iron*
(Redrawn from Bothwell et al.[31])

by the mucosal cells the Fe must be in a soluble form at the site of absorption and since this site is in the upper duodenum the pH is likely to be between 3 and 7 as a result of the pancreatic secretion of bicarbonate. At this pH very few ferric ions will be in solution and at pH 8 the maximum concentration is only 10^{-18} M, compared with 1.6×10^{-2} M for ferrous iron. Food Fe can, however, interact with many potential ligands in food and it is those ligands which form soluble monomeric complexes with Fe and thus prevent its precipitation and polymerization which tend to promote its absorption.

The model for Fe absorption proposed (Figure 2) still has many areas of uncertainty associated with it:

(1) How is Fe (both haem and non-haem) transferred from the intestinal lumen into the mucosal cell?
(2) Is all the haem Fe in meat released into the lumen with the Fe locked in the porphyrin ring and absorbed intact or is some of the porphyrin degraded and the Fe absorbed as non-haem Fe?
(3) What range of chemical forms of non-haem Fe are absorbed and is there competition between them?
(4) What enzyme systems are involved in the Fe absorption (both haem and non-haem Fe) in normal and Fe-deficient states?
(5) How does the control mechanism for transfer of Fe from the mucosal cell into the blood operate?

These and many other areas of uncertainty still remain to be elucidated in this field. In practical nutritional terms, however, what seems to be well established is that when present together in a mixed diet the contribution made by the haem Fe to the total Fe absorbed (*i.e.* the total Fe bioavailability) is much greater than that of the non-haem Fe. In one now classical study[13] while the haem Fe represented only 5.7% of the total Fe intake it formed 29.0% of the total Fe absorbed. It is quite clear from these studies that haem Fe, even when it forms only a small proportion of the total dietary Fe content, is of great nutritional significance. The reason for this finding is said to be that the pathway for haem Fe absorption is almost unaffected by potential inhibitors of Fe absorption in foods such as phytic acid, tannins (polyphenols), oxalates, *etc.* and may even be stimulated by the presence of some foods or food components, *e.g.* soya[14] and meat.[15] Conversely the non-haem Fe pathway, while it is capable of absorbing large amounts of Fe if presented in sufficiently large amounts in a highly available form (reduced, soluble, and low molecular weight), is much more susceptible to both inhibitory and enhancing influences from other foods and food components and in most mixed diets with low levels of meat, poultry, and fish the inhibitory influences usually outweigh the enhancing ones. The enhancing influence on non-haem Fe absorption exerted by the presence of meat, which is 'dose dependent' is known as the 'meat effect' or 'meat factor'. The 'meat factor' results from the presence of a variety of meat (muscle) products but eggs, milk, cheese, and ovalbumin either have no effect or an inhibitory one on non-haem Fe absorption.[16,17]

The exact mechanism for the 'meat effect' has not been fully explained, but is probably related to the release of amino acids and/or polypeptides during proteolytic digestion which chelate dietary non-haem Fe, thereby facilitating its absorption.[18,22] Since the solubility of the iron in the lumen of the gastrointestinal tract is the first prerequisite for its absorption a considerable number of *in vitro* studies have been performed looking at this solubilization process, with a view to using it as a predictor for the bioavailability of both food and fortification Fe *in vivo* in man.

The following methods have been used to try to measure the soluble Fe from foods which is potentially available for absorption:

(1) Incubation with pepsin/HCl mixtures or gastric juice.[19,20]
(2) Pepsin/HCl digestion followed by neutralization with either sodium hydroxide or sodium bicarbonate (solid) and further digestion at pH 7 with pancreatic enzymes with or without emulsifying agents.[21,22]
(3) Pepsin/HCl digestion followed by simulated pancreatic digestion inside a dialysis membrane to allow separation of low molecular weight (dialysable) from high molecular weight Fe compounds.[23]

Widely varying proportions of food Fe are solubilized by such procedures, the figures lying in the range 20–70%. Although soluble not all of this Fe is ionizable and when these foods are ingested the absorption of Fe is almost always a good deal less than might be expected from the ionizable Fe figures (Figure 3). There is a number of possible reasons for such discrepancies but the most important is probably that the released Fe is vulnerable to the various ligands in the diet and in gastrointestinal secretions such as bile and pancreatic juices since its fate is largely determined by the nature of the particular complexing agent with which it binds. *In vitro* studies on how much Fe can be rendered ionic by gastric juice therefore only provide an index of the maximum amount of Fe available for absorption in a particular foodstuff.

4 *In Vitro* Studies On Meat Digestion and Fe Release

Some of the first detailed work on the chemical nature of Fe released from meat during simulated physiological digestion was performed by Hazell *et al.*[22] using radiolabelled rat muscle and unlabelled beef. Their studies showed that following pepsin/HCl digestion and prolonged pancreatic digestion followed by fractionation by gel filtration the Fe compounds released were predominantly low molecular weight non-haem Fe compounds and not haem compounds as might be expected (Figure 4). When injected into loops of rat small intestine *in vivo* these low molecular weight compounds were very rapidly absorbed compared with whole blood (haemoglobin primarily) or an aqueous muscle extract (containing soluble

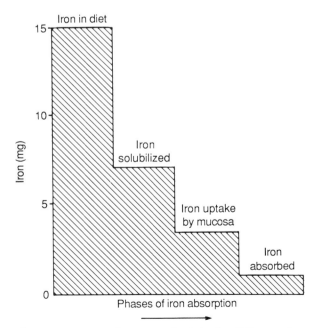

Figure 3 *The relative quantities of iron involved in the different phases of the daily absorption of dietary iron*
(Redrawn from Bothwell *et al.*[31])

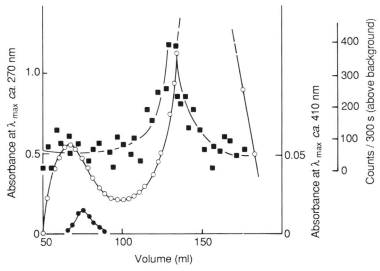

Figure 4 *Chromatogram of supernatant from ^{59}Fe labelled rat meat digest following pepsin and pancreatic digestion at pH 7.0 and 35 °C for 18 h and fractonation on Sephadex G-50:* ○——○ *haemoproteins;* ●——● *protein fragments;* ■——■ ^{59}Fe *activtiy*
(Redrawn from Hazell et al.[22])

haemoproteins and other meat proteins) but excluding actin and myosin (Table 3).

The studies by Hazell et al.[16] questioned the widely held assumption that the high Fe absorption from meat was due to the intact absorption of haem Fe but rather indicated that low molecular weight non-haem Fe compounds, which must partly originate from the cleavage of the haem Fe in the lumen of the gastrointestinal tract, were the Fe compounds most readily (and speedily) absorbed.

This view that part of the haem Fe from meat is cleaved before

Table 3 ^{59}Fe *absorption from rat duodenal loops* in vivo *after injection of a dialysate of digested rat muscle, an aqueous extract of undigested rat muscle, or rat blood all labelled* in vivo *with ^{59}Fe (values are means ± SE of at least two observations)*

	% ^{59}Fe absorbed from loop into blood	
	Zero time*	60 min
Dialysed muscle digest	19.3 ± 0.3	34.2 ± 0.8
Aqueous muscle extract	6.2 ± 0.3	28.3 ± 5.7
Whole blood	4.2 ± 0.3	16.8 ± 2.4

*Zero time was *ca.* 30–60 s after administration. From ref.16.

absorption has found little support in the literature although it is now becoming more widely recognized that the haem Fe content of meat can be significantly reduced by processing (cooking and curing) and therefore the actual amount of haem Fe present in meat products just prior to consumption cannot be inferred from measurements made of the haem Fe levels in raw (unprocessed) meat (Table 2).

The effect of cooking meat would appear to cause its Fe absorption to decrease in both human[24] and rat[25] *in vivo* studies, the explanation being that the haemoproteins suffer a loss in bioavailability due to their co-precipitation with other proteins to yield the insoluble cooked meat haemoproteins.[26] These complexes can be solubilized by strong acid but *in vivo* the stomach may not be able to do this very efficiently; thus throughout the digestion period all the haem Fe is not readily accessible. Undoubtedly there is still confusion into the exact mechanism whereby haem Fe in meat is absorbed and particularly the role of haem-splitting enzyme in the intestinal mucosa. There can be no doubt, however, that the level of haem Fe in foods and diets is very important for Fe absorption whether as a fortificant for weaning food[27] or taken as meat in a mixed diet.

5 Mechanism of the 'Meat Effect'

There is little doubt that meat promotes the absorption of both haem and non-haem Fe but the mechanism is not yet completely understood.

Politz and Clydesdale[18] demonstrated that a dilute salt-insoluble fraction of chicken muscle was able to improve the solubility of added $FeCl_3$ during simulated physiological digestion and that the fractions which solubilized most Fe were those in the lowest molecular weight range (6200–2500). Solubilization capacity did not correlate to binding by free sulfhydryl groups.

Kapsokefalou and Miller[28] produced evidence that meat (raw and cooked), haemoglobin, and red blood cells when mixed with $FeCl_3$ and subjected to simulated physiological digestion produced large increases in the level of dialysable Fe^{II} and that this may be part of the mechanism for the 'meat effect', *i.e.* a reducing effect. The nature of the binding of digestion products to Fe released from meat and other foods is clearly important also in the context of the pH changes occurring in the small intestine.

At the near-neutral or weakly acid pH of the duodenum Fe would normally exist in aqueous solution in the form of Fe hydroxides.

The presence of protein digestion products, however, produced following *in vitro* digestion of bovine serum albumin (Shears *et al.*[29]) produced such a high affinity for the Fe^{II} that the formation of hydroxides was prevented. With Fe^{III} the formation of hydroxides was delayed from pH 3 in the absence of the digest to pH 7 in its presence. In addition Shears *et al.* showed that the only sites for the complexation of the Fe were the carboxyl groups despite the presence of other chelatable groups (amino and sulfhydryl groups). This result using bovine serum albumin differs somewhat from that of Taylor et al.[30] who provide evidence that the sulfhydryl groups of cysteine are responsible, through Fe complexation, for the enhancing effect of meat on non-haem Fe absorption.

Further work is undoubtedly needed to understand this effect more precisely in chemical terms and to understand why only meat and not other animal or plant foods enhances non-haem Fe absorption.

5 References

1. 'The Dietary and Nutritional Survey of British Adults: a Survey of the Dietary Behaviour, Nutritional Status, and Blood Pressure of Adults Aged 16 to 64 Living in Great Britain', J. Gregory et al., HMSO, 1990.
2. E.R. Monsen, L. Hallberg, M. Layrisse, M. Hegsted, J.D. Cook, W. Mertz, and C.A. Finch, Am. J. Clin. Nutr., 1978, **31**, 134.
3. M.S. Buchowski, A.W. Mahoney, C.E. Carpenter, and D.P. Cornforth, J. Food. Sci., 1988, **53**, 43.
4. B.R. Schricker, D.D. Miller, and J.R. Stouffer, J. Food Sci., 1982, **47**, 740.
5. T. Hazell, J. Sci. Food Agric., 1982, **33**, 1049.
6. J.O. Igene, J.A. King, A.M. Pearson, and J.I. Gray, J. Agric. Food Chem., 1979, **27**, 838.
7. B.R. Schricker and D.D. Miller, J. Food Sci., 1983, **48**, 1340.
8. O. Jansuittivechakul, A.W. Mahoney, D.P. Cornforth, O.G. Hendricks, and K. Kangsadalampai, J. Food Sci., 1985, **50**, 407.
9. C.V. Moore and R. Dubach, Trans. Assoc. Am. Physicians, 1954, **64**, 245.
10. C. Martinez-Torres and M. Layrisse, in 'Clinics in Haemotology', Vol.2, ed. S. T. Callender, W. B. Saunders & Co., London, Philadelphia & Toronto, 1973, pp.339–352.
11. I. Kaldor, Australas. Ann. Med., 1957, **6**, 244.
12. R. Sanford, Nature (London), 1960, **185**, 533.
13. E. Bjorn-Rasmussen, L. Hallberg, B. Isaksson, and B. Arvidson, J. Clin. Invest., 1974, **53**, 247.
14. S.R. Lynch, S.A. Dassenko, T.A. Morck, J.L. Beard, and J.D. Cook, Am. J. Clin. Nutr., 1985, **41**, 13.
15. A.L. Turnball, F. Cleton, and C.A. Finch, J. Clin. Invest., 1962, **41**, 1898.
16. J.D. Cook and E.R. Monsen, Am. J. Clin. Nutr., 1976, **29**, 859.
17. M. Layrisse, C. Martinez-Torres, and M. Roche, Am. J. Clin. Nutr., 1968, **21**, 1175.
18. M.L. Politz and F.M. Clydesdale, J. Food Sci., 1988, **53**; 1081.
19. H.V. Hart, J. Sci. Food Agric., 1971, **22**, 354.
20. O. Bergeim and E.R. Kurch, J. Biol. Chem., 1949, **177**, 591.
21. B.S. Narasinga Rao and T. Prabhavathi, Am. J. Clin. Nutr., 1978, **31**, 169.
22. T. Hazell, R.J. Neale, and D.A. Ledward, Br. J. Nutr., 1978, **39**, 631.
23. B.R. Schricker, D.D. Miller, R.R. Rasmussen, and D. Van Campen, Am. J. Clin. Nutr., 1981, **34**, 2257.
24. C. Martinez-Torres, I. Leets, P. Taylor, J. Ramirez, M. Camacho, and M. Layrisse, J. Nutr., 1986, **116**, 1720.
25. F.E. Bogunjoko, R.J. Neale, and D.A. Ledward, Br. J. Nutr., 1983, **50**, 511.

26. D.A. Ledward, *J. Food Sci.,* 1971, **36**, 883.
27. E. Calvo, E. Hertrampf, S. Pablo, N. Amar, and A. Stekel, *Eur. J. Clin. Nutr.*, 1989, **43**(4), 237.
28. M. Kapsokefalou and D.D. Miller, *J. Food Sci.*, 1991, **56**, 352.
29. G.E. Shears, D.A. Ledward, and R.J. Neale, *Int. J. Food Sci. Technol.*, 1987, **22**, 265.
30. P.G. Taylor, C. Martinez-Torres, E.L. Romano, and M. Layrisse, *Am. J. Clin. Nutr.*, 1986, **43**, 68.
31. T.H. Bothwell, R.W. Charlton, J.O. Cook, and C.A. Finch, 'Iron Metabolism in Man', Blackwell, Oxford, 1979.

Chemistry of Lipids

J.B. Rossell

LEATHERHEAD FOOD RESEARCH ASSOCIATION, RANDALLS ROAD,
LEATHERHEAD, SURREY KT22 7RY, UK

Introduction

This chapter is about the chemistry of lipids in foods based on muscles. It is not about the chemistry of muscle lipids, and as most foods contain depot fats as well as intramuscular lipids both sources of lipid must be taken into account. However, since depot fats occur in much larger quantity than the intramuscular lipid, they form a far greater proportion in the final food. Furthermore, there is much more information available about the composition and chemistry of depot fats, and this review will therefore mainly consider the chemistry of depot lipids from fatty tissue.

This chapter will also consider the composition of oils in fatty fish, but in this case little distinction is made between the intramuscular lipid and lipids from fatty tissue.

The lipids in muscle-based foods are essentially triglycerides, or triacyl-glycerols, in which a variety of fatty acids is combined with the glycerol 'backbone' of the molecule. The different properties of lipids from different animal muscles are related to the compositions of the constituent triglycerides and through these to the fatty acids. These can vary with regard to the position of the muscle within the body, the diet of the animal, its weight or fattiness, age and sex, and, more importantly, to the animal species. Lipids from muscles nearer the skin tend to be more unsaturated, while diet is of importance with non-ruminant animals. Animals with higher amounts of body fat tend to have more unsaturated lipids - a generalization that helps rationalize, in part, the influences of age, body weight, and sex.

Beef, mutton, and pork fats do not differ greatly in degree of unsaturation, pork tending to be more unsaturated, and mutton somewhat less saturated than beef fat. Chicken fat is less saturated than mutton, pig, or beef, while horse flesh, not favoured in the UK, but appreciated on the European continent, has far less saturated fat, with a comparatively high concentration of linolenic acid.

In comparison with these land animals, fish oils constitute a completely different category of lipid being liquid at normal ambient temperatures, and having high concentrations of some very unsaturated fatty acids.

In dietary terms, the concentrations of myristic and linoleic acids have been of some interest, as has the polyunsaturated/saturated (P/S) ratio. The physical properties, however, are influenced not only by the fatty acid composition but also by the distribution of the fatty acids in the triglyceride molecules. The proportion of saturated fatty acid at the triglyceride 2-position could also be of dietary importance.

As muscle tissues age the constituent fatty acids may become liberated, generating free fatty acids, or oxidized to form initially peroxides and subsequently aldehydes and ketones. These minor components cause off-flavours or rancidity. Other minor components are sterols such as cholesterol, carotene, and tocopherols. Carotenes and tocopherols have vitamin A and E potencies, respectively.

The actual levels of oils and fats in muscle-based foods are important. Processed foods, such as tinned meat and fish, sometimes contain a lipid-type material, which is soluble in hexane, but does not yield fatty acid methyl esters when treated in the normal way, as the unprocessed flesh does. The nature of this material is still being studied but its presence leads to doubts about true contents of normal fats in tinned products, especially tinned fish.

It is of interest to review the world production of animal fats and fish oil in comparison with those of vegetable oils, as illustrated[1] in Table 1. In this Table, the data refer to free fat produced for world trade, not the fat consumed, for example, as part of a meat product. Table 1 shows that lard, tallow, and fish oils have shown steady growth, while production of mammalian butter fats has decreased. However, the growth in animal fats has been far outweighed by the much larger growth in the production of vegetable oils since 1983.

Of course, the production of food from muscular tissue does not usually involve separation of the fat, unless of course we use the fat to make pastry in a food such as a pork pie. Table 2 therefore shows the fat content

Table 1 *World production/trade of oils and fats (1000 tonnes)*[1]

	1983/84	1985/86	1988
Vegetable oils	43 300	52 995	62 376
Animal fats:			
Butter	6 255	6 260	6 114
Lard	5 185	5 430	5 445
Tallow	6 110	6 270	6 735
Total animal fats	17 540	17 960	18 294
Fish oils	1 265	1 520	1 466
World total	62 105	72 475	82 136

Table 2 *Typical fat contents (m/m%)*

	Mutton	Beef	Pig	Chicken	Fatty fish
Lean meat	9	5	7	4	–
Lean meat with all visible fat removed	4	2	–	–	–
Fatty tissue	72	67	72	–	18

of some typical meats. Lean mutton, beef, or pig flesh usually has between 5 and 10% fat, while chicken usually has a lower level of *ca.* 4%. If the meat is trimmed to remove all visible fat, and the lipid content is then determined, beef cuts may have as little as 2% lipid,[2,3] while mutton has a somewhat higher level of *ca.* 4%. In these carcasses the fatty tissue normally contains *ca.* 70% triglyceride fat. Fatty fish have not been studied in the same depth, but normally contain *ca.* 18% lipid.

2 Fatty Acid Compositions

The lipids are of course mainly triglycerides of fatty acids, and the major compositional characteristic of the fat is its fatty acid composition. Table 3 shows some typical fatty acid compositions of a variety of muscle-based fats. Mutton tallow has *ca.* 21% palmitic acid (C16:0) and 28% stearic acid

Table 3 *Fatty acid compositions of land animal fats: typical values* (m/m%)

Fatty acid	Mutton	Beef	Pig	Chicken	Horse
14:0	2.0	2.5	1.5	1.3	6.0
14:1	0.5	0.5	0.5	0.2	1.0
15:0	0.5	0.5			0.5
16:0	21.0	24.5	24.0	23.2	23.0
16:1	3.0	3.5	3.0	6.5	10.5
17:0	1.0	1.0	0.5	0.3	0.5
18:0	28.0	18.5	14.0	6.4	3.5
18:1	37.0	40.0	43.0	41.6	29.0
18:2	4.0	5.0	9.5	18.9	5.0
18:3		0.5	1.0	1.3	17.5
20:0	0.5	0.5	0.5		0.5
20:1	0.5	0.5	1.0		0.5
Others	2.0	2.5	1.5	0.3	2.5
P/S Ratio	0.07	0.11	0.25	0.64	0.66
IV*	42.6	48.7	60.3	78.3	90.4

*Iodine value by calculation from fatty acid composition.

(C18:0). The oleic and linoleic acids together constitute 41%. On this basis mutton tallow is regarded as predominantly saturated fat. This is also the case for a typical beef fat, which shows a fair similarity to mutton tallow. These two fats have P/S ratios of *ca.* 0.1, and iodine values in the 40–50 unit range. Pig fat, or lard, has about the same level of palmitic acid, but generally a lower level of stearic and a higher level of linoleic acid. This results in somewhat higher P/S ratio of 0.25, and higher iodine value of *ca.* 60. To some extent, this may account for the view that lard has a lower resistance to oxidation than beef fat.

Chicken fat has *ca.* 20–25% palmitic acid, a lower level of stearic acid (5–10%), and a higher level of the unsaturated acids oleic (40–45%) and linoleic (C18:2) (*ca.* 20%). This gives chicken fat a P/S ratio of *ca.* 0.6, and an iodine value of 75–80.

Horse flesh is not much favoured in the United Kingdom, but is often eaten on the mainland of Europe. Its composition is of interest, in that it contains a higher level, *ca.* 6%, of myristic acid (C14:0), again *ca.* 20–25% palmitic acid, but a higher level of palmitoleic acid (C16:1) (10.5%). It has lower concentrations of stearic and oleic acids than the other muscular fats considered here. This is compensated by the significantly higher concentration of linolenic acid (C18:3) in horse fat, a typical value being 17.5%. This results in horse fat having a P/S ratio of *ca.* 0.65, similar to that of chicken fat, but a higher iodine value of *ca.* 90 units.

Fish oils are significantly different from animal fats, in that they have very much higher iodine values. Capelin oil has an iodine value of *ca.* 145, while menhaden oil has an iodine value of *ca.* 186. These oils have high P/S ratios of 2.3 and 1.05. Herring and mackerel fats, as shown in Table 4, again have *ca.* 8% myristic acid and *ca.* 16% palmitic acid, but high levels of polyunsaturated acids with 20 and 22 carbon atoms.[4] These are thought to have significant dietary properties.

At this stage it is useful to consider the variation in properties of fat from different parts of the carcass. Table 5 shows the composition of fats from different parts of a sheep's carcass.[2,3,5] In most cases, the myristic acid level is *ca.* 2%, but in the intramuscular fat (lean meat with all visible fat removed) the myristic acid concentration rises to 5%. The palmitic acid level varies from 17 to 26% depending on the meat cut analysed. The most significant variation, however, is in the stearic acid level, as this varies from 3% of the fat in legs and ears, to 11% in rump fat, 21% in fat from the chest, and 33–34% in mesenteric (between intestines) and perinephric (kidney) fats. This is commonly thought to be because the animal needs to mobilize its fat, which must therefore have a lower melting point for fats near the skin, but a higher melting point for internal fats which must have some structural rigidity. These variations in the stearic acid level are reflected in corresponding variations in the oleic acid concentrations. In general, levels of linoleic and linolenic acids range from 1 to 2%, with a somewhat higher level of 4% in the intramuscular fat, presumably associated with the higher concentration of linoleic acid in cell wall membranes.

Table 4 Fatty acid compositions of some fish oils: typical values (m/m%)[4]

Fatty acid	Anchovy	Capelin	Pilchard	Sardine	Herring	Mackerel	Menhaden
C14:0	6.9	7.0	7.8	8	9.9	7.2	9.0
C15:0	–	–	0.4	trace	–	–	0.5
C16:0	20.3	10.5	15.7	19	16.3	12.7	20.0
C16:1	9.4	9.0	8.5	10	5.6	4.5	12.0
C16:2	–	–	2.0	–	–	–	
C16:3	–	–	2.0	–	–	–	5.0
C16:4	–	–	3.2	–	–	–	
C17:0	–	–	0.8	trace	–	–	0.9
C18:0	3.7	1.5	3.7	3	1.3	2.3	3.5
C18:1	13.7[a]	14.0	9.3	14	10.2	13.6	10.2
C18:2	1.0	1.5	1.5	1	1.1	1.6	1.2
C18:3	–	1.0	1.1	trace	0.6	1.7	1.5
C18:4	–	7.0	2.2	3	0.8	4.9	2.0
C20:1	3.5[b]	8.5	2.5	2	15.2	11.7	–
C20:3	–		1.7	–	–	–	–
C20:4	0.8	14.5	0.8	1	0.3	–	–
C20:5	19.6		19.3	22	1.7	6.5	–
C22:1	2.6		3.1	trace	27.0	16.3	1.0
C22:2	–		–	–	0.6	–	
C22:5	1.3	12.5	2.4	2	1.3	–	26.0[c]
C22:6	9.3		6.5	4	1.1	9.1	
Others	7.9	4.0	5.5	11	6.5	7.9	7.2
IV	180–198	145	182	159–192	ca. 140	ca. 147	186

[a] Combined C18:1 and C16:4 acids.
[b] Combined C20:1 and C18:4 acids.
[c] Combined C20 polyunsaturated + C22 polyunsaturated.

Table 5 Composition of fats from sheep carcass

	Major fatty acids (mass %)					
	C14:0	C16:0	C18:0	C18:1	C18:2	C18:3
Perinephric (kidney)	2	26	34	30	1	1
Mesenteric (between intestines)	2	24	33	32	1	1
Chest	2	26	21	42	1	1
Rump	2	23	11	53	1	2
Leg	1	17	3	69	1	1
Ear	1	17	3	64	2	2
Intramuscular (all visible fat removed)	5	24	14	38	4	2

As mentioned previously, muscle-based lipids are mainly triglycerides. In these, different fatty acids are normally combined at the 2-position in comparison with the 1- and 3-positions. Table 6 shows some typical overall and 2-position acids for three meat fats, although the comparisions vary

Table 6 *Overall and 2-position acids*

	Overall			2-position		
	Beef	Sheep	Pig	Beef	Sheep	Pig
16:0	27	20	29	26	16	72
18:0	17	34	18	17	11	4
18:1	33	38	41	33	30	14
18:2	5	4	8	5	41	3
Others	18	4	4	19	2	7

somewhat depending on the part of the carcass, the animal's diet, and the other factors discussed here.[2] In beef fat, there is only a slight variation of palmitic acid at the 2-position in comparison with overall concentrations. In sheep lipids there is a contrasting situation in that both the palmitic and stearic acid levels are lower at the 2-position in comparison with the overall concentration. The linoleic acid is at a higher concentration at the 2-position.

The composition of pig fat, or lard, is most significant in that it has a very much higher concentration of palmitic acid, and a much lower concentration of stearic acid, at the 2-position in comparison with the overall concentration, and with the other fats. The 2-position unsaturated acids are also at much lower levels. This provides an interesting possibility for further development in the identification of pig fat contamination of beef tallows, and may be of interest to food analysts in Jewish and Muslem communities. The composition of fatty acids at the 2-position probably has important nutritional consequences, an aspect as yet inadequately researched.[6-14]

3 Oxidation of Animal Fats

All fats are prone to atmospheric oxidation, some being more resistant to oxidation than others. Although animal fats have relatively low concentrations of unsaturated fatty acids, and relatively low iodine values, they are still prone to oxidation, mainly because they do not contain natural antioxidants. Furthermore, manufacture of muscle-based food may lead to liberation of iron from haemoglobin, which leads to significant levels of iron in the fatty materials. Iron is a very powerful pro-oxidant, and is probably one of the causes of rancidity in meat products. Haemoproteins themselves may also catalyse lipid oxidation, as discussed in this volume by Dr Gray. In some foods, the oxidation state of the iron is closely linked to the colour of the meat and the rancidity characteristics of the fat, as explained by Rankin.[15] The very unsaturated fatty acids in fish oils make them much more susceptible to oxidation, and food products manufactured from fatty fish must be very carefully processed in order to avoid oxidation and the development of unsatisfactory off-flavours.

Oxidation of oils and fats naturally leads to the production of aldehydes, which are responsible for the off-flavours of rancid products. Table 7 shows the flavour thresholds of some unsaturated aldehydes produced by this route, illustrating the very low level at which they can be detected.[16]

Food manufacturers often wish to evaluate different fats with regard to their resistance to oxidation. A common apparatus used in this application is the Rancimat. The principle by which the Rancimat apparatus works is that air is passed through fat held at constant temperature of, say, 100 °C, causing oxidation of the fat. The air stream is then passed through a tube of distilled water, volatile components generated by the oxidation being carried over by the air stream and dissolving in the water. These volatile components change the conductivity of the water which is monitored by an electrical recording system. Initially, the fat is resistant to oxidation, and virtually no volatile products are carried over. After a while, however, the resistance to oxidation of the fat becomes exhausted, and oxidation sets in. As this is autocatylic, there is a very sharp increase in the rate of oxidation, in the evolution of oxidation breakdown products, and in the conductivity of the distilled water. The time taken for this change to occur is recorded, and is called the 'induction period'. Fats with a longer induction period have a greater resistance to oxidation. The operation and use of the Rancimat apparatus has been fully described elsewhere.[16]

Table 8 shows some typical induction periods of lard and tallow samples.[16] In this case, the lard has a greater resistance to oxidation than the tallow, and in this respect may have been an untypical sample. Alternatively, and more probably, it shows that accelerated tests at high temperatures, such the Rancimat, are not always a reliable guide to the flavour stability of fat. Table 8 also shows induction periods of a sample of tallow treated with synthetic antioxidants. The Rancimat results show slight improvement, whereas an alternative method of evaluation, namely the FIRA/Astell evaluation, shows much greater influence of the antioxidants. This illustrates a problem in the use of the Rancimat apparatus, as the synthetic antioxidants are volatile at the temperature of measurement, and are carried over in the passing air stream. This shows that the Rancimat results are not a reliable guide to the influence of antioxidants on the keeping quality of foods treated with antioxidants.[16]

Table 7 *Off-flavours of some aldehydes*

Aldehyde	Threshold ($mg\,kg^{-1}$ *oil*)	Flavour
trans-Hex-2-enal	0.5	'Green'
Deca-*trans*-2-*trans*-4-dienal	0.1	Stale frying oil
Nona-*trans*-2-*trans*-6-dienal	0.02	Cucumber
Nona-*trans*-2-*cis*-6-dienal	0.0015	Beany
Non-*trans*-6-enal	0.00035	'Hydrogenation'

Table 8 *Induction periods* (h) *of animal fats and fish oils*

Sample	Rancimat test	FIRA/Astell test
Lard	10.6	8.2
Beef tallow A	9.5	6.8
Beef tallow B	6.3	3.8
Tallow A + 100 mg kg^{-1} BHT	12.3	29.6
Tallow A + 100 mg kg^{-1} BHA	17.6	36.9
Tallow A + 200 mg kg^{-1} BHA	23.5	86.7
Hydrogenated fish oil	26.1	29.5

4 Minor Components

The most significant minor component in muscle-based lipids is the free fatty acid content. Fatty acids are liberated by hydrolysis of the triglyceride, to form a partial glyceride and a free acid. Many processes of curing or processing meat lead to the generation of free fatty acids, which contribute to the desirable flavour of the finished foods. If the free fatty acid contents rise above *ca.* 4% by weight, (as a proportion of the fat), however, the flavour becomes too strong and leads to unpalatability. High free fatty acid levels are not immediately noticeable in most foods, but do lead to an unpleasant aftertaste, usually in the form of a burning sensation at the back of the tongue.

Another minor component is of course cholesterol. Table 9 shows cholesterol contents of some meat and fish products in comparison with those of vegetable oil and eggs. It is not always appreciated that vegetable oils do contain low levels of cholesterol, as these pale into insignificance in comparison with the very much higher levels in meat, fish, and egg products. It is, however, curious that some people prefer vegetable-based frying oils even when eggs, lamb chops, or bacon are being cooked.

Animal fats also contain low levels of tocopherol and carotene, as illustrated in Table 10. These are mainly derived from dietary sources. High levels of carotene in animal fats cause the fat to have a deep yellow colour, which is often a disadvantage with regard to quality perception by the public for whole cuts of meat, but is of little consequence for manufactured meat products where the colour of the fatty tissue is obscured by the changes incurred during the manufacturing process.

Table 9 *Cholesterol contents* (mg kg^{-1})

Whole fish	500–7000
Fish oil	2000–6000
Chicken, lamb chop	*ca.* 1000
Beef tallow	1000–1200
Egg yolk	*ca.* 12 600
Vegetable oils	10–100

Table 10 *Minor components in animal fats and fish oils: typical values* (mg kg^{-1})

	Carotene (vitamin A)	Tocopherol (vitamin E)
Beef	2	2
Pig	0.75	1
Sheep	3	No data
Anchovy[4]	60	
Herring[4]	Spring 355/Autumn 200	Summer 37/Winter 12
Capelin[4]	Summer 45/Winter 25	(Typical range 30–77) (Mean *ca.* 50)
Palm oil (for comparison)	500–800	600–1000

5 Extraction of Fat from Manufactured Food Products

Some recent work at the Leatherhead Food Research Association[17] has shown that the extraction of fat from manufactured food products is not always straight forward. We have found that this is a particular problem with tinned fish. We purchased fresh mackerel, extracted the fat from this, and analysed the extracted lipid. We found that the fresh fish contained *ca.* 8% fat. When the fatty acid composition of this was determined against an internal standard, the fatty acids recovered responded to *ca.* 95% of the total lipid.

The fresh fish was gutted, brined, and canned. The canned fish was stored for a few weeks, after which the cans were opened, and the lipid was again extracted. The recovery of total lipid was *ca.* 8%, but when this was analysed for fatty acid composition, again against an internal standard, only *ca.* 65% of the extracted lipid could be recovered as fatty acids.

Our first thought was that the highly unsaturated fatty acids might have become oxidized, and therefore not chromatographed in the normal way during fatty acid analysis. However, calculations about the loss of fatty acids showed that we would have needed to have extremely high levels of oxidation before this mechanism could explain the losses found. Furthermore, the canned fish still tasted quite wholesome and did not display any of the attributes associated with a high degree of oxidation.

Work is currently progressing on this topic, and at the moment we think that partly oxidized fatty acids might have reacted with protein in the fish to form a lipoprotein which is then extracted together with other lipids. However, it is possible that the lipoprotein does not yield fatty acid during the analytical procedure. If this theory is correct, it would have implications with regard to the fatty acid, and amino acid, content of the fish, and probably also implications with regard to the appropriate form of food labelling.

6 Acknowledgements

I wish to express thanks to the management of the Leatherhead Food Research Association for permission to attend and speak at this Symposium and to publish this chapter, to Ann Pernet for editing, and Lynn Rajack for typing.

7 References

1. International Association of Seed Crushers, Proceedings Buenos Aires Conference, September 1989.
2. A.J. Sinclair and K. O'Dea, *Food Technol. Aust.*, 1987, **39**, 228.
3. A.J. Sinclair and K. O'Dea, *Food Technol. Aust.*, 1987, **39**, 232.
4. F.V.K. Young, *Fish Oil Bull.*, 1986, No.18, International Association of Fish Meal Manufacturers, Potters Bar, Herts., UK.
5. M. Enser, 'Analysis of Oilseeds, Fats and Fatty Foods', ed. J.B. Rossell and J.L.R. Pritchard, Elsevier Applied Science, London and New York, 1991, p.329.
6. L.D. Lawson and B.G. Hughes, *Lipids*, 1988, **23**, 313.
7. R.M. Tomaselli and F.W. Bernhart, US Patent 3 542 560, 1970.
8. W. Droese, E. Page, and H. Strolley, *Int. J. Pediatrics*, 1976, **123**, 277.
9. D. Kritchevsky, S.A. Tepper, D. Vesselinovitch, and R.W. Wissler, *Atherosclerosis*, 1971, **14**, 53.
10. D. Kritchevsky, S.A. Tepper, D. Vesselinovitch, and R.W. Wissler, *Atherosclerosis*, 1973, **17**, 225.
11. J.J. Myher, L. Marai, A. Kuksis, and D. Kritchevsky, *Lipids*, 1977, **12**, 775.
12. F. Manganaro, J.J. Myher, A. Kuksis, and D. Kritchevsky, *Lipids*, 1981, **16**, 508.
13. Food and Agriculture Organisation (FAO), 'Dietary Fats and Oils in Human Nutrition - Report of an Expert Commission', FAO, UNO, Rome, 1980.
14. J.M. Muderhwa, C. Dhuique-Mayer, M. Pina, P. Galzy, P. Grignac, and J. Graille, *Oleagineaux* 1987, **42**, 207.
15. M. Rankin, in 'Rancidity in Foods', 2nd Edn., ed. J.C. Allen and R.J. Hamilton, Elsevier Applied Science, London and New York, 1989.
16. J.B. Rossell, in 'Rancidity in Foods', 2nd Edn., ed. J.C. Allen and R.J. Hamilton, Elsevier Applied Science, London and New York, 1989.
17. P.N. Gillatt, Leatherhead Food Research Association, personal communication, 1990.

Structural Aspects of Processed Meat

D.F. Lewis

LEATHERHEAD FOOD RESEARCH ASSOCIATION, RANDALLS ROAD, LEATHERHEAD, SURREY KT22 7RY, UK

Meat processing generally involves a combination of mechanical action, addition of salts and other additives, and heat processing. The objective of the process is to produce a stable and consistently acceptable product. An understanding of the structural changes produced by these processes is an important part of controlling and optimizing the product. At the most basic level the effect of salt addition to meat is to help reduce the weight loss on cooking and in structural terms this is related to the extent to which the myofibrillar proteins are dispersed from their natural position in the meat structure. The pH of the system is a crucial feature in controlling the action of salt on meat and much of the beneficial effects of polyphosphates can be related to their pH buffering ability.

Where mechanical treatment is involved the purpose is often to improve the meat-to-meat binding or to help incorporate fat in a stable form. In both these cases the ability to produce stable products is critically related to the extent to which proteins are dispersed from the meat structure; hence the availability of salt and water at the time of mechanical treatment is most important. Meat is highly variable and the underlying structure of the meat and its post mortem history can greatly influence the way that a particular piece of meat behaves.

The effects of other additives may be simply water holding or may be to supplement the natural stabilizing effects of the dispersed meat proteins. The time of addition is often crucial; for example, some proteins do not disperse well in the presence of salt and these should be incorporated before the salt to give the best effect. Other dry additives, such as rusk, may increase the amount of damage to fat cells if the fat and additive are bowl chopped in a dry state and this can lead to excessive fat loss during cooking.

Virtually all meat products are eaten after cooking and it is important to consider the effects of cooking when developing mechanisms to explain meat behaviour during processing. It is possible to incorporate relatively large amounts of fat or water in a raw product but it is clearly important to be able to retain the integrity of the product during cooking.

Lipids: Nutritional Aspects

R.C. Cottrell

LEATHERHEAD FOOD RESEARCH ASSOCIATION, RANDALLS ROAD, LEATHERHEAD, SURREY KT22 7RY, UK

A great deal of confusion has arisen in the professional and public discussion of the nutritional and toxicological implications of dietary lipids. Misunderstandings have been particularly common in respect of muscle-based foods because of the widespread ignorance of nutritionists and dietitians of the distinction between depot fat and the structural lipid of muscle. It is not commonly appreciated that lean meat has a radically lower lipid content than marbled meat and, moreover, that the lipid of the cell wall is significantly different in composition from that in adipose tissue.

However, this confusion can be readily disposed of by appropriate information.

What remains an issue is the inadequacy of current nutritional and toxicological data on the positive and adverse effects of the consumption of different lipids at different intake levels. The available data are sufficient to demonstrate that certain lipids have undesirable effects if consumed in substantial amounts, and that others are dietary essentials. There is little clear information, however, on dose–response relationships in either animals or man, and interactions between individual lipids and between lipids and non-lipid components of the diet (such as antioxidants) are largely unresearched. A significant research effort is required to resolve these issues. Fortunately funds are being made available by both the UK Government and the European Community.

Chemistry of New and Existing Technologies

Surimi from Fish

I.M. Mackie

TORRY RESEARCH STATION, 135 ABBEY ROAD, ABERDEEN AB9 8DG, UK

1 Introduction

Surimi is the Japanese term for minced fish which has been washed free of the water-soluble or sarcoplasmic proteins. This enriched myofibrillar protein fraction is the starting material for the traditional fish gel or kamaboko products of Japan[1] and, in recent years, shellfish substitutes or analogues.[2-6] Of these latter products the best known are crabsticks which seem to feature almost universally in fish counter displays throughout Europe and North America. Traditionally the manufacture of fish gel products in Japan was one way of extending the shelf life of fish. It involved adding salt to washed fish mince and heating the paste, usually by steaming to form heat set gels. A wide range of products based on this process continue to be produced in Japan today. In the traditional process, surimi, because of its perishable nature, was processed directly into kamaboko products but in today's industrial processes it is invariably made into a frozen 'intermediate' product which is manufactured into kamaboko or other gel products at a later date and very often in plants far removed geographically from where it was produced from the raw fish.

The growth of the surimi industry to its present 1 million tonnes per annum scale of production worldwide was made possible by the discovery of Nishiya *et al.*[7] in the early 1960s of the cryoprotective effect of low molecular weight carbohydrates such as sorbitol and sucrose on the myofibrillar proteins of fish flesh during frozen storage. By adding these carbohydrates at concentrations of *ca.* 4% of each to fish muscle mince washed free of sarcoplasmic proteins, they showed that it was possible to store surimi in the frozen state over many months without incurring a significant loss in its ability to form kamaboko gels. By being able to stabilize surimi in this way the industry was no longer dependent upon the variability of supplies of fresh fish and it could embark on the production of surimi as a separate commodity. Fish gel manufacturing plants were able to extend their base beyond local supplies and to draw on fish resources which had until then been unavailable because of distance.[4] The Alaska pollack (*Theragra chalcogramma*) of the North Pacific Ocean was then a largely unexploited resource but it was available in the quantity required to sustain the huge expansion in the fish gel industry of Japan in the 1960s and 1970s.[5,8]

To a large extent the history of the surimi industry is that of the Japanese fish processing industry but relatively recent developments in the USA and elsewhere have taken the industry beyond that of Japan. In the USA, in particular, the industry has drawn on the Japanese experience and is now well established.[5,6,8] Unlike the Japanese surimi industry it is concerned primarily with the production of shellfish analogues.[5,6,8]

While the surimi industry is still based largely on Alaska pollack, it is well recognized that this resource is under pressure and that other species of fish will have to be utilized even to maintain the present level of production. These developments are now taking place throughout the world often in association with Japan, *e.g.* in Thailand, and a number of species of fish are currently being used for surimi production. In the main, the most suitable species for surimi are those with white flesh and low contents of fat - the white fish species. However, underexploited species tend to be the fatty species which have a higher proportion of red muscle and a higher content of lipid in their flesh, both undesirable characteristics in raw material for surimi production.[9-11]

Resources of Fish Available

In any consideration of raw material for surimi production the species of fish must not only meet the required standards of quality but it has to be available preferably throughout the year and in sufficient quantities to sustain a manufacturing process.[8] It must also be available at a relatively low price. Most species of fish which have comparable gelling properties to Alaska pollack are excluded immediately on the grounds of cost as they are likely to be fully exploited as fish for direct human consumption, *e.g.* cod, haddock, hake, *etc.* The underexploited species of the world available in the largest quantities are mainly the small pelagic fish - anchovy, sardine, *etc.*, a large proportion of which are currently reduced to fish meal.[12] Efforts are being made particularly in Japan to develop acceptable surimi from those species and, it would appear from recent reports, with some success.[9]

Observations made at Torry Research Station in a comparative study on kamaboko prepared from fish available in the North East Atlantic have shown that surimi from fatty species is in general inferior to that from the white flesh species.[13] In addition to being darker in colour, it has a lower water holding capacity and less elasticity than surimi from white fish species (Table 1). That is not to say that surimi from fatty species will not have a role in the food manufacturing industry but today it would appear that outlets and uses have still to be identified. Recent approval from the Department of Agriculture in the USA to allow mixing of surimi and meat products may encourage further developments in this area.[14]

Table 1 Analyses of kamaboko from various species of fish[13]

Species/treatment	Moisture (%)	Water loss (%)	Folding test score	Correx score (gf)	Puncture test Breaking force (gf)	Puncture test Breaking distance (mm)	Puncture test Gel strength (gfXcm)	Compression test Firmness (gf)	Compression test Springiness (dimensionless)
Alaska pollack	73.5	2.39	7	651	663	7.0	467	7837	0.875
Cod	76.46	2.96	7	280	454	11.8	536	3552	0.896
Fresh whiting (May 1987)	76.81	2.59	7	221	332	11.1	368	2625	0.909
Fresh whiting (July 1987)									
Fillet	76.65	2.96	7	253	292	8.7	255	3322	0.903
Trimmings	75.71	4.26	6.5	245	262	7.5	198	3490	0.897
Skeletons*	76.92	6.12	2	<100	47	2.0	9	1907	0.844
Fresh haddock									
Fillet	74.98	2.08	7	465	698	11.4	805	5415	0.887
Trimmings	73.72	3.81	7	467	504	8.2	416	6327	0.886
Skeletons*	74.57	6.08	3.5	231	168	3.9	67	4530	0.862
Commercial haddock skeletons*	73.80	6.17	2	294	202	3.3	67	5132	0.837
Fresh saithe	75.35	6.06	6	293	498	11.2	559	3937	0.890
Commercial plaice skeletons*	75.07	17.10	1	<100	39	1.1	4	992	0.650
Fresh Clyde herring	72.66	2.28	6	257	284	8.3	237	3590	0.887
Fresh North Sea herring*									
With bicarbonate*	67.60	3.51	3	225	156	3.5	55	3522	0.834
Without bicarbonate*	64.22	4.21	3	398	310	4.7	146	6795	0.834
Light flesh only*	67.98	3.14	3	584	367	3.8	141	8095	0.831
Frozen herring*	63.35	8.51	2	265	215	3.6	77	5130	0.812
Frozen horse mackerel*	67.46	1.75	3	409	370	4.6	169	7390	0.859
Frozen mackerel*	64.61	11.80	2	336	250	4.0	100	5570	0.815
Fresh dogfish*	68.62	1.64	2	116	48	0.7	3	2845	0.856
Least-significant difference	0.04	2.13							0.018
Least-sigificant ratio			1.085	1.122	1.116	1.230	1.087		

* Puncture test curves did not have the shape of those from gels with high elastic properties. The breaking point was assigned to the point of maximum load at the end of each smooth compression curve.

The situation today is that Alaska continues to be the main raw material for surimi production in Japan and the USA. Other species which are used (Table 2) are not available in quantities which would enable the industry to return to the production levels of 1973 and 1974 when 3 million tonnes were caught by Japan specifically for surimi production.[5,15] The Alaska pollack fishery, however, remains one of the major fisheries in the world even with heavy exploitation. Currently over 6 million tonnes of Alaska pollack are caught annually but, because of the extension of territorial waters in the USA and the USSR in 1977, the proportion caught by Japan has decreased steadily; in 1989 it was 1.1 million tonnes.[16] These figures should be related to the total fish landings in Japan of 11.8 million tonnes and in the UK of 916 713 tonnes, out of a total world catch of 99.5 million tonnes.[16]

The Structure of Fish Muscle

Like all forms of contractile muscle that of fish has the common general arrangement of thick and thin filaments responsible for contraction within the cells. Structurally, however, there are fundamental differences in the arrangement of the muscle cells and in their attachment to connective tissue; fish muscle cells are short (10–30 mm depending upon the size of the fish) and interspersed between the ends of the block of cells are sheets of connective tissue - the myocommata).[17] Fish muscle always consists of two types of muscle cells, white and red, the relative proportions of which vary depending upon the swimming behaviour of the fish. In the pelagic species such as herring, mackerel, and tuna for example, red muscle is well developed. It penetrates deeply into the white muscle structure but when present in low amounts as in cod and Alaska pollack it usually lies

Table 2 *Japan's catch of fish species which are used primarily for the production of surimi*[a]

		1980–84 average	
Species		1000 tonnes	%
Alaska pollack		1554	87.2
Atka mackerel		93	8.2
Croaker		29	1.6
Sharp toothed eel		14	0.8
Lizard fish		19	1.1
Cutlass fish		36	2.0
Shark		37	2.1
	Total	1782	100.0

[a] S.C. Sonu, NOAA Technical Memorandum, National Marine Fisheries Service, SWR013, 1986, p.37.

immediately under the skin along the lateral line. White muscle is not normally required for swimming but it can be used for 'flight' or 'escape' when bursts of speed are required for short periods of time. It has an anaerobic metabolism which depends upon glycogen for energy in contrast to red muscle which receives a copious supply of blood. Red muscle is rich in haemoglobin, myoglobin, and mitochondria while muscle is well supplied with glycolytic enzymes which permit intense contraction during the brief period of flight.[17]

Chemical Composition of Fish Muscle

The main constituents of fish flesh (Table 3) are as would be expected for the flesh of any animal: water, protein, and fat. Their concentrations are, in the main, within the range found in mammals but, in fatty species in particular, there is a wide fluctuation in the lipid content. In those species, the energy reserves of lipid are stored in the flesh rather than in the liver as in white flesh species. As a consequence, the proximate analyses of the flesh of the former species show wide variation depending upon the season. In mackerel for example the lipid content of the flesh can be as high as 35% when food is abundant but it can fall to levels below 1% prior to spawning in late spring. Even in non-fatty species there is a well defined seasonal variation in the contents of lipid and water. Lipids of fish are characterized by their high concentrations of polyunsaturated fatty acids which are extremely susceptible to oxidation during handling, processing, and storage.[19] The fatty species with much higher concentrations of lipids thus develop rancid and other off-flavours more readily than white flesh species and, primarily for this reason, they have remained largely underutilized.

Table 3 *Percentage chemical composition of fish flesh*[a]

Species	Water	Fat	Protein
Blue whiting (*Micromesistius poutassou*)	79–80	1.9–3.0	13.8–15.9
Cod (*Gadus morhua*)	78–83	0.1–0.9	15.0–19.0
Hake (*Merluccius capensis*)	79–84	0.2–1.4	15.2–18.6
Herring (*Clupea harengus*)	60–80	0.4–22.0	16.0–19.0
Pilchard (*Sardina pilchardus*)	60–80	2.0–18.0	17.0–20.0
Mackerel (*Scomber scombrus*)	60–74	1.0–23.5	16.0–20.0

[a] J. Murray and J.R. Burt, 'The Composition of Fish', Torry Advisory Note, No. 38, HMSO, Edinburgh.

The protein content (usually expressed as nitrogen content × 6.25) also shows variation with season but the fluctuation in levels is much less than for lipids. Most protein contents lie between 15 and 19% of the wet weight of muscle but it should be remembered that *ca.* 10% of this protein equivalent arises from non-protein sources such as free amino acids, peptides, nucleotides, trimethylamine oxide, ammonia, and other volatile bases.[18]

The proteins of muscle can be divided into three main groups depending upon their solubilities in water and strong salt solutions (Table 4). The main group of proteins, the myofibrillar proteins, are extractable in 0.5 M NaCl while the sarcoplasmic proteins are water-soluble. The latter proteins are mainly enzymes which are required for the anaerobic metabolism within the muscle cells while the myofibrillar proteins are the structural proteins responsible for muscular contraction and movement of the animal. In fish muscle, the latter group accounts for 65–75% of the total protein content compared with 52–56% in mammals. The connective tissue protein content of fish muscle is considerably lower than that of mammals (3–10% compared with 10–15%), reflecting the differences in the structural arrangements of their contractile mechanisms.[18]

The myofibrillar proteins myosin and actin and their product of interaction, actomyosin, are by far the major proteins of skeletal muscle, myosin being the main protein component of the thick filaments while actin, together with tropomyosin and the troponins, constitute the thin filaments. Myosin has a molecular weight of *ca.* 500 000 but it differs in detail from one species of animal to another. It comprises two heavy chains of molecular weight 200 000 and four light chains of molecular weight *ca.* 20 000. The two heavy chains exist in a supercoiled helical transformation (Figure 1) and at the head end of the molecule both chains are folded into a globular structure. The twin globular heads contain the site involved in the contractile mechanism and in ATPase activity. Fish myosins, particularly those of white fish species from temperate waters, are the least stable

Table 4 *Main groups of skeletal muscle protein in the flesh of fish and mammals*

Protein group	Percentage composition	
	Fish[a]	Mammals[b]
Sarcoplasmic	20–35	30–35
Myofibrillar	65–75	52–56
Connective tissue	3–10	10–15

[a] I.M. Mackie, in 'Developments in Food Proteins-2' ed. B.J.F. Hudson, Applied Science Publishers Ltd, London, 1983, p.228.
[b] D.E. Goll, in 'Food Proteins', ed. J.P. Whitaker and S.R. Tannenbaum, AVI Publishing Company Inc. Westport, Connecticut, 1977, p.123.

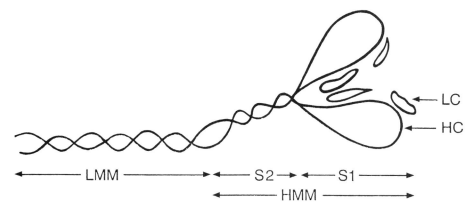

Figure 1 *Diagramatic presentation of the myosin molecule*

of all myosins and are in general more susceptible to the action of proteolytic enzymes. Actin from fish appears to be identical to that from mammals. As the monomer it has a molecular weight of 42 000 and it exists in the thin filament as the polymeric form, F actin, a helical double strand of monomer units.

Denaturation of Myofibrillar Proteins during Frozen Storage

The greater susceptibility of fish myosin to denaturation on heating and to breakdown by proteolytic enzymes also manifests itself in its instability during frozen storage. Again, the myosins of gadoid species are the least stable of all fish myosins and as a result the flesh of those species toughens more rapidly on frozen storage than that of flat fish species,[18] for example. As a general rule, gadoid species must be stored at −30 °C to avoid significant loss in quality if the period of storage is over 6 months whereas beef can be stored at −20 °C over the same period. The deteriorative changes in the muscle proteins are progressive and depend upon the time and temperature of frozen storage. When the flesh is minced or comminuted they are markedly accelerated owing, it is believed, to the release and dispersion of enzymes and pro-oxidants when the muscle cells are ruptured. These effects are even more pronounced when the flesh is contaminated with viscera or organs such as kidneys, as can happen when flesh is separated mechanically from whole or eviscerated fish and inadequately washed.

The nature of the transformation taking place in the myofibrillar proteins and particularly in myosin during frozen storage is not well understood. It has been postulated that formaldehyde released by the action of the enzyme trimethylamine oxide demethylase (TMAO ase) interacts with amino groups of protein to form inter- and intra-chain methylene cross-linkages.[20] Carbonyls or hydroperoxides resulting from oxidation of the fish lipids have also been suggested as likely reactants. Some —S—S— bond cross-linking might also take place on oxidation of SH groups. It would seem, however, on the evidence presently available, that non-covalent rather than covalent bonds are likely to be involved in protein denaturation on frozen storage.[18,19]

In the preparation of surimi, the water-soluble proteins are removed by washing, and in this process many of these agents including lipids are also largely removed.[8,11,15] It is still necessary, however, to add cryoprotectants to the washed mince in the form of low molecular weight carbohydrates such as sucrose and sorbitol to inhibit the denaturation process substantially.[7,8] The stabilizing effect of these compounds on the myofibrillar proteins can be explained by their ability to increase the surface tension of water as well as the amount of bound water. As a result, the proportion of water withdrawn on freezing is reduced and the native structure of the protein is stabilized. The action of the cryoprotectants as solutes in reducing the freezing point of water must also be a contributory factor in inhibiting protein denaturation.[21]

Gelling Properties of Fish Myofibrillar Proteins

It is generally understood that to produce gel products with the required elasticity and firmness, two basic requirements must be met.[1-4,8,11,15]

(1) The myofibrillar proteins must initially be dissolved in a salt solution.
(2) On heating to form a gel the proteins must be denatured in such a way that they form a regular network structure capable of immobilizing the water present in the uncooked surimi.

To meet the first requirement, proteins must be in their native, undenatured state in the surimi if they are to be fully dissolved in the salt solution and the temperature must be kept below 5 °C to avoid denaturation during the blending operation. For the second requirement it is necessary that the myofibrillar proteins are free of sarcoplasmic proteins and that adequate time is allowed for the transition from a native to a denatured state as the temperature is raised. During this heat-induced transition - setting or 'suwari' - the actomyosin chains unfold and interact with one another to form an initial network which immobilizes a large part of the free water.[21-23] On heating to 100 °C the network is further strengthened without incurring aggregation and loss in the water-retaining properties of the gel. The nature of the bonds formed on heating is poorly understood but it is probable that during low-temperature gelation ionic, hydrogen, and hydrophobic bonds are involved while at the higher temperatures —S—S— and hydrophobic bonds predominate.[21,23]

The low temperature setting generally takes place between 40 and 50 °C but it can proceed at 0 °C if sufficient time is given for the protein interactions to take place. At higher temperatures between 60 and 70 °C the gel structure can be weakened by the 'Modori' phenomenon, which is believed to be due to heat-stable proteases acting on the myofibrillar proteins in this temperature range. As a result the final gel can be soft and unacceptable as a product. Proteases can be present in the tissue or they can arise from contamination by the viscera during the preliminary processing operations. There also seems to be considerable variation in the severity of the 'Modori' phenomenon depending upon the species. However, it is possible, in practice, to minimize the effect by adding wide-spectrum protease inhibitors such as bovine $\alpha 2$ macroglobulin to the raw surimi.

2 Preparation of surimi

A flow diagram of the surimi process is shown in Figure 2. The main concern throughout the manufacture of surimi is to maintain the gel forming characteristics of the myofibrillar proteins close to those in freshly killed fish. It is well established that these properties decrease as the fish spoils and that to obtain surimi of the highest quality it is very important that fresh fish is used initially and that the temperature is held between 0 and 5 °C during storage and processing.[1-4] When fish is in *rigor mortis* it is difficult to fillet mechanically and for this reason it is generally preferable

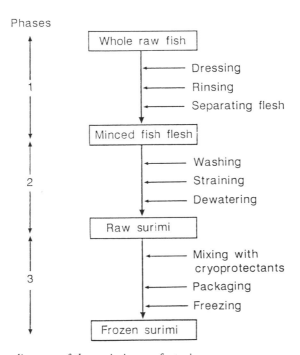

Figure 2 *Flow diagram of the surimi manufacturing process*

to process the fish in shore-based operations in the post-*rigor* condition. With sea-board operations, however, fish is usually processed in the pre-*rigor* state. As the catch will be reasonably uniform in size, machine filleting is likely to be used and it is also likely that evisceration will have been carried out immediately after the fish are brought on board. The removal of viscera as soon as practically possible is desirable as intestinal proteases can penetrate the flesh and act on the myofibrillar proteins during storage.

The alternative process to filleting is to remove the head and viscera only and to pass the remainder of the fish through a deboning machine. However, although a higher yield is obtained, a lower grade surimi, darker in colour and with poorer gel forming properties, is usually obtained.[15] Minced fish flesh produced by either route must then be passed through a deboning machine of Baader (Model 694), Bibun, or Yanagiya types, the capacities of which vary from 200 kg to 8 tonnes per hour. In the deboning operation the fish is compressed between a stainless steel perforated drum (3–5 mm hole size) and a rubber belt and extruded as a mince, free of bones, while the skin and bones are retained. The comminuted flesh thus obtained is washed with water either continuously or batch-wise to remove the water-soluble or sarcoplasmic proteins. Generally three washes using ratios of water to mince of 3:1 each over a 10 minute period are sufficient to remove the sarcoplasmic proteins, although factors such as the species of fish, the degree of comminution of the flesh, and the extent and duration of agitation influence the efficiency of the process.[9,11,21,22] The first wash is usually carried out with water containing 0.5% $NaHCO_3$ which has the effect of neutralizing the pH of the mince; the second wash is with water only and the third with water containing 0.1–0.5% NaCl, which has the effect of reducing osmosis and the water-holding capacity of the mince. In general, a decanting or centrifugation (rotary screens) is carried out between each wash. Excessive agitation or long washing times often have the effect of increasing the water uptake by the muscle making it more difficult to remove during the later dewatering operation. Also, too much salt in the final step may cause solubilization of myofibrillar proteins leading to a lower yield of surimi and premature setting of the protein gel prior to heating to form the final gel product.

After the third wash the yield of concentrated myofibrillar proteins is *ca.* 20% from lean fish and *ca.* 15% from fatty fish (the yield being the weight % of the washed compressed pulp expressed as a percentage of the weight of whole fish). It is now necessary to carry out a final purification step by extruding the pulp through a screw press, 'a refiner', to remove from it matter such as small bones and scales, residual fragments of skin, intestinal membranes, connective tissue, *etc.*, which would otherwise detract from the appearance and texture of the final fish gel product. Finally, it is necessary to reduce the water content to less than 85% so that, after the addition of 8–9% of cryoprotectants, a concentration between 75 and 79% will be obtained. This concentration of water in surimi is close to that of untreated fish muscle tissue and is considered to be the optimum for the gel forming process. The dewatering is almost universally carried out on a screw press which purges water from the fish muscle slurry by squeezing the product into a progressively reducing chamber with the aid of a rotating screw while allowing the pressurized water to escape through tiny drain holes in the chamber wall.

The purified pulp thus obtained is finally mixed with cryoprotectants, usually with a silent cutter-type mixer. Typically 4% sorbitol, 4% sucrose, and 0.2–0.3% sodium polyphosphate are added. This addition has the effect of reducing the protein content to *ca.* 16% and the water content to *ca.* 76%. It is important that chopping is done intermittently (*e.g.* at 15 s intervals) to avoid excessive rise in temperature. The surimi is then frozen in 10 kg blocks in polyethylene bags, transferred to cartons, and stored frozen until required. It is generally accepted that surimi from white fish species such as Alaska pollack is stable for up to a year when stored at −20 °C. At higher temperatures, however, the shelf life is considerably less.

3 Preparation of Products from Surimi

For the manufacture of the gel products, the blocks of surimi are allowed to thaw in air until the temperature rises to *ca.* 0 °C (Figure 3). The pulp is then thoroughly mixed with salt at a concentration of 2.5–3.5% to give a gel or sticky paste of the myofibrillar proteins using either a cutter or a mechanical stone mortar.[1,4,21,22] Throughout this operation is is essential that the temperature is maintained at *ca.* 5 °C to avoid or at least minimize any denaturation of the myofibrillar proteins. For reasons of good hygienic practice also it is important that the temperature of the surimi is held low to prevent the growth of fish spoilage micro-organisms. At this stage other ingredients are often introduced depending upon the product and grade

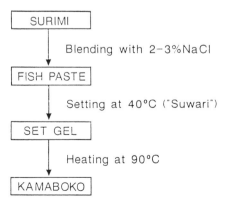

Figure 3 *Flow diagram of the kamaboko manufacturing process*

being manufactured, *e.g.* starch, egg white, colouring, and flavouring. For shellfish analogues, vegetable oil and appropriate flavours for the product are added. The salt content influences the solubility of the myofibrils, which in turn affects the texture of the final gel. Gels prepared from surimi to which 3% NaCl has been added are more elastic and firmer than those with 2.3% NaCl, the minimum concentration required for satisfactory gel formation. The pH is also a factor in determining the degree of solubilization of the myofibrillar proteins, 6.0 being considered the minimum value required to produce acceptable textures in the gel product.

The second step in the formation of gels is the setting process, which often requires the paste to be held at 40–50 °C for one hour. The extent of setting will be determined by a number of factors such as the species of fish and its freshness. For some processes, however, particularly those involving extrusion, premature gel formation can be a disadvantage.

The third step is the thermo-coagulation or gelling process which takes place when the surimi is heated to 100 °C. At this stage the paste will have been pumped through a die or into moulds to give it its final shape. Steaming is a widely used method of heating and is usually for 20–90 minutes depending upon the sizes of the portions of surimi. When the internal temperature reaches 80–85 °C the gelling process is considered to be complete and at the same time spoilage micro-organisms are destroyed. The final product is normally packed in sterile transparent heat-shrink plastic film followed by pasteurization. The sealed product is finally cooled in cold air and stored either under refrigerated conditions (temperature <5 °C) for up to three weeks or frozen for longer term storage.[8]

Products of Surimi

Traditional Products. A wide range of traditional products is manufactured in Japan and these continue to be the main outlets for surimi. These kamaboko products are very diverse and are available in many variants of shape, flavour, texture, and colour. They include itatsuki kamaboko - half cylinder shapes of fine textured white gel prepared by steaming surimi on thin wooden slabs, fried kamaboko - satsúmaage - a wide range of products fried in vegetable oils, and chickuwa, a tubular product, made by wrapping surimi round a bamboo or brass rod prior to cooking (Figure 4).

Shellfish Analogues. The essential requirement for these products is that they simulate the typical firm fibrous structure of crustacean and shellfish flesh such as that of crab, clam, lobster, scallop, and shrimp.

Shellfish and crustacean analogue products are manufactured by two main procedures, either by a continuous extrusion system or in moulds of the shape of the flesh being simulated. The former extrusion process is used for the manufacture of crab sticks and requires the paste to be

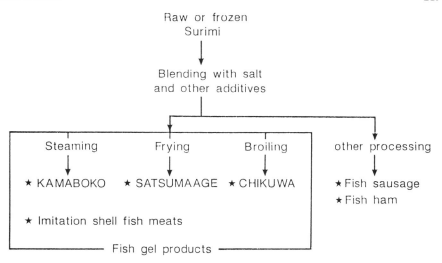

Figure 4 *Surimi-based fish gel products*

extruded continuously onto a transporter belt to give a thin sheet (1–3 mm high and 200 mm wide). The extruded sheet is partially heat set and scarified lengthwise into parallel filaments by rotating blades. The scored sheet is then rolled and folded in over itself mechanically to form a continuous cylinder. This 'rope' is finally coloured, wrapped, and cut into the desired length by a wrapping machine. The sticks are then heated further to ensure pasteurization and cooled quickly to *ca.* 5 °C.[15] There are many variants of the extrusion process, included in which are the use of dies in the shapes of longitudinal sections of crustacean flesh portions. The paste extruded can be of uniform consistency or it can contain pre-cut lengths of previously formed kamaboko gels to give improved textural qualities to the final product. As the surimi is extruded it is cut transversely at regular intervals and the portions are then carried by conveyer belt for thermo-coagulation and packaging.

Moulded products are produced by pumping the surimi paste into moulds of the desired shape, *e.g.* lobster tails. Again, as for the extrusion systems many variants of the composition of the paste have been developed to give a wide range of textures of the final product. These range from a homogenous paste to a paste containing thin strings of kamaboko gel prepared either as for crab sticks or by finely slicing previously manufactured gels. This type of product is claimed to give a better 'bite' than the variety made from a simple homogeneous paste which often tends to be rubbery and uniform in texture.[15]

Such manufacturing processes are generally continuous, highly automated procedures from the intitial preparation of the surimi paste right through to the final packaged product.

General Conclusions

To make gels of the required strength and elasticity for kamaboko and shellfish analogues it is necessary to remove the water-soluble or sarcoplasmic proteins from the raw fish flesh.

The myofibrillar protein fraction free of sarcoplasmic proteins must be heat-processed directly or preserved in the frozen state by the addition of cryoprotectants such as sorbitol and sucrose.

The mechanism of the heat gelling process is poorly understood but it is well established that the proteins must dissolve in salt solution during the initial blending operation. When the gel or paste is heated the myofibrillar proteins, still largely in their native state, unfold to form a regular network of denatured chains possibly interacting through hydrogen and ionic bonds. It is important that this unfolding process - setting or 'suwari' - takes place slowly so that water is immobilized within the gel structure. On heating to 100 °C the outline structure is reinforced by further bond formation including possibly —S—S— and hydrophobic bonds. The balance between aggregation of the denatured protein chains with loss of water and gelation with immobilization of water depends upon many factors such as species of fish, quality of surimi, pH, temperature, and concentration of salt.

The strength and elasticity of fish gels have been attributed to the greater instability of fish myosins which favours the initial formation of the gel structure at temperatures below those required for aggregation. It would appear that the fish with the least stable myosin, Alaska pollack, can produce the strongest and most elastic gels. The gels obtained are white and elastic - essential requirements for the traditional kamaboko products of Japan and the newer shellfish analogues. Surimi can be made from pelagic species but it is darker in colour and softer in texture. It is not yet produced to any significant extent commercially. Although outlets for surimi from pelagic species have yet to be identified it would seem that provided the problem of oxidation can be overcome they could be blended with the more traditional surimi to give a wider range of textures than presently available and which could have more general application in the food processing industry. The cold setting phenomenon for example makes it possible to reform raw flesh of animals without cooking.

Surimi production is, however, an essentially wasteful process. It can only be economically viable when the raw material is relatively cheap and available in sufficient quantities to sustain the industry. Today Alaska pollack meets those requirements and it would appear that for the foreseeable future it will continue to be the main raw material for surimi. Pelagic species meet the requirements on cost and quantity but it has yet to be established that surimi from these species can be marketed successfully.

Surimi is now well established as a product of the fish industry and has added to the variety of foods available from the sea. If present trends continue surimi-based products will continue to diversify.

It should not be forgotten, though, that most species of fish will continue to be consumed as they are without the need for this sophisticated form of processing.

5 References

1. T. Suzuki, 'Fish and Krill Protein Processing Technology', Applied Science Publishers, London, 1981, p.62.
2. T.C. Lanier, *Food Technol.*, 1986, **40** (3), 107.
3. C.M. Lee, *Food Technol.*, 1986, **40** (3), 115.
4. T. Akahane and J.C. Cheftel, *Ind. Agro-Aliment.*, 1989, 881.
5. M. Okada, in 'Engineered Seafood Including Surimi', ed. R.E. Martin and R.L. Collette, Noyes Data Corporation, Park Ridge, New Jersey, 1991, p.30.
6. D. Hamann, in 'Chilling and Freezing of New Fish Products', International Institute of Refrigeration, Commission C2, Aberdeen (UK), Paris, 1990-3, p.19.
7. K. Nishiya, F. Takeda, K. Tamoto, O. Tanaka, and T. Kubo, Hokkaido Fisheries Research Laboratory, Fisheries Agency, Japan, 1960, Report 21, p.44.
8. Anon., 'Surimi - It's American Now', Alaska Fisheries Development Foundation, Inc., 1987, p.1.
9. F. Nishioka, T. Kokunaga, T. Fujiwara, and S. Yoshioka, Ref.6, p.123.
10. H. Roussel and J.C. Cheftel, *Int. J. Food Sci. Technol.*, 1988, **23**, 607.
11. Y. Shimizu, R. Machida, and S. Takenami, *Bull. Jpn. Soc. Sci. Fish.*, 1981, **47** (1), 95.
12. M. Windsor and S. Barlow, 'Introduction of Fishery By-Products' Fishing News Books Ltd, Farnham, Surrey, 1981, p.5.
13. R. J. Hastings, J.N. Keay, and K.W. Young, *Int. J. Food Sci. Technol.*, 1990, **25** (3), 281.
14. Luc de Franssu, *Infofish Int.*, 1991, **1**, 20.
15. S.S. Sonu, 'Surimi', National Oceanic and Atmospheric Administration, National Marine Fisheries Services No. AA-TM-NMFS-SWR-013, 1986, p.1.
16. Anon., 'FAO Yearbook of Fisheries Statistics, Catches and Landings', 1989, p.85.
17. R.M. Love, 'The Chemical Biology of Fisheries', Academic Press, 1974, p.1.
18. I.M. Mackie in 'Developments in Food Proteins - 2', ed. B.J.F. Hudson, Applied Science Publishers, London, 1983, p.215.
19. R.G. Ackman, in 'Advances in Fish Sciences and Technology', ed. J.J. Connell, Fishery News Books Ltd, Farnham, Surrey, p.86.
20. H.O. Hultin, *J. Chem. Educ.*, 1984, **61**, 289.
21. C.M. Lee, *Food Technol.*, 1984, **38** (11), 69.
22. J.K. Babbit, *Food Technol.*, 1986, **40** (3), 97.
23. E. Niwa, Ref.5, p.136.

Red Meat and Poultry Surimi

M.K. Knight

GRIFFITH LABORATORIES, COTES PARK ESTATE, SOMERCOTES, DERBYSHIRE
DE5 4NN, UK

1 Introduction

Some food commodities such as cereals and oilseeds must be refined before use as foods. Other commodities, *e.g.* vegetables, fruits, and milk, may be refined or concentrated before entering the food chain. Flesh foods have, traditionally, not been refined prior to consumption in western societies; however, there is a major example of this in eastern cultures.

A traditional Japanese method of preparing fish flesh is described in the previous chapter and is to mince the flesh and then wash it with water (Figure 1).[1] This process refines the flesh by washing out the smaller, largely water-soluble proteins (the sarcoplasmic proteins) to produce a material enriched in salt-soluble fibrous proteins (the myofibrillar proteins).[2] This enriched material is known as surimi and is the basic ingredient in a range of foods known as kamaboko products.[3] The textures of these products can be quite different from that of the original fish flesh. The most common examples of kamaboko products in western societies are 'crab' sticks, in which the fish flesh has a texture similar to that of crab meat.[1]

The successful development of the fish surimi process and increasing market share of surimi-based fish products throughout the world have prompted interest in applying the surimi process to red meat and poultry. The meat industry has traditionally included low-grade meat raw materials, such as mechanically recovered meat (MRM) and trimmings, into manufactured meat product formulations. Some of these products, for example sausages and beefburgers, have gained a poor image over recent years. Perhaps the main concern about these red meat products is the amount of saturated fat they contain. The industry has responded to this concern by developing a range of low-fat products. However, the fat content of certain types of MRM and trimmings is high and prevents their use at high levels in some low-fat products.

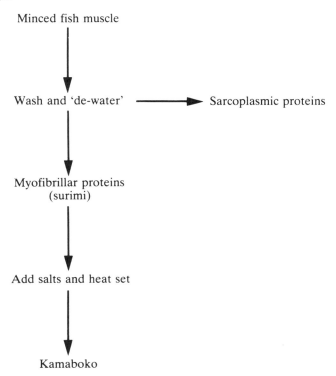

Figure 1 *Surimi production - the refinement of fish muscle protein*

2 What is Surimi?

Surimi is a Japanese term for mechanically deboned mince, usually fish mince, that has been refined by water-washing and mixed with cryoprotectants for frozen storage.[3] The water-soluble proteins, together with enzymes, blood, metal ions, and fat, are separated from the fish mince during the refining process. The dewatered fish mince therefore contains a higher concentration of the more functional salt-soluble myofibrillar proteins. The resultant material is not only more functional but also more stable, owing to lower concentrations of enzymes, of haem iron, and also of the sugars that are major substrates for microbial growth.[4]

Surimi is commonly made from white-fleshed, non-oily fish such as Alaska pollack (*Theragra chalcogramma*), although a wide range of other species is used as surimi raw materials, such as cod (*Gadus morhua*), croaker (*Scianidae* spp.), mackerel (*Scomber* spp., *Pneumatophorus* spp., or *Rastrelliger* spp.), menhaden (*Brevoortia tyrannus*), and blue whiting (*Micromesistius pontassou*).[5,6] In the southern hemisphere, surimi is made from Chilean mackerel and New Zealand hoki.

The consistency of surimi is close to that of mashed potato and it is an almost white, odourless, and tasteless material. The proteins myosin and actin are the major components and these salt-soluble fish proteins have the ability to form a strong, highly elastic gel at relatively low temperatures (*ca*. 40 °C).[7]

Surimi can be formed, extruded, shaped, and cooked to produce a wide range of what in Japan are called kneaded products. Surimi is therefore an intermediate product for further processing into a range of fish products. In Japan the kneaded products are grouped into kamaboko (steamed), chikuwa (broiled), and hanpen (boiled).[6] Surimi-derived products include fish and ham sausage, fabricated crab legs, and shrimp, lobster, and scallop analogues.[8-11]

Cryoprotectants such as sucrose, sorbitol, and polyphosphates are added to surimi to prevent a loss of functional properties, particularly gelation, during frozen storage.[1]

A list of terms used in the production of surimi is given in Table 1.

3 History of Surimi Production

Commerical production of fish surimi began in Japan in the 19th century on a small scale using locally caught fish. The development of the kamaboko industry started at the beginning of the present century, when the Japanese fishery industry introduced western trawl fishing.

In 1940 kamaboko production was 185 000 tons and this had risen from only a few thousand tons in 1910.[3] World War II greatly depressed the kamaboko industry and production was almost zero in 1945. After a gradual recovery following the war, a fish sausage, made from surimi, was introduced into the market in 1953. The fish sausage industry developed very rapidly and its production increased from 2000 tons in 1954 to 101 000 tons in 1960 and 188 000 tons in 1965. The traditional kamaboko industry also increased in production from 268 000 tons in 1954 to 408 000 in 1960.

A significant growth in the kamaboko industry occurred in the 1960s and 1970s. Total production of kamaboko and fish sausage increased from 509 000 tons in 1960 to 796 000 tons in 1965, with a further rise to 1 187 000 tons in 1973.

Table 1 *Terminology in surimi production*[12]

Surimi	Mechanically deboned fish flesh that has been washed with water and mixed with cryoprotectants.
Minced fish	Mechanically separated fish flesh that has not been washed.
Kamaboko	Products made from surimi that are mounted on a wooden plate and steamed or broiled.
Chikuwa	Products made from surimi that are broiled.
Tempura	Products made from surimi that are fried.
Kaen-surimi	Surimi containing salt.
Muen-surimi	Surimi without salt.

In recent years the interest in surimi and surimi-based products has increased enormously in the West. Phenomenal growth has been experienced in the USA, where sales grew from virtually zero in 1980 to over 68 000 tonnes in 1987.[13] More modest success has been achieved in Europe, with UK consumption estimated at 3200 tonnes in 1988. World surimi production in 1987 was estimated to be between 350 000 and 400 000 tonnes. Most of this was produced by Japan, where the Japanese Fisheries Association estimated production at between 200 000 and 250 000 tonnes, divided equally between at-sea and on-shore processing plants.[13]

As a result of the growth in surimi production numerous articles have been written describing the opportunities for further exploiting the raw material in processed products.[9,14,15]

4 Surimi Technology

Traditionally, Japanese surimi was prepared from fresh fish and immediately processed into kamaboko products. During the 1960s, a mechanical surimi process was developed in Japan,[3] which allowed surimi to be frozen without destroying its gelling ability. This process led to the industrial production of surimi on a large scale on land as well as on-board fishing vessels.

The industrialized surimi process, as it evolved in Japan over 25 years, was essentially based on mechanization of the traditional manual production methods. This process has been described by Swafford and coworkers,[16] and is illustrated in Figure 2. After the fish were headed, filleted, and deboned, the fish mince was washed two or three times with water and the wash water was strained through large rotary screens. The amount of washing required at each stage was judged subjectively by the operators. The washed mince then passed through a refiner to remove blade skin fragments. The refiner commonly used was a rotating drum of the Beehive design, also used in the production of mechanically recovered red meat and poultry.

A screw press removed a significant amount of water before final admixture of cryoprotectants and block freezing. This process produced good quality surimi in accordance with Japanese quality standards. However, the process has the following disadvantages:

(i) very high water consumption;
(ii) equipment requires large surface area;
(iii) difficult and time-consuming cleaning procedures;
(iv) considerable loss of myofibrillar protein in the screen effluents.

In the early 1980s a number of workers began to investigate alternative processing methods. In particular, Alfa Laval Ltd successfully introduced a process based on the use of a decanter centrifuge (G. Søbstedt, Alfa Laval Ltd, Denmark - personal communication). This process resulted in substantial improvements in the yield of surimi because of the greater recovery of

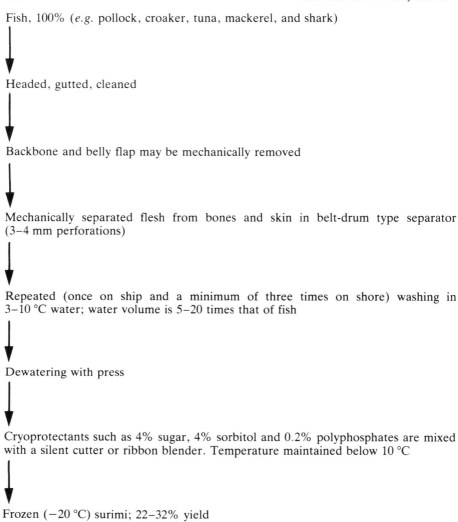

Fish, 100% (*e.g.* pollock, croaker, tuna, mackerel, and shark)

Headed, gutted, cleaned

Backbone and belly flap may be mechanically removed

Mechanically separated flesh from bones and skin in belt-drum type separator (3–4 mm perforations)

Repeated (once on ship and a minimum of three times on shore) washing in 3–10 °C water; water volume is 5–20 times that of fish

Dewatering with press

Cryoprotectants such as 4% sugar, 4% sorbitol and 0.2% polyphosphates are mixed with a silent cutter or ribbon blender. Temperature maintained below 10 °C

Frozen (−20 °C) surimi; 22–32% yield

Figure 2 *Conventional surimi process*
(Source: Ockerman and Hansen[12])

solid material during centrifugation compared with the traditional screen techniques. The Alfa Laval process, illustrated in Figure 3, was developed after a series of trials in an Alaskan surimi plant. The process is based on the washing of the deboned fish mince by means of an in-line mixer, a holding cell and a decanter centrifuge replacing the two- or three-stage screen wash systems. Similar systems to that shown in Figure 3 now operate in both on-shore and off-shore surimi plants throughout the world.

The design of the Alfa Laval decanter centrifuge is shown in Figure 4 and this illustrates the separation of solid and liquid components of the fish

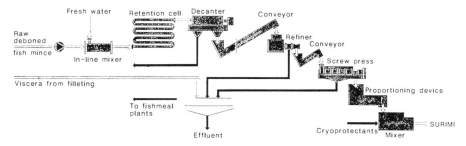

Figure 3 *Alfa Laval surimi plant*
(Courtesy: Alfa Laval Ltd)

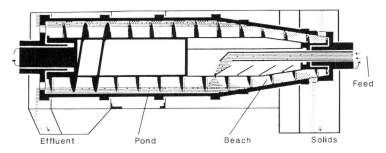

Figure 4 *Decanter centrifuge design*
(Source: Swafford et al.[16])

mince after water-washing. The centrifuge exerts a maximum force of 3000 g and continuously discharges the solid and liquid components. The introduction of centrifugation into the surimi process was not studied systematically and therefore optimum process conditions have never been established.

There is some doubt about requirement for large volumes of water during the washing stages. Hastings,[17] for example, examined the influence of the number of washing stages on the rheological properties of heat-set surimi gels. Gels made from surimi, previously washed three times in water (fish:water ratio 1:3) were soft and very flexible; gels from unwashed fresh cod mince were firmer and less elastic, but both gels gave a maximum score on the folding test (see below for folding test description). A single wash made the gels firmer and slightly more elastic than those from unwashed mince. Sensory assessment of the gels found no difference in firmness between the different washing procedures and small differences in elasticity and toughness. The colour of gels derived from cod mince was off-white, compared with the white surimi-derived gels.

A number of other workers have investigated the effects of varying the process conditions on the properties of surimi gels.[18,19] Generally, these

Table 2 *Properties used or proposed for grading surimi*[12]

Chemical and visual
Moisture level
pH
Whiteness - Hunter colour meter
Impurities - black skin and bones
Physical properties
Expressible drip - pressed
Viscosity - in 3/5% NaCl solution
Gel-forming ability (constant moisture)
Gel strength - plunger
Folding test - crack when folded
Firmness - sensory
Chewiness - energy used with repeated compressions
Elasticity - tensile force to break sheet
Water binding - slope of gel strength versus moisture
Frozen storage
Freeze-thaw cycles - pressed fluid

studies have looked superficially at a number of selected process conditions, but a fuller analysis of all the factors affecting surimi yield and composition has not been carried out.

5 Why Red Meat and Poultry Surimi?

Interest in producing red meat and poultry surimi has been stimulated to make better use of raw materials such as MRM and trimmings by upgrading to produce high-quality meat ingredients for further processing.

The potential benefits of using red meat and poultry surimi to the meat manufacturer are:

(i) lower fat and cholesterol content than other manufacturing meats;
(ii) reduced risk of rancidity development;
(iii) reduced risk of microbial spoilage;
(iv) bland-testing raw material to which any flavour can be added;
(v) almost colourless raw material for incorporation into a range of products;
(vi) improved rheological properties compared with other manufacturing meats;
(vii) a base raw material that can be used as the major component of a product, therefore providing a wider product range than is possible from other manufacturing meats.

Early attempts to extract haem pigments from meat, for analytical purposes, suggest that a single water-wash treatment is not effective.[20,21] DeDuve[22] extracted 70–95% of the haem pigments from deboned turkey meat with 0.01 N acetate buffer, pH 4.5. A popular method used by many analysts to extract haem pigments from meat and meat products is that used by Hornsey.[23] This involves extracting into a 4:1 mixture of acetone:water.

Since the early 1980s a limited amount of research has been conducted on pigment and fat extraction from red meat poultry for the production of surimi-like raw materials. This work is summarized below together with other relevant studies.

6 Pork Surimi

Only two studies on pork surimi have been reported in the literature. The first was by Lee and co-workers,[24] who examined the effects of pH, ionic strength, temperature, and number of washings on the colour and yield of surimi. An outline of the method they used to prepare the surimi is shown in Figure 5. The type of minced pork used as raw material was not

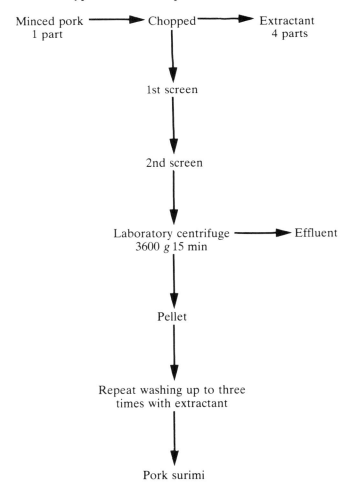

Figure 5 *Outline of pork surimi preparation used by Lee et al.*[24]

specified and neither was the homogenization method. Their results show that, as the number of washings increased, the fat and salt-soluble protein content decreased, but the water content and colour lightness values (L, a, b scale) increased. Increasing the pH value above 5.5 increased the water content of the surimi. The maximum colour lightness values were obtained from pork treated between pH 5.0 and 5.5, but the salt-soluble protein content of the surimi was lowest in this pH range, compared with higher values (pH 7.5). This is not surprising since the isoelectric point of myosin is in the range pH 5.0–5.5.

The effect of increasing the ionic strength from zero to 0.25 M sodium chloride was to reduce the amount of salt-soluble protein in the surimi produced. Consistent effects of different wash temperatures (2, 10, and 20 °C) were not observed. The only significant interaction observed was between yield and pigment content.

The protein concentration of the samples prepared by Lee et al.[24] varied from 8 to 18%, but the fat content was generally less than 0.5%. The authors concluded that pork could be used as a raw material for a protein product, similar in many respects to fish surimi. Of the variables examined, pH had the largest effect on the properties of the pork surimi.

The second study on pork surimi was by McKeith et al.,[25] who made surimi from lean pork chopped with five parts water. The resulting slurry was filtered through a metal screen with 2 mm mesh to remove connective tissue. The filtrate was centrifuged at 2000 g for 15 min and the pellet collected and rewashed twice with water.

The yield (wet weight basis) of pork surimi was ca. 45% and the authors attributed this low yield to the removal of substantial amounts of myofibrillar protein, along with connective tissue, on the metal screen.[25] The pork surimi had a fat content of ca. 0.1%, compared with 5.2% fat in the lean pork before processing, and the protein content of the surimi was generally 1% higher than that of the lean pork at 22.5%.

A compression test was used to compare the textural properties of pork surimi with those of commercial fish surimi (the grade of surimi was not specified). The 'hardness' of the pork surimi gels was lower than that of the fish surimi gels up to 50 °C. Between 50 and 80 °C a substantial increase occurred in the hardness of the pork surimi gels, whereas the fish surimi gels remained relatively soft. The hardness of fish surimi approximately doubled over the range 40–80 °C, while that of pork surimi increased ten-fold. The salt-soluble protein content of fish surimi was reported to be 3.6% and that of pork surimi to be 7.9%.

7 Beef Surimi

In addition to their study of pork surimi, McKeith et al.[25] also examined the properties of surimi-like materials made from lean beef and beef heart, weasand meat, head meat, and tongue. The process conditions were identical to those stated above for pork surimi, except that the heart, weasand meat, head meat, and tongue were not chopped.

When cooked, the beef surimi from all the raw materials was significantly darker than commercial fish surimi. The textural characteristics of surimi made from lean beef, measured by a compression test, were similar to those of the pork surimi. However, at equivalent protein concentrations, beef surimi gels were slightly 'harder' than the pork surimi gels.

The salt-soluble protein content of surimi made from lean beef was 7.8%. Texture data and salt-soluble protein contents of beef surimi made from heart, weasand meat, head meat, and tongue were not reported.

8 Mutton Surimi

A group at the Meat Industry Research Institute of New Zealand (MIRINZ) began working on the production of mutton surimi in 1987 (O.A. Young, MIRINZ, New Zealand - personal communication, 1988). The group has used the name 'Myobind' to describe the mutton surimi. Torley et al.[26] reported some of the group's results and pointed out that it is not possible simply to apply the fish surimi process to mutton, for the following reasons:

(i) Sheep meats have much higher collagen contents than fish and the quality of the collagen differs between the two species.
(ii) Sheep meats are much redder than fish and may therefore require a greater number of washings. Water-soluble mutton flavours are also removed by washing on several occasions.
(iii) Sheep meats also have a higher fat content than fish and the levels of unsaturation in mutton fat are far lower than in fish oils. The removal of solid mutton fat to make mutton surimi may require a different approach to the removal of liquid oil to make fish surimi.

Torley et al.[26] described two methods for mutton surimi production. The first is a 'coarse mince' method, which involves suspending the mince in water, allowing the fat to float and to be removed prior to centrifugation. This method reduces the mutton fat content from 30% to 5–10% and the mutton surimi has a moisture content of 75%. The second method is a 'fine particle' method, which involves chopping the mutton prior to mixing with water. The fat content of material produced in this way is only 1% and the moisture content is ca. 80%.

For both processes, the majority of myoglobin is removed in the first two washes. The moisture content of the fine particle product increases significantly with the number of washes, until the mutton cannot be dewatered adequately. Details of the centrifugation method used were not provided. Frozen storage trials were conducted over a 15 week period and the results show that there is an initial fall in the first week of 20% in gel strength, even when sucrose and sorbitol are used as cryoprotectants. However, after this period, there is little change in gel strength.

9 Poultry Surimi

Increased demand for processed turkey and chicken pieces and for poultry white meat has created a need for new ways to process the over-supply of necks, backs, thighs, and drumsticks.[27] One way of processing these parts is by the use of mechanical deboning to produce meat raw materials for use in emulsion-type products and in reformed and restructured products. However, several constraints have limited the use of mechanically deboned poultry meat:

(i) It has a dark colour.
(ii) The particle size is generally small, which results in poor textural properties.[28]
(iii) The storage life is short.[29]
(iv) The high levels of unsaturated fatty acids in poultry fat result in a greater chance of rancidity development compared with red meats.[30]

In recent years there has been a growing interest, particularly in America, in the application of the surimi process to mechanically deboned poultry. Much of the published work was conducted at North Carolina State University, Rayleigh, North Carolina. An Alfa Laval pilot-scale decanter centrifuge has been used in the trials conducted there.

Froning and Johnson[31] reported that much of the pigment in mechanically deboned chicken meat is loosely held and can be removed relatively simply by water-washing and centrifugation processes. Warris[32,33] extracted haem pigment with phosphate buffer and concluded that maximum extraction was achieved by buffers having a pH above 6.8. Various extracting media were investigated by Ball and Montejano,[34] including tap water, sodium bicarbonate (pH 8.45), and sodium acetate (pH 5.25). These solutions were used to extract pigment from the thigh meat of broilers. The largest amount of pigment was removed by the bicarbonate treatment, with an 88% reduction, and the other treatments removed *ca.* 70–75% of the pigment. Hydrogen peroxide, sodium metabisulphite, and ascorbic acid have been suggested as treatments for the removal of the dark colour of mechanically deboned poultry meat.[35]

Hernandez and co-workers demonstrated that turkey meat pigments were more effectively removed by raising the pH value of the washing medium.[27] Phosphate buffers at pH 6.4, 6.8, 7.2, and 8.0 were examined for their efficiency of pigment extraction, which was measured by the Hunterlab Colour Difference meter. The pH 8.0 buffer produced the lightest-coloured turkey meat after washing and pressing to de-water twice.

Dawson *et al.*[36] showed that bicarbonate buffer (pH 8.5) was more effective in removing pigment from mechanically deboned chicken than washing with acetate buffer (pH 4.5) or neutral tap water. However, there was no difference in the amounts of fat removed from the chicken meat by each of the extractants. The moisture content of the chicken surimi was, of course, influenced by the pH of the extraction medium, with higher moisture contents produced at higher pH values. In a subsequent investigation, these authors evaluated a decanter centrifuge in the production of chicken surimi.[37] An outline of the pilot-scale process they designed is shown in Figure 6. This includes two separate mixing vats and two decanter centrifuge units. The process reduced the pigment content to an acceptable level in the raw surimi, but a grey colour was noticed when this was cooked. The fat content was reduced from 12.8% in the deboned chicken to 1.5% in the surimi obtained after centrifugation. The polyunsaturated:saturated (P/S) ratio of the chicken fat was unaffected by the process. Ball described the development of shrimp and chicken Cordon Bleu analogues containing chicken surimi made by the process illustrated in Figure 6.[38]

10 Defining the Quality of Surimi

The tests that are used to grade surimi depend upon the intended use of the material. Ockerman and Hansen[12] summarized the properties that have been used or proposed for assessing surimi quality, and these are listed in Table 2. In addition to these physical and chemical tests, the microbiological status of the surimi should be assessed as well as the chemical stability of the fat. Dawson *et al.*[36,37] have reported obtaining high thiobarbituric acid (TBA) values from chicken surimi produced in a decanter-centrifuge based process. The amount of fat remaining in the surimi will, however, influence the extent to which rancid flavour development will occur in both red meat and poultry surimi. Verrez and co-workers[39] and Babbitt and Reppond[40] suggest that vacuum chopping of surimi,

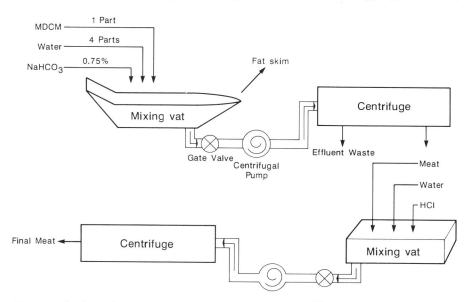

Figure 6 *Outline of surimi process used by Dawson* et al.*[37]*

with ingredients such as egg white and potato starch, has a positive effect on the textural quality of fish surimi. The use of vacuum chopping methods is also likely to reduce the chances of rancidity development.

11 Test Methods

The following methods have been used to assess the quality of red meat and poultry surimi:

Compression Testing

Three 5 g samples of each material are tested as follows:

The sample is packed into a plastic cylinder, to produce a plug of surimi of length 28 mm and diameter 24 mm, cut from a syringe barrel. The sample is contained by two rubber syringe bungs, the upper of which has a hole through it to allow air to escape.

When packed, the cylinders are fixed in a metal rack, which is placed in a water bath at 80 °C such that the upper bung is just above the water level. The samples are then heated for 30 min, removed from the bath, and left to stand at ambient temperature before transferral to a refrigerator at 4 °C for a further 10 min. The samples are then removed from the holders and tested for compressibility using a Stevens CR Analyser to record the force required to compress the sample by 6 mm. The Stevens Analyser is filled with a flat plate plunger (5 cm diameter) and the sample is compressed at a rate of 2 cm min^{-1}.

Cooking Loss Measurements

Cooking loss measurements are made in conjunction with compression tests. The weight of surimi placed in the cylinder is measured before cooking. The weight of the surimi plug removed after cooking is then measured and the loss calculated. Care is taken to remove any fragments of surimi that have become attached to the plastic during cooking and to weigh these with the plug.

Tensile Adhesive Strength (TAS) Testing

Beef rump muscle (*semitendinosus*) is vacuum-packaged and blast-frozen to −40 °C and stored at −18 °C before use. Prior to use the meat is tempered to 5 °C before slicing into 1 cm slices.

For the TAS test, two 3 cm squares of meat are cut from the slices such that the meat fibres are parallel to the smallest dimension, then 0.5 g of surimi is spread evenly onto one face of each square with a small spatula. Care is taken to ensure that the surimi is spread evenly over the whole surface. The covered surfaces are then pressed together to form a meat-surimi-meat sandwich. The sandwich is blast-frozen at −40 °C for

45 min, vacuum-packed, and then immersed in a water bath at 80 °C for 30 min. Upon removal from the hot water bath, the sample is cooled in iced water for 10 min and then further cooled in a refrigerator at 4 °C for 10 min.

After removal from the vacuum bag the sample is gently dried using tissue and fixed to two metal templates with Superglue, as illustrated in Figure 7. The assembled test piece is attached to an Instron Universal Test Machine. The meat block is pulled apart at a speed of 10 cm min^{-1} and the force–time curve for each sample recorded. This shows when the sample fractured and measures the force required to fracture and the total work done. For each sample of surimi, three separate tests are performed.

Storage Modulus, Loss Modulus, and Dynamic Viscosity

Samples of the materials are tested using a CarriMed CS Rheometer set in oscillatory mode. The measurement system consists of 4.0 cm diameter parallel plates with a gap of 2000 μm and an inertia of 63.60 dyne cm s^{-2}.

The following instrument settings are used:

Pre-shear time	0 dyne cm
Pre-shear time	0 s
Equilibration time	10 s
Stress	79.58 dyne cm^{-2}
Frequency	0.7 Hz

At these settings the instrument is used in temperature-sweep mode over the range 25–80 °C. Measurements of the following parameters are recorded at 24 timed points during heating:

G' The storage modulus - a measure of the energy stored during deformation of the sample and recovered when the stress is removed, *i.e.* a measure of elasticity.

G'' The loss modulus - a measure of the energy lost during deformation (dissipated as heat), *i.e.* a measure of fluidity.

π^* poise The dynamic viscosity - numerically equal to $(G' + G'')/$ (frequency of oscillation). This is a measure of the total resistance of the system to the stress applied.

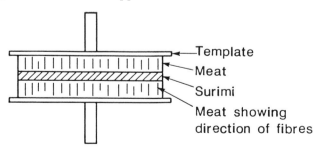

Figure 7 *Tensile adhesive strength (TAS) assembly*

These measurements are also made during cooling of the sample from 50 to 25 °C and during subsequent standing at 25 °C. Observations of the above parameters are made over the range of frequencies of oscillation from 0.05–10.00 Hz.

Microscopy Methods

Small amounts of the samples are placed in boiling tubes and heated in a water bath for 1 hour at either 45 or 80 °C. Unheated and heated samples are frozen in liquid nitrogen on suitable stubs and frozen sections (*ca.* 15 μm) are cut using a Slee Cryostat set at −25 °C.

The sections are stained as follows:

(*a*) with 0.1% Toluidine Blue in water;[41]
(*b*) with Picro-Sirius Red;[42]
(*c*) with 0.1% osmium tetroxide followed by treatment with acid Fuchsin/Ponceau de Xylidine.[41]

12 Effect of pH, Ionic Strength, and Cryoprotectants

Experiments at the Leatherhead Food Research Association on the effects of salt concentrations and pH value have been carried out in accordance with a statistically designed pattern. The experiments formed a one-third fraction of a 3^4 factorial experiment. In this plan only one-third of the experimental variables are actually investigated. These data provide most of the information on the main effects and on second-order interactions but at the expense of possible higher-order interactions that cannot be investigated.

The factors under investigation were as follows:

pH	5.0, 5.5, and 6.8
Sodium chloride	0–2%
Sodium tripolyphosphate	0–0.3%
Calcium chloride	0–2%
Sorbitol	0–4%
Polydextrose	0–8%
Sucrose	0–4%
Soya isolate	0–5%
Sodium caseinate	0–5%
Potato starch	0–5%

The compositions of samples prepared to investigate the effects of pH and ionic strength are listed in Table 3.

The effects of polydextrose and glycerol on the frozen storage life of surimi were investigated over a 6 month period. The effects of soya isolate, sodium caseinate, and potato starch on the textural characteristics of the surimi were also investigated.

Table 3 *Compositions of the experimental samples used to examine the effects of pH and ionic strength*

Experiment block no.	Sample no.	pH	Composition (%)			Composition (g in 500 g surimi)			
			NaCl	CaCl$_2$	STP[a]	NaCl	CaCl$_2$	STP[a]	Surimi
I	1	5.50	1.25	1.25	0.15	6.25	6.25	8.75	486.75
	2	6.80	2.00	2.00	0.30	10.00	10.00	1.50	478.50
	3	6.80	1.25	0.50	0.00	6.25	2.50	0.00	491.25
	4	6.80	0.50	1.25	0.15	2.50	6.25	8.75	490.50
	5	5.50	0.50	2.00	0.30	2.50	10.00	1.50	486.00
	6	5.00	2.00	1.25	0.15	10.00	6.25	0.75	483.00
	7	5.00	1.25	2.00	0.30	6.25	10.00	1.50	482.25
	8	5.50	2.00	0.50	0.00	10.00	2.50	0.00	487.50
	9	5.00	0.50	0.50	0.00	2.50	2.50	0.00	495.00
II	10	6.00	2.00	0.50	0.15	10.00	2.50	0.75	486.75
	11	5.00	1.25	0.50	0.15	6.25	2.50	0.75	490.50
	12	5.50	2.00	1.25	0.30	10.00	6.25	1.50	482.25
	13	5.50	1.25	2.00	0.00	6.25	10.00	0.00	483.75
	14	5.00	0.50	1.25	0.30	2.50	6.25	1.50	489.75
	15	5.50	0.50	0.50	0.15	2.50	2.50	0.75	494.25
	16	6.80	1.25	1.25	0.30	6.25	6.25	1.50	486.00
	17	6.80	0.50	2.00	0.00	2.50	10.00	0.00	487.58
	18	5.00	2.00	2.00	0.00	10.00	10.00	0.00	480.00
III	19	6.80	0.50	0.50	0.30	2.50	2.50	1.50	493.50
	20	5.00	0.50	2.00	0.15	2.50	10.00	0.75	486.75
	21	5.50	1.25	0.50	0.30	6.25	2.50	1.50	489.75
	22	5.50	0.50	1.25	0.00	2.50	6.25	0.00	491.25
	23	5.50	2.00	2.00	0.15	10.00	10.00	0.75	479.25
	24	6.88	2.00	1.25	0.00	10.00	6.25	0.00	483.75
	25	5.00	2.00	0.58	0.38	10.00	2.50	1.58	486.00
	26	5.00	1.25	1.25	0.00	6.25	6.25	0.00	487.50
	27	6.80	1.25	2.00	0.15	6.25	10.00	0.75	483.00

[a] STP = sodium tripolyphosphate.

13 Composition of Beef Surimi

The composition of the starting material and of the resultant beef surimi for four trial runs conducted at Leatherhead Food Research Association is shown in Table 4. These results demonstrate the variability of the starting material (beef MRM), with the fat content ranging from *ca.* 28 to 42% and protein from *ca.* 11 to 14%. The results also demonstrate the efficiency of the production process in greatly reducing fat, reducing the ash content by about one-third, and concentrating collagen to increase its concentration by about one-third.

The composition of the refined material (beef surimi) largely varied in the protein and water content. Protein contents ranged from 15.8 to 19.4%

Table 4 *Composition (% m/m) of beef MRM and the resultant beef surimi*

Analysis	1		2		3		4		5	
	MRM	Surimi	MRM	Surimi	MRM	Surimi	MRM	Surimi	MRM	Surimi
Protein (N × 6.25)	14.4	19.4	11.1	19.2	10.7	16.6	11.1	18.1	13.6	15.8
Water	57.8	81.3	47.7	80.3	50.3	82.6	47.8	80.6	53.3	81.8
Water: protein	4.0	4.2	4.3	4.2	4.7	4.9	4.3	4.4	3.9	5.2
Fat	27.8	0.6	41.2	0.7	40.1	0.5	4.2	1.5	32.9	1.8
Ash	1.1	0.7	–	–	–	–	–	–	–	–
Collagen (hydroxy proline × 8)	–	–	–	–	–	–	–	–	1.0	2.9

and water from 80.3 to 82.6%, giving a range of water:protein ratios of 4.2–5.2.

Microscopical Structure of Beef Surimi

Observations were made of the following Food Research Association produced materials when raw, when heated to 45 °C, and when heated to 80 °C:

(*a*) beef MRM;
(*b*) beef surimi;
(*c*) beef surimi chopped for 1 min alone;
(*d*) beef surimi chopped for 1 min with 2% sodium chloride;
(*e*) beef MRM chopped for 1 min with 2% sodium chloride and 0.15% sodium tripolyphosphate.

Observations were made of the effects of heating to 45 °C and 80 °C on the gross appearance and texture of the MRM and surimi. On heating to 45 °C no marked change in texture was noted in either the MRM or the surimi; both remained 'crumbly'. Some degree of firming or gelation was noted in the surimi chopped with salt and the surimi chopped with salt and phosphate. Some free liquid was separated in both the MRM and surimi samples. On heating to 80 °C the MRM showed fat separation but there was no fat separation in the surimi samples. The surimi chopped with either salt or with salt and phosphate produced smooth, springy gels that were more sliceable than the surimi-only samples. A small amount of liquid had separated from all the surimi samples and this presumably contained gelatin, since the liquid formed a thermo-reversible gel that set upon cooling.

The beef MRM, as seen in the light microscope, had a coarse particulate structure with a fair amount of fat (Figure 8a). Muscle fragments with intact myofibrillar structure were also seen, together with large fragments of connective tissue (Figure 8b). Beef surimi showed much less fat than the

MRM (Figure 9a) and there was an increased concentration of connective tissue (Figure 9b). The myofibrillar structures were more disrupted in the surimi and less easily identifiable, but small areas of banded structure could still be seen.

Chopped surimi showed a more continous structure with even more finely divided and dispersed fat and connective tissue (Figure 10a). Some

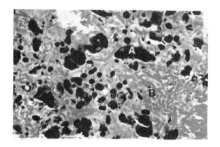

Figure 8a *Uncooked MRM showing fat* (A) *and protein* (B) *(× 20)*

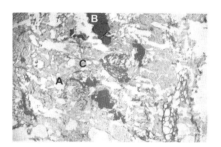

Figure 8b *MRM chopped alone and cooked to 80 °C showing myofibrillar proteins* (A), *collagen* (B), *and gelatin* (C) *(× 20)*

Figure 9a *Uncooked surimi showing fat* (A) *and protein* (B) *(× 20)*

Figure 9b *Surimi chopped alone and treated to 80 °C showing myofibrillar protein* (A) *and connective tissue* (B) *(× 48)*

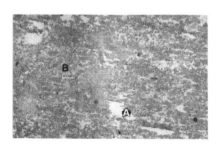

Figure 10a *Uncooked surimi chopped showing fat* (A) *and protein* (B) *(× 20)*

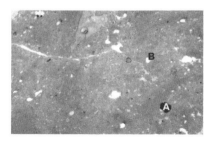

Figure 10b *Uncooked surimi alone chopped with salt and phosphate showing fat* (A) *and protein* (B) *(× 20)*

muscle myofibrills were still seen within the fibres. Surimi chopped with salt and phosphate showed a dense continuum with very finely divided and dispersed fat and very little muscle structure. The myofibrillar structure was highly disrupted and the protein more dispersed compared with the surimi chopped alone (Figures 9a and 10b).

The changes in the structures seen between the raw MRM and surimi samples were mainly in the following features:

(i) The level of fat was much lower in the surimi.
(ii) The level of connective tissue was higher in the surimi.
(iii) The degree of comminution and dispersion of muscle protein was greater in the surimi, particularly on addition of salt and polyphosphate.

After heating to 80 °C, areas of free fat were seen in the MRM (Figure 11a) and to a much lesser extent in the surimi (Figure 11b). In the chopped surimi (Figure 12a) and particularly the surimi chopped with salt and phosphate (Figure 12b) the samples heated to 80 °C were much more continuous, with very finely dispersed fat.

Comparison of the cooked samples showed the formation of a continuous matrix in the surimi, particularly after addition of salts. Staining with Picro-Sirius Red (Figures 8b and 13) showed the presence of pink areas,

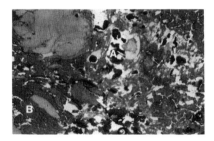

Figure 11a *MRM cooked to 80 °C showing fat (A) and protein (B) ($\times$ 20)*

Figure 11b *Surimi cooked to 80 °C showing fat (A) and protein (B) ($\times$ 20)*

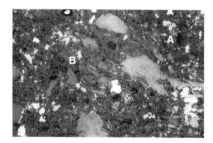

Figure 12a *Surimi chopped alone and cooked to 80 °C showing fat (A) and protein (B) ($\times$ 20)*

Figure 12b *Surimi chopped with salt and phosphate and cooked to 80 °C showing fat (A) and protein (B) ($\times$ 20)*

indicating the formation of gelatin. The amount of gelatin appeared to increase with addition of salts.

The formation of an increasingly connected fine network could also be seen and was most marked in the surimi with salt and polyphosphate. This appears as lilac with Toluidine Blue staining (Figure 14) and red with Picro-Sirius Red (Figure 10b), indicating that the network is collagen-based. The large pieces of connective tissue also appeared to be affected, particularly in the surimi with salts at 80 °C. The pieces appeared to have more holes, and gelatin areas were often associated with their edges, indicating possible extraction of material (Figure 13).

A general increase in the dispersion of myofibrillar proteins was seen in the surimi samples together with changes in the connective tissue and this also appears to form a network. This is most easily seen in the cooked samples (Figures 13 and 14).

Figures 8–14 therefore illustrate some aspects of the changes induced on MRM by processing into surimi and also by treatment with salt, phosphate, and heat. None of the samples were homogeneous but major structural differences were observed as a result of each treatment and these have been summarized in Figure 15.

15 Effects of pH and Ionic Strength on Beef Surimi Rheology

Effects of Compressibility

The values obtained for compressibility of surimi sample prepared at the Food Research Association are listed in Table 5. The compression tests were carried out on the twenty-seven compositional types of beef surimi listed in Table 3 and three replicate measurements of compressive strength were made for each compositional type.

The values found for MRM were considerably lower than those for the surimi samples, probably reflecting the high level of fat (*ca.* 28% - see

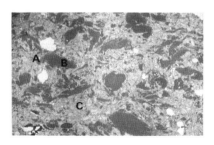

Figure 13 *Surimi chopped with salt and phosphate and cooked to 80 °C showing myofibrillar protein (A), collagen (B), and gelatin (C) (× 20)*

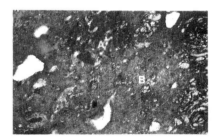

Figure 14 *Beef surimi chopped with salt and phosphate and phosphate and cooked: to 80 °C showing myofibrillar protein (A) and connective tissue (B) (× 48)*

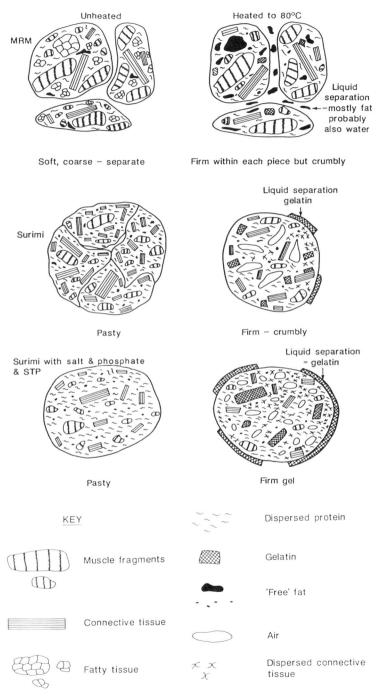

Figure 15 *Major trends in the structure of MRM and surimi and the effects of salt, phosphate, and heat*

Table 5 Effect of pH and ionic strength on the compressive strength of beef surimi

Sample no.[a]	pH	Salt concentration (%)			Compressive strength (g)			
		NaCl	CaCl$_2$	STP[b]	1	2	3	Mean
1	5.50	1.25	1.25	0.15	1985	1960	2740	2228
2	6.80	2.00	2.00	0.30	1975	1740	1590	1768
3	6.80	1.25	0.50	0.00	3050	2695	2645	2797
4	6.80	0.50	1.25	0.15	2850	3005	3300	3052
5	5.50	0.50	2.00	0.30	3525	2795	3030	3117
6	5.00	2.00	1.25	0.15	3705	3250	2560	3172
7	5.00	1.25	2.00	0.30	2765	3290	4135	3397
8	5.50	2.00	0.50	0.00	3020	4305	2930	3418
9	5.00	0.50	0.50	6.00	2950	2440	2015	2468
10	6.80	2.00	0.50	0.15	2335	2555	2340	2410
11	5.00	1.25	0.50	0.15	2485	3005	2775	2755
12	5.50	2.00	1.25	0.30	2530	2720	2055	2435
13	5.50	1.25	2.00	0.00	2660	3030	3035	2908
14	5.50	0.50	1.25	0.30	3345	3860	3915	3707
15	5.00	0.50	0.50	0.15	2650	2220	2550	2807
16	6.80	1.25	1.25	0.30	2740	2445	2630	2605
17	6.80	0.50	2.00	0.00	2630	3120	2605	2785
18	5.00	2.00	2.00	0.00	3175	4095	3115	3462
19	6.80	0.50	0.50	0.30	1490	1785	2210	1828
20	5.00	0.50	2.00	0.15	1980	1425	1010	1472
21	5.50	1.25	0.50	0.30	1865	2225	1070	1720
22	5.50	0.50	1.25	0.00	2790	3340	3185	3105
23	5.50	2.00	2.00	0.15	2615	2725	2285	2542
24	6.80	2.00	1.25	0.00	1495	2025	1825	1782
25	5.00	2.00	0.50	0.30	2940	1880	2890	2570
26	5.00	1.25	1.25	0.00	2145	3190	2305	2547
27	6.80	1.25	2.00	0.15	1550	1665	1580	1598

[a] See Table 3.
[b] STP = sodium tripolyphosphate.

Table 4) in the MRM. The triplicate samples of surimi showed considerable variation in compressive strength. However, analysis of variance (ANOVA) carried out on the mean values showed that there were significant effects of sample composition (*i.e.* pH and ionic strength) on compressive strength and this is shown in Table 6. The analysis shows that pH had a significant influence on compressibility, producing softer gels with increasing pH value. The results also indicate that sodium tripolyphosphate causes a small reduction in compressibility and that there can be interactive effects between pH and sodium chloride and between pH and sodium tripolyphosphate. In these interactive effects there appears to be a pattern. At low salt concentrations the highest compression values were

Table 6 *Statistical analysis of the effects of pH and ionic strength on compressive strength of beef surimi*

Factors and interactions tested	Probability (%) of a significant effect on compressive strength[a]
pH	1.7
Sodium chloride	51.0
Sodium tripolyphosphate	6.7
pH × sodium chloride	5.3
pH × sodium tripolyphosphate	5.3
Sodium chloride × sodium tripolyphosphate	12.3
Calcium chloride	25.2

[a] $> 5\%$ = not significant; $< 5\%$ = significant; $< 1\%$ = highly significant.

generally found at pH 5.5, *i.e.* near the isoelectric point of myosin. At the higher salt concentrations compressibility decreased as pH was increased (*i.e.* as pH was adjusted further away from the isoelectric point of myosin, which is lowered through salt addition). The concentrations of sodium and calcium chlorides did not have a significant effect on the compressibility.

Effects on Adhesion

The results of TAS measurements on the twenty-seven different treatments (Table 3) of pH and ionic strength are listed in Table 7. Considerable difficulty was experienced in conducting these tests owing to frequent detachment of the sample at the meat–metal interface. When tension was applied, rather than fracturing either in the meat block or at the meat–surimi junction, *ca.* 40% of samples fractured at the meat–metal interface. Hence there are some missing values in Table 5 because of insufficient amounts of sample to repeat certain treatments. However, the statistical design was such that it could tolerate some missing values and ANOVA on these results is shown in Table 8.

The most significant factor affecting the TAS values was pH, with a small increase in TAS with increasing pH. The results generally showed no significant effects of the factors tested on TAS values. This could be due to the difficulties found in conducting these measurements. The test is dependent upon a good junction having formed between the meat blocks, where the surimi should act as the adhesive. Care was taken to ensure that the cut surfaces were smooth and that meat fibre direction was constant for each sample. However, some replicate samples of the same treatment fractured at the meat–surimi junction and others fractured some distance from the junction. Therefore, the test did not in every case measure the strength of the meat–surimi junction.

Table 7 *Effects of pH and ionic strength on the tensile adhesive strength (TAS) values of beef surimi*

Sample no.[a]	pH	Salt concentration (%)			TAS (g)			
		NaCl	CaCl$_2$	STP[b]	1	2	3	Mean
1	5.50	1.25	1.25	0.15	363	301	238	301
2	6.00	2.00	2.00	0.00	238	454	350	247
3	6.80	1.25	0.50	0.00	352	419	365	379
4	6.80	0.50	1.25	0.15	238	419	454	370
5	5.50	0.50	2.00	0.30	238	269	334	280
6	5.00	2.00	1.25	0.15	*	421	437	429
7	5.00	1.25	2.00	0.30	354	197	284	278
8	5.50	2.00	0.50	0.00	*	321	300	311
9	5.00	0.50	0.50	0.00	259	*	489	374
10	6.80	2.00	0.50	0.15	314	338	*	326
11	5.00	1.25	0.50	0.15	440	425	388	418
12	5.50	2.00	1.25	0.30	431	444	351	409
13	5.50	1.25	2.00	0.00	427	326	413	389
14	5.00	0.50	1.25	0.30	179	258	328	255
15	5.50	0.50	0.50	0.15	*	353	318	336
16	6.80	1.25	1.25	0.30	*	648	*	648
17	6.80	0.50	2.00	0.00	*	*	*	*
18	5.00	2.00	2.00	0.00	351	285	173	270
19	6.80	0.50	0.50	0.30	136	450	532	373
20	5.00	0.50	2.00	0.15	351	156	*	254
21	5.50	1.25	0.50	0.30	299	400	204	301
22	5.50	0.50	1.25	0.00	330	345	252	309
23	5.50	2.00	2.00	0.15	303	349	247	300
24	6.80	2.00	1.25	0.00	367	*	*	367
25	5.00	2.00	0.50	0.30	245	235	297	359
26	5.00	1.25	1.25	0.00	289	343	*	316
27	6.80	1.25	2.00	0.15	353	264	*	309

[a] See Table 3.
[b] STP = sodium tripolyphosphate.
* Not determined.

Effects of Storage Modulus, Loss Modulus, and Dynamic Viscosity

Gel systems vary in their elements of solid-like and liquid-like characteristics. In strong gels G' (storage modulus) is normally substantially greater than G'' (loss modulus) and the absolute values of both moduli show little dependence upon the frequency of oscillatory deformation. In addition the dynamic viscosity, π^* poise, decreases steeply with increasing frequency.

Examples of plots of G', G'', and π^* poise against frequency of oscillation are shown in Figures 16 and 17 for MRM and in Figures 18 and

Table 8 *Levels of significance of the effects of pH and salt concentration on TAS*

Factor and interaction tested	Probability (%) of a significant effect on TAS[a]
pH	7.5
Sodium chloride	46.6
Sodium tripolyphosphate	85.2
pH × sodium chloride	61.4
pH × sodium tripolyphosphate	19.2
Sodium chloride × sodium tripolyphosphate	41.3
Calcium chloride	26.6

[a] $> 5\%$ = not significant; $< 5\%$ = significant; $< 1\%$ = highly significant.

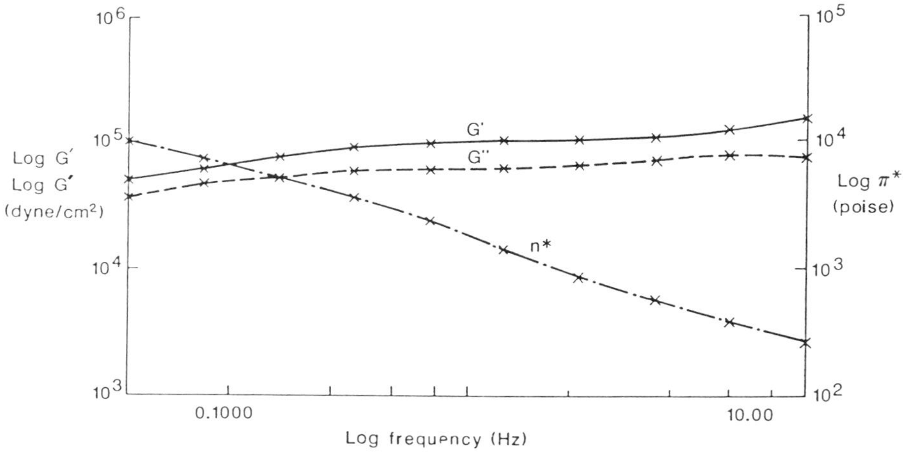

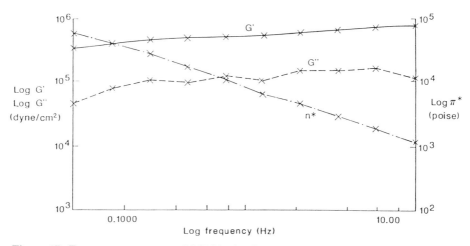

Figure 16 *Frequency spectrum of MRM at 25 °C before heating to 80 °C*

Figure 17 *Frequency spectrum of MRM after heating to 80 °C and cooling to 25 °C*

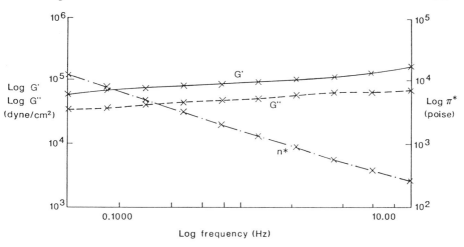

Figure 18 *Frequency spectrum of beef surimi at 25 °C before heating to 80 °C*

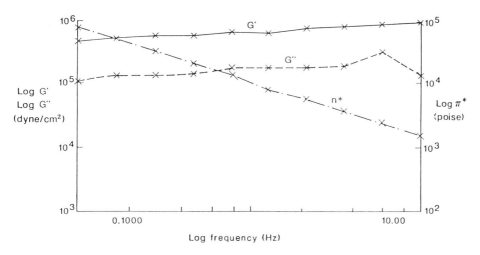

Figure 19 *Frequency of beef surimi after heating to 80 °C and cooling to 25 °C*

19 for surimi. It can be seen that beef MRM and beef surimi show the characteristics of strong gels - both before and after heating to 80 °C and cooling to 25 °C.

The raw beef MRM and beef surimi show very similar frequency sweep patterns, with surimi showing very slightly lower values than MRM. The cooked and cooled samples of MRM and surimi were also similar to one another, each showing higher values of G' and G'' compared with the raw samples. After cooking, surimi showed a ten-fold increase in the value of G', but a lesser increase in G''. A similar pattern was seen for MRM but with lesser increases in the values of G' and G''.

The three rheological parameters, G', G'', and η^* poise, were also monitored at a fixed frequency of oscillation (0.7 Hz) during heating of the samples from 25 to 80 °C and during subsequent cooling to 25 °C. The general forms of the thermorheological profiles obtained for beef MRM and for beef surimi are shown in Figure 20. The relationship of G' to temperature for MRM shows a considerable decrease from 25 °C to a temperature of ca. 56 °C (probably initially due to fat melting), when G' began to increase. The rate of increase accelerated at ca. 58 °C and G' then rose linearly with temperature before levelling off at ca. 76 °C.

In beef surimi G' was more than an order of magnitude greater than that for the MRM from which it was produced. The beef surimi also showed a decrease in G' as temperature increased from 25 °C but much less so than for MRM (probably reflecting the lower fat content of surimi). This decline was reversed at a higher temperature (63 °C) than that seen for MRM, and G' continued to increase in surimi up to 80 °C. In both MRM and surimi, G' at 80 °C was less than its value at 25 °C, indicating a slight decrease in elasticity of the materials during cooking.

The repeatability of the thermorheological profile of surimi was examined and two sets of results are illustrated in Figures 21 and 22. These results were considered acceptable.

Thermorheological profiles were obtained for each of the twenty-seven samples of different pH values and salt concentrations listed in Table 3.

The following values were taken from the profiles:

T_1 the temperature at which G' began to increase;
T_2 the temperature at which the rate of increase in G' began to decline;

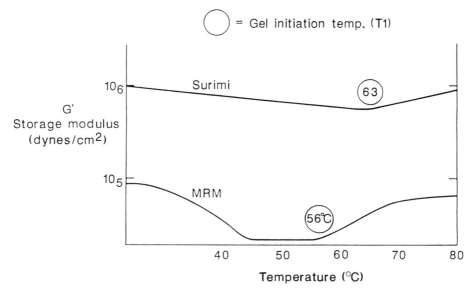

Figure 20 *Thermorheological profiles of beef MRM and beef surimi*

$G'T_1$ the value of G' at T_1;
$G'T_2$ the value of G' at T_2;
G'_{25} the value of G' after the heated material had cooled to 25 °C.

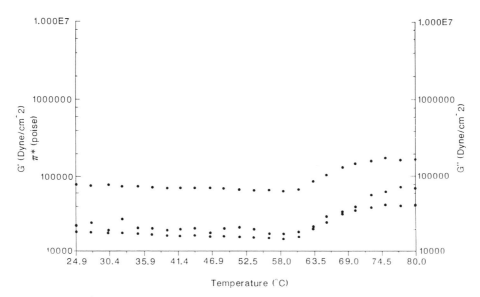

Figure 21 *Thermorheological profile of beef surimi sample 2 (see Table 4 for composition)*

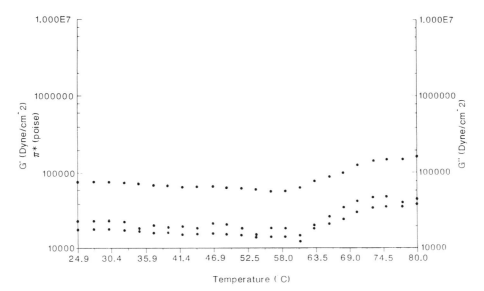

Figure 22 *Thermorheological profile of beef surimi sample 4 (see Table 4 for composition)*

These values are shown in Table 9 together with the composition of the samples. The following parameters were examined by ANOVA and linear regression analysis to determine whether they were influenced by pH and salt concentration:

T_1; T_2; $G'T_1$; $G'T_2$; G'_{25}; $G'T_2 - G'T_1$; $T_2 - T_1$;

$(G'T_2 - G'T_1)/(T_2 - T_1)$

The results of these analysis showed that only T_1 was significantly

Table 9 *Temperature and storage modulus values for the 27 samples of beef surimi differing in pH and salt concentration*

Sample no.[a]	pH	Salt concentration (%)			Temperature (°C)		Storage modulus (dyne cm^{-2})		
		NaCl	CaCl$_2$	STP[b]	T_1	T_2	$G'T_1$	$G'T_2$	G'_{25}
1	5.50	1.25	1.25	0.15	54.0	66.0	100	227	764
2	6.80	2.00	2.00	0.30	54.0	66.0	103	198	596
3	6.80	1.25	0.50	0.00	58.5	66.0	87	199	592
4	6.80	0.50	1.25	0.15	57.0	62.6	95	150	640
5	5.50	0.50	2.00	0.30	52.0	60.6	95	170	457
6	5.00	2.00	1.25	0.15	58.5	69.0	133	240	686
7	5.00	1.25	2.00	0.30	58.5	64.5	185	250	776
8	5.50	2.00	0.50	0.00	58.0	69.0	270	475	1,960
9	5.00	0.50	0.50	0.00	55.9	69.0	290	470	1,636
10	6.80	2.00	0.50	0.15	61.3	66.9	145	205	273
11	5.00	1.25	0.50	0.15	64.0	76.5	137	220	550
12	5.50	2.00	1.25	0.30	55.0	77.5	22	204	423
13	5.50	1.25	2.00	0.00	52.2	63.6	79	125	332
14	5.00	0.50	1.25	0.30	52.5	70.4	221	389	1,061
15	5.50	0.50	0.50	0.15	60.5	80.0	160	280	438
16	6.80	1.25	1.25	0.30	56.0	66.2	117	278	491
17	6.80	0.50	2.00	0.00	54.5	64.0	86	148	235
18	5.00	2.00	2.00	0.00	58.2	69.7	92	125	375
19	6.80	0.50	0.50	0.30	55.9	64.6	52	128	264
20	5.00	0.50	2.00	0.15	49.6	72.0	26	62	100
21	5.50	1.25	0.50	0.30	54.5	80.0	40	102	164
22	5.50	0.50	1.25	0.00	51.0	67.5	60	109	377
23	5.50	2.00	2.00	0.15	56.2	80.0	83	202	311
24	6.80	2.00	1.25	0.00	57.1	63.4	56	83	144
25	5.00	2.00	0.50	0.30	60.7	68.2	60	104	275
26	5.00	1.25	1.25	0.00	59.3	64.8	144	214	473
27	6.80	1.25	2.00	0.15	52.7	63.5	62	105	170

[a] See Table 3.
[b] STP = sodium tripolyphosphate.

influenced by pH and salt concentration. The levels of significance of these results are listed in Table 10.

The temperature T_1 was significantly influenced by pH value, sodium chloride concentration, and calcium chloride concentration. There was also a significant interactive effect between sodium chloride and pH value. No effect was found for sodium tripolyphosphate nor did this salt exhibit any significant interactive effect with sodium chloride or pH value. The direction of these influences was also indicated by this analysis. Increasing the level of sodium chloride increased the value of T_1, whereas increasing the level of calcium chloride decreased T_1. The variation at T_1 with pH showed minimum value of pH 5.5, close to the isoelectric point of myosin.

The increase in G', which begins at temperature T_1, indicates the initiation of the formation of a different type of network within the system, capable of storing more energy. This is a fundamental change in the system and the question arises of whether the change in initiation temperature, influenced by pH, sodium chloride, and calcium chloride, was related to other rheological characteristics of the material. The possible relationships between T_1 and other rheological characteristics were therefore tested.

The following values derived from the twenty-seven thermorheological profiles were analysed statistically and plots derived of the following parameters against the T_1 value: $G'T_1$, $G'T_2$, G'_{25}, $G'T_2 - G'T_1$, $(G'T_2 - G'T_1)/(T_2 - T_1)$

Also plotted was the ratio of percentage calcium chloride to percentage sodium tripolyphosphate. In this plot a constant value was added to the concentration of STP to overcome the difficulty of an infinite value due to zero sodium tripolyphosphate concentrations. Examples of these plots are shown in Figures 23 and 24. In none of the plots was there a pattern indicating a close relationship of the parameters to T_1. The most important feature of the results is the lack of relationship between the temperature at which the gelation process begins and any of the other factors that describe

Table 10 *Level of significance of the effects of pH and salt concentration on the gel initiation temperature (T_1) for beef surimi*

Factor and interaction tested	Probability (%) of a significant effect of gel initiation temperature[a]
pH	2.2
Sodium chloride	0.8
Sodium tripolyphosphate	9.3
pH × sodium chloride	1.4
pH × sodium tripolyphosphate	22.4
Sodium chloride × sodium tripolyphosphate	63.2
Calcium chloride	3.0

[a] $> 5\%$ = not significant; $< 5\%$ = significant; $<1\%$ = highly significant.

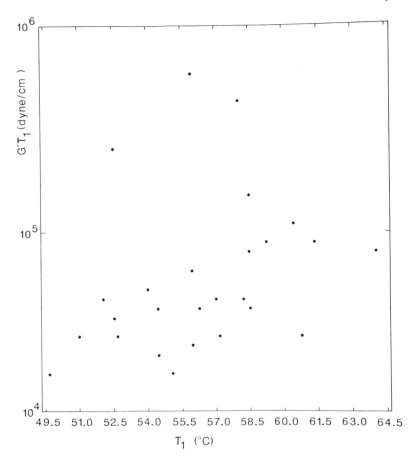

Figure 23 *Plot of $G'T_1$ against T_1 for beef surimi*

the process. It appears that the temperature at which the new network is initiated has little effect upon the course of the reaction or the final value of G'.

16 Cryoprotectant Effects on Beef Surimi

Several cryoprotectants that are used to protect the functional properties (*i.e.* protein solubility, water-holding capacity, and gelation) of fish surimi during frozen storage have been investigated at the Leatherhead Food Research Association for their effects on beef surimi. The cryoprotectant compositions of the samples examined are shown in Table 11.

When beef surimi was vacuum-packaged, blast-frozen, and transferred to −18 °C overnight before thawing, the thawed material was paler in colour than fresh surimi. It also had a 'bitty' or loose feel to the touch and felt

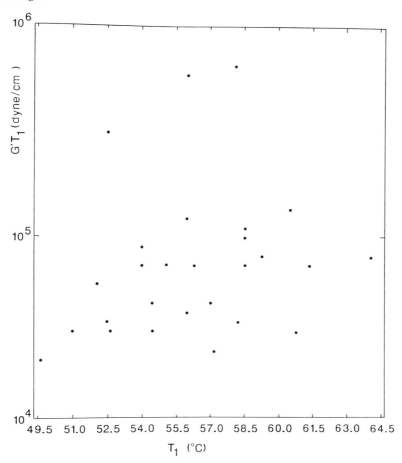

Figure 24 *Plot of $G'T_2$ against T_1 for beef surimi*

Table 11 *Cryoprotectant composition[a] (%) of beef surimi samples*

Sample code	Sodium chloride	STP[b]	Calcium chloride	Sorbitol	Sucrose	Poly-dextrose
A	2.0	0.15	0	0	0	0
B	2.0	0.15	0.5	0	0	0
C	2.0	0.15	0	0	0	8.0
D	2.0	0.15	0	4.0	4.0	0
E	2.0	0.15	0	0	4.0	0
F	2.0	0.15	0	4.0	0	0

[a] Based on levels used in fish surimi.[7]
[b] STP = sodium tripolyphosphate.

less cohesive than the fresh material. When salt and phosphate were chopped into the frozen–thawed surimi it became more cohesive and doughy. It also took on a light-tan colour.

Effects on Cooking Performance

The cooking losses of samples containing cryoprotectants were determined after 1 week of frozen storage (in vacuum packs) from triplicate 5 g samples heated at 80 °C for 30 min. The results are shown in Table 12.

Beef surimi alone showed the greatest loss and the addition of salt and phosphate reduced this to 32%. Only the further incorporation of 8% polydextrose gave lower losses, substantially reducing the loss to less than half of that found for beef surimi alone. The addition of calcium chloride did not reduce cooking losses much below those found for surimi alone. Addition of sorbitol and sucrose together and sucrose alone (samples D and E) to surimi containing salt and phosphate had no substantial influence on cooking losses. Addition of sorbitol alone, however, appeared to raise cooking losses to the level found from surimi alone. The reason for this effect is unclear.

Effects on Compressibility

The compressibilities of the samples were measured as previously described, using a Stevens CR Analyser. The results of triplicate measurements are shown in Table 13. These values show considerable variation between the triplicate observations of each sample. When the mean values are considered, only in sample C (containing 8% polydextrose) was a higher force required to compress the sample compared with the others.

Table 12 *Cooking losses for beef surimi samples containing cryoprotectants*

Sample code[a]	Mean wt loss (g) from 5 g of surimi sample	Observed loss (%)	Adjusted loss[b] (%)
Beef surimi alone	1.66	33	42
A	1.26	25	32
B	1.56	31	40
C	0.73	15	20
D	1.23	25	34
E	1.33	27	35
F	1.53	30	41

[a] See Table 11.
[b] Adjusted for water content of sample (*i.e.* calculated as % total water in the sample).

Table 13 *Effects of cryoprotectants on the compressive strength of beef surimi*

Sample code[a]	Compressive strength (g)			
	1	2	3	Mean
Surimi alone	2605	1810	2240	2218
A	2775	1775	2015	2182
B	2350	2045	2750	2382
C	2430	2885	2885	2733
D	2590	1630	2170	2130
E	2115	1785	2420	2106
F	2100	2050	2105	2218

[a] See Table 11.

Effects on Adhesion

The tensile adhesive strength (TAS) of the samples was measured by the method previously described for surimi except that in those tests surimi alone was used in place of the meat–surimi–meat sandwich. Considerable difficulty was encountered in this series of tests, with frequent splitting of the material from the metal plate. Triplicate results were obtained for only three of the samples (A–F listed in Table 11) and these are shown in Table 14.

There was considerable variation in the values of the triplicate determinations, which, together with the difficulty in mounting the test samples, indicated that this method is unsuited to the material. No further attempts were made to complete these experiments. The values shown in Table 14 were within the range found previously for samples of varying salt composition (Table 7).

Effects of Freezing on Storage Modulus

The storage modulus G' is a measure of the elastic nature of a material. This parameter might be expected to vary if denaturation or other changes

Table 14 *Tensile adhesive strength (TAS) of beef surimi containing cryoprotectants*

Sample code[a]	TAS (g)			
	1	2	3	Mean
B	72.3	192.2	440.0	234.8
C	144.6	196.3	309.9	215.9
D	239.7	314.0	417.6	323.6

[a] See Table 11.

occur in the proteins during freezing, frozen storage, and thawing. This phenomenon does occur in fish surimi and a range of cryoprotectants are used to ameliorate freeze denaturation. To examine the effects of cryoprotectants on beef surimi the value of G' was measured in fresh samples containing the various protectants and then monitored with time for the frozen stored material. G'_1, G'', and π^* poise were recorded during heating to 80 °C and cooling to 25 °C. These values paralleled one another and therefore only values of G' are shown. The values of G' at 80 °C and when the sample had cooled to 25 °C are shown in Figures 25 and 26, respectively, for fresh and for frozen-thawed surimi.

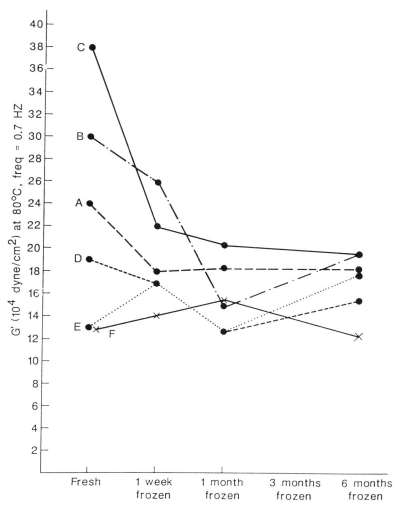

Figure 25 *Effect of cryoprotectants on the storage modulus (G') at 80 °C of fresh and frozen–thawed beef surimi*

Values of G' at 80 °C (Figure 25)

The fresh materials showed a considerable range of values - ca. $13–38 \times 10^4$ dyne cm^{-2}. The sample containing only sodium chloride and sodium tripolyphosphate (A) was in the centre of the range. The further addition of polydextrose (C) or of calcium chloride (B) increased the value of G', whereas the addition of sucrose (E), sorbitol (F), or a combination of these (D) decreased the value from that obtained for salt and phosphate alone.

After freezing for a week, the range of values had narrowed to ca. $14–26 \times 10^4$ dyne cm^{-2}. The values for salt and phosphate alone (A) were

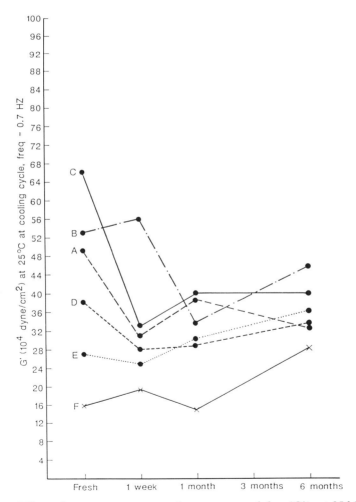

Figure 26 *Effect of cryoprotectants on the storage modulus (G') at 25 °C of fresh and frozen in fresh and frozen–thawed beef surimi*

lower but still in the middle of the range. Polydextrose (C) and calcium chloride (D) still produced higher values than salt and phosphate alone, and the other samples showed lower values.

After freezing for one month the range of values had further narrowed to *ca.* $13\text{–}21 \times 10^4$ dyne cm^{-2} and the value for the sample containing calcium chloride (B) had fallen below that for sample A at 1 month. The range of G' values after 6 months frozen storage was similar to that at 1 month. In particular, the differences between the values for samples A, B, and C were small.

Values of G after Cooling from 80 °C to 25 °C (Figure 26)

The values of G' for fresh surimi determined at 25 °C were higher than those found at 80 °C but also showed a wider range - *ca.* $16\text{–}66 \times 10^4$ dyne cm^{-2}, compared with $13\text{–}38 \times 10^4$ dyne cm^{-2} for fresh surimi at 80 °C. In their order of ranking they paralleled the values found for the fresh samples measured at 80 °C.

After one week of frozen storage two samples (B and F) had shown a rise in the value of G', whereas the others showed a drop. At 1 month this position had been reversed, with samples B and F showing a fall in the value of G', whereas all of the others had shown a rise. At 6 months the range of G' values had narrowed and samples B, C, D, and E had G' values higher than sample A, which contained salt and phosphate only. The results of this 6 month trial demonstrate that the addition of cryoprotectants has a considerable effect upon the storage modulus of fresh beef surimi and, more importantly, that freezing and frozen storage modify the rheology of the material. The addition of polydextrose or calcium increased the G' value of surimi in the fresh state in the initial period of frozen storage. Addition of sucrose and/or sorbitol tended to decrease the G' value of the surimi. Overall, the samples containing calcium chloride (B), polydextrose (C), and sucrose (E) had higher G' values, after 6 months frozen storage, than surimi containing salt and phosphate only.

Effects on Rancidity

The effect of frozen storage on rancidity development was followed by measurement of thiobarbituric acid (TBA) values. TBA values are sometimes used as an index of rancidity development. The method measures the content of malondialdehyde and high levels of aldehydes in meat products are usually associated with a rancid taste. Threshold values above which rancid tastes may be detected are difficult to set; therefore, it is usual to follow TBA values over the period of storage and to look for changes over the life of the product.

The TBA values for MRM and surimi are listed in Table 15. Values for

Table 15 *TBA values of beef MRM and surimi after frozen storage*

Sample	Time frozen	TBA values
MRM	1 week	1.78
Surimi	1 week	1.08
Sample A[a]	1 month	0.52
	3 months	0.61
	6 months	1.22

[a] See Table 11.

surimi after up to 6 months frozen storage are all lower than the value for MRM after just 1 week. This indicates that the surimi was not rancid after 6 months frozen storage and that there is less risk of rancidity development in the surimi compared with MRM.

17 Effects of Non-meat Proteins and of Starch on Beef Surimi

A range of samples of beef surimi was prepared at the Leatherhead Food Research Association to examine the effects of soya isolate, sodium caseinate, and native potato starch on the compressive strength and cooking losses of beef surimi. The compositions of the samples and the results obtained are shown in Table 16. The values shown are for samples frozen for 3 months. Values for fresh samples are shown in parentheses. Each value is the mean of three determinations.

The results for the frozen–thawed material in Table 16 show that the inclusion of non-meat proteins and of starch can markedly alter the properties of beef surimi. Surimi alone showed a high compressive strength and also high cooking losses. Mixing with salts did not change this behaviour but chopping with salts for 1 minute appeared to decrease the compressive strength. The cooking losses of surimi with salts were similar to those from surimi alone. The addition of 5% soya isolate increased the compressive strength compared with the value for surimi chopped with salts and considerably reduced cooking losses, whilst starch reduced both compressive strength and cooking losses. Casein also reduced the strength of surimi but had much less effect than either soya or starch in reducing cooking losses.

In addition to these individual effects there appeared to be considerable interactive effects of the non-meat ingredients. In combinations of each at 5% m/m, casein and starch virtually eliminated cooking losses whilst soya and casein produced the highest compressive strength with cooking losses characteristic of soya used alone. A combination of all three ingredients also showed a very low cooking loss value but had a lower strength than that for the combination of soya and casein.

Table 16 *Effects of soya isolate, sodium caseinate, and native potato starch on the compressive strength and cooking losses of beef surimi*

Sample composition	Compressive strength (g)	Cooking loss (% m/m)
Surimi alone	2400	28
	2330	33
Surimi + salts[a]	2360	32
Surimi + salts, chopped[b]	2090	33
Surimi + salts + 5% soya, chopped	2250 (1670)[c]	12 (17)
Surimi + salts + 5% starch, chopped	1600 (1420)	10 (22)
Surimi + salts + 5% casein, chopped	1490 (1340)	26 (28)
Surimi + salts + 5% each of soya, casein, and starch, chopped	2110 (2070)	1.0 (6.0)
Surimi + salts + 5% each of casein and starch, chopped	1790 (1760)	1.0 (6.0)
Surimi + salts + 5% each of soya and casein, chopped	2860 (2510)	11 (12)
Surimi + salts + 4% PD[d]	1520	30
Surimi + salts + 4% PD + 5% soya, chopped	2380 (1890)	19 (12)
Surimi + salts + 4% PD + 5% starch, chopped	1280 (1150)	11 (11)
Surimi + salts + 4% PD + 5% casein, chopped	1280 (1330)	29 (26)
Surimi + salts + 8% PD	1520	35
Surimi + salts + 8% PD + 5% soya, chopped	2280 (2000)	15 (14)
Surimi + salts + 8% PD + 5% starch, chopped	1310 (1040)	10 (15)
Surimi + salts + 8% PD + 5% casein, chopped	1780 (1520)	30 (28)

[a] 2% sodium chloride + 0.15% sodium tripolyphosphate mixed in a Kenwood Chef mixer for 1 min.
[b] as above, but chopped for 1 min in a Robot Coupe instead of a mixed.
[c] ( ) values for fresh surimi.
[d] PD = polydextrose.

Overall, the results indicate that soya had the greatest influence in increasing compressure strength whilst potato starch had the greatest influence in decreasing cooking loss. The results obtained for the fresh materials generally showed lower compressive strengths and higher cooking losses than those found for the frozen–thawed material, indicating that freezing and thawing changed the nature of these materials. When polydextrose was included in these formulations at 4% m/m, the compressive strength of the sample with soya isolate was little changed whereas that of samples containing potato starch and casein was decreased. The cooking losses for all of the samples were slightly increased. With 8% polydextrose a similar picture was seen except that the sample containing casein showed a considerably increased strength compared with that with 4% polydextrose.

In the controls for the experiments with polydextrose, it was found that this material did not reduce the cooking losses of surimi. This is in contradiction to earlier results (Table 12), in which distinct reduction in losses was observed from 33% down to 15% for the fresh material. This requires further investigation.

18 General Discussion

The majority of examples given in the text above relate to beef MRM and surimi. The choice of beef MRM as the meat on which to carry out this work was made on the basis of 'the most difficult case'. Beef fat contains a high proportion of saturated fatty acids and therefore has a relatively high melting point. Beef MRM has a high level of fat and the fat is difficult to remove because of its high solid content at chill temperatures. Therefore, if surimi can be produced from beef then other species should present fewer problems.

The transition from beef MRM to beef surimi is reflected in the proportions of protein, water, fat, ash, and collagen listed in Table 4, the major differences being the virtual removal of fat, with a concomitant increase in protein concentration. The water:protein ratio was not greatly altered under the conditions used here, although this parameter can be manipulated by varying certain processing conditions.

The observed microscopic structural changes during the transition from MRM to surimi reflect not only the reduction of fat and increase in protein but also the increase in the relative proportion of connective tissue. Although collagen appears to constitute a relatively small proportion of the material (Table 4), it appears on microscopic observation to be a widespread and important part of the matrix formed when surimi is chopped with salts and heated. Knight[43] also found connective tissue to be an important component of a meat gel matrix. These observations suggest that the gel network in meat surimi is a co-gel of connective tissue (collagen) and gelatin (a cold-setting system), with dispersed myofibrillar protein (a heat-setting system) similar to some fish systems.[44] Such a structure would explain the high cooking losses (> 30%) found for this material even when the rigorous cooking régime is taken into account. The cooking losses were determined immediately after cooking and after allowing the sample to drain. At this point the myofibrillar protein would be set (*i.e.* solid), but the gelatin and its associated water would still be liquid and would drain from the mass. If the samples had been allowed to cool in their own juices substantially lower cooking losses would be expected, as found by Evans for mixtures of meat and fat.[45]

This co-gel structure would also explain the substantial reductions in cooking losses produced by soya and starch (Table 16), both of which bind water and set on heating. Caseinate, which does not set on heating, had a much smaller effect in reducing cooking losses.

The use of a small deformation, non-destructive technique to study the dynamic rheological behaviour of meat emulsions was demonstrated previously by Poulson and Tung[46] to provide information on the molecular changes occurring during cooking. They found that for a pork meat emulsion (10% protein, 57% water, 29% fat, and 2.5% sodium chloride), the temperatures at which the storage modulus (G') began to increase (60 °C) and then plateau (75 °C) corresponded closely to the characteristic differential scanning calorimetry (DSC) profile for muscle protein comminuted with salt. DSC analysis shows a single endothermic peak with onset at *ca.* 60 °C and a maximum at 72 °C. Poulson and Tung[46] proposed that protein denaturation as observed by DSC was followed very closely by the formation of a strong cohesive meat gel, as indicated by an increase in G'. The values of $G'T_1$ and $G'T_2$ for the meat emulsion were similar to those found for beef MRM in the Leatherhead Food Research Association work reported here. In their interpretation of the mechanism of meat protein gel formation, Poulson and Tung noted that the denaturation temperature of the most abundant myofibrillar proteins is similar to T_1 - the temperature at which G' begins to increase. In the Food Research Association study it was observed that this temperature varied with pH and with the type and concentration of salt. Wright and Wilding[47] found a similar dependency of myosin gelatin on pH and salt concentration using DSC. Our own observations of the effects of sodium chloride, calcium chloride, and pH value suggest *prima facie* that they do this via their influence on the distance of molecular separation. Sodium ions induce structural changes in myosin molecules, which tend to increase T_1, whereas calcium ions, which can form divalent cation bridges between molecules, tend to decrease T_1. In addition, T_1 was minimal at pH 5.5, close to the isoelectric point of muscle proteins, at which the molecules occupy minimal volume.

The lack of relationship between T_1 and other parameters of the thermorheological profiles suggests that once the proteins are denatured the formation of the new network is a random or chaotic process, rather than an ordered process. This may indeed be the case for a given rate of increase of input energy. In these experiments a uniform heating rate of 3 °C min^{-1} was applied. However, there is evidence from fish surimi studies that the nature of the final gel network and texture found from fish proteins can be greatly modified by the rate of heating. This is shown by the process known as 'suwari', in which holding the surimi at sub-gelation temperatures, *e.g.* 40 °C for 1 h,[7] produces a more ordered gel network formation and a homogeneous and more elastic gel on subsequent cooking than does immediate cooking. Arkahare and Cheftel[48] have indicated that fish protein showed these effects of 'pre-gelation' to a greater extent than proteins from warm-blooded species (*e.g.* rabbit). This is probably due to the higher denaturation temperatures of proteins from warm-blooded species, which would be expected to require higher pre-gelation temperatures.

It seems possible that there is a range of gel networks that could be formed, differing in structure and texture according to the sequence in which the molecules are permitted to react to form linkages. There appear to be at least three factors over which we can exercise control: (i) the intermolecular distance through salt composition and pH; (ii) the amplitude of molecular vibration through the input energy (heating); and (iii) the probability of a reaction via the time at a given energy level (the heating rate). In this way it seems possible that alignments of molecules will be made through the formation of weaker bonds (*e.g.* hydrogen and van der Waals bonds) prior to the fixation at the structure of higher energy levels by covalent, disulfide, and hydrophobic bonds that require higher energies for their formation. If this is the case then the cooking of mixtures of proteins could become more than just a means of denaturing proteins and inactivating biological structures such as enzymes and bacteria. It would also become a means of influencing structure and texture by the control of the molecular interactions of proteins and other ingredients.

19 Acknowledgements

The author would like to thank MAFF for supporting much of the work carried out on red meat and poultry surimi. The author is also indebted to staff at the Leatherhead Food Research Association for their valued assistance, including Mrs Kathy Groves for the microscopical work, Dr David Rose for assistance with statistical analyses, Dr Robert Hart for assistance and advice on rheological aspects of this work, and Bee Khim Choo and Nick Church for conducting the experiments. Finally, this work would not have been carried out without the determination and enthusiasm of the late Dr John Wood.

20 References

1. C.M. Lee, *Food Technol.*, 1984, **38**, 69.
2. J.N. Olley, in 'Advances in Fish Science and Technology', ed. J.J. Connell, Fishing News (Books) Ltd, London, 1980, Chapter 1.
3. M. Okada, 'The history of surimi and surimi based products in Japan', in Proceedings of the International Symposium on Engineered Seafood including Surimi, Seattle, 1985, p.30.
4. D. Green and T. Lanier, 'Fish as the soybean of the sea', in Ref.3, p.42.
5. J.N. Keay, 'Surimi: a European perspective' in Ref.3, p.619.
6. A. Sribhibhadh, 'Prospects in developing countries on production of surimi products', in Ref.3, p.8.
7. E. Niwa, 'Functional aspects of surimi', in Ref.3, p.141.
8. Y. Kammuri and T. Fujita, 'Surimi-based products and fabrication processes', in Ref.3, p.254.
9. Anon, *Food Manuf.*, 1987, November, 48.
10. Anon, *Meat Process.*, 1988, February, 40.

11. G.A. MacDonald, *Food Technol. NZ*, 1988, August, 29.
12. H.W. Ockerman and C.L. Hansen, 'Animal By-product Processing', Ellis Horwood, Chichester, 1988.
13. J.N. Young, 'Surimi - A Significant Protein Food of the Future', Information Group Services & Market Information Section, Leatherhead Food Research Association, 1988.
14. Anon, *Food Eng.*, 1985, June, 73.
15. J.N. Keay, 'Surimi: a European perspective', in Ref.3, p.619.
16. T.C. Swafford, J. Babbitt, K. Reppond, A. Hardy, C.C. Riley, and T.K.A. Zetterling, 'Surimi process yield improvements and quality contribution by centrifuge', in Ref.3, p.483.
17. R.J. Hastings, *Int. J. Food Sci. Technol.*, 1989, **24**, 93.
18. C.M. Lee and J.M. Kim, 'Texture and freeze–thaw stability of surimi gels in relation to ingredients and formulation', in Ref.3, p.168.
19. J. Opstredt, 'A National programme for studies on the value for 'surimi' production of industrial fish species in Norway', in Ref.3, p.218.
20. W.E. Poel, *Am. J. Physiol.*, 1949, **156**, 44.
21. H.P. Flemming, T.N. Blumer, and H.B. Craig, *J. Anim. Sci.*, 1960, **19**, 1164.
22. C.A. DeDuve, *Acta Chem. Scand.*, 1984, **38**, 264.
23. H.C. Hornsey, *J. Sci. Food Agric.*, 1956, **7**, 534.
24. M.-Y. Lee, F.K. McKeith, J. Novakofski, and P.J. Bechtel, *J. Anim. Sci.*, 1987, **57**, (Suppl. 1), 283.
25. F.K. McKeith, P.J. Bechtel, J. Novakofski, S. Park, and J.S. Arnold, 'Characteristics of surimi-like material from beef, pork and beef by-products', in Proceedings of International Congress of Meat Science and Technology, Brisbane, 1988, p.325.
26. P.J. Torley, D.H. Reid, O.A. Young, and R.D. Archibald, *Food Technol. NZ*, 1988, June, 51.
27. A. Hernandez, R.C. Baker, and J.H. Hotchkiss, *J. Food Sci.*, 1986, **51**, 865.
28. J.C. Acton, *J. Food Sci.*, 1972, **93**, 240.
29. D.V. Vahedra and R.C. Baker, *Food Technol.*, 1970, **24** (7), 776.
30. L.E. Dawson and R. Gartner, *Food Technol.*, 1983, **37** (7), 112.
31. G.W. Froning and F. Johnson, *J. Food Sci.*, 1973, **38**, 279.
32. P.D. Warris, *Anal. Biochem.*, 1976, **72**, 104.
33. P.D. Warris, *J. Food Technol.*, 1979, **14**, 75.
34. H.R. Ball and J.G. Montejano, *Poult. Sci.*, 1984, **63** (Suppl. 1), 60.
35. K. Thompson, *Meat Ind.*, 1984, **31** (12), 23.
36. P.L. Dawson, B.W. Sheldon, and H.R. Ball, Jr., *J. Food Sci.*, 1988, **53** (6), 1615.
37. P.L. Dawson, B.W. Sheldon, and H.R. Ball, Jr., *Poult. Sci.*, 1988, **67**, 73.
38. H.R. Ball, Jr., *Poult. Sci.*, 1988, **67**, 67.
39. V. Verrez, D.J. Gallant, B. Bouchet, and L. Han-Ching, 'Surimi and kamaboko processing as seen under the microscope', Abstracts from Food Microstructure meeting held at Reading University, UK, SEM Inc., USA, 1988.
40. J.K. Babbitt and K.D. Reppond, *J. Food Sci.*, 1988, **53**, 965.
41. D.F. Lewis, 'Manual of Microscopical Methods', Leatherhead Food Research Association, 1978.
42. F.O. Flint and K. Pickering, *Analyst (London)*, 1984, **109**, 1505.
43. M.K. Knight, 'Interactions of meat proteins in meat products', PhD Thesis, University of Nottingham, 1990.

44. M. Tülsner and F. El Bedawey, *Lebensm.-Ind.*, 1980, **27**, 175.
45. G.G. Evans and M.D. Ranken, *Leatherhead Food RA Res. Rep.*, 1975, No. 219.
46. A.T. Poulson and M.A. Tung, *Can. Inst. Food Sci. Technol.*, 1989, **22** (1), 80.
47. D.J. Wright and P. Wilding, *J. Sci. Food Agric.*, 1984, **35**, 357.
48. T. Arkahare and J.C. Cheftel, *Ind. Aliment. Agric.*, 1989, **106** (10), 881.

Cured Meat Products and Their Oxidative Stability

L.H. Skibsted

RVAU CENTRE FOR FOOD RESEARCH, ROYAL VETERINARY AND AGRICULTURAL UNIVERSITY, THORVALDSENSVEJ 40, 1871 FREDERIKSBERG C, DENMARK

1 Introduction

The meat curing process and the developed technology has served mankind in producing safe meat products with good keeping properties. The curing involves the treatment of meat with sodium chloride and sodium nitrite, often in combination with ascorbic acid. Sodium nitrite is a strong oxidant, and together with added ascorbate it produces nitric oxide (NO). Nitric oxide is a gaseous molecule with an odd number of electrons and is very reactive towards radicals and oxygen. Nitric oxide holds the key as to why the treatment of meat with the strong oxidant sodium nitrite in combination with the notorious pro-oxidant sodium chloride results in products with improved oxidative stability and prolonged shelf-life.

The 'purely inorganic' compound NO (Figure 1) does not seem to be very biological or physiological, and if meat curing had not been discovered by serendipity in ancient time but had been the result of scientific design and needed to be approved by legal authorities, meat curing would be facing serious problems.[1] However, current medical research seems to have identified nitric oxide with the so-called endothelium-derived relaxing factor (EDRF) as a positive health factor since it plays a crucial role in controlling human blood pressure through vasodilation.[2] Direct evidence

Bond order: $2\frac{1}{2}$

Figure 1 *Nitric oxide is a so-called odd molecule with a three-electron bond. Compared with the nitrosonium ion (NO^+), the bond is elongated as the odd electron is added to an antibonding orbital: $(\sigma_1)^2(\sigma_1{}^*)^2(\sigma_2)^2(\pi)^4(\pi^*)^1$. In the Lewis formalism, the unpaired electron hides between the nitrogen atom and the oxygen atom, in effect making nitric oxide a slow-reacting free radical*

has been presented for nitric oxide biosynthesis in man; and it has furthermore been shown that endogenous nitric oxide is an inhibitor of platelet aggregation in the blood stream.[3] Nitric oxide may be concluded to be a 'biomolecule', and the chemistry of meat curing finds surprising parallels in the chemistry behind two important aspects of human well-being.[4]

The chemistry of meat curing has, in view of its great economical importance and the health concerns related to dietary intake of nitrite, been the subject of numerous investigations and reviews.[5–8] The present account is not intended to cover all of the many aspects of the curing process but focuses on the chemistry of the process and on the oxidative stability of the resulting products during subsequent storage.

2 The Chemistry of Meat Curing

Curing is most significantly recognized by the colour changes of the meat system, by the absence of warmed-over flavour for heated products and by the development of unique flavours. The chemistry behind the colour change is central to our understanding of the improved oxidative stability as it involves transformation of the meat pigments myoglobin, oxymyoglobin, and metmyoglobin into nitrosylmyoglobin and nitrosyl myochrome. The initial step in this transformation involves the reaction of nitrite with various endogenous or added reductants such as ascorbic acid.

Reaction of Nitrite with Ascorbate

The conjugate acid of nitrite, nitrous acid, has a pK_a of 3.22, and only a minor fraction of nitrite added to meat is converted into the more reactive free acid.[9] Nitrous acid is unstable in solution and decomposes in a reversible reaction with an equilibrium constant K of $10 \, \text{mol} \, \text{l}^{-1}$, as calculated from standard potentials:[10]

$$3HNO_2 \rightleftharpoons H^+ + NO_3^- + 2NO + H_2O \qquad (1)$$

Nitrous acid behaves in solution both as a reductant and as an oxidant, and the standard reduction potentials for nitrate and nitrous acid are very similar.[10] In meat systems, nitrite primarily acts as an oxidant:

$$NO_2^- + H_2O + e^- \rightarrow NO + 2OH^- \qquad E^\ominus_{red} = +0.46 \, V \qquad (2)$$

and decreasing pH increases the oxidizing capability. It should, however, be noted, that nitrite is partly converted into nitrate during curing and storage, indicating that part of the nitrite is eventually oxidized:[8]

$$NO_2^- + 2OH^- \rightarrow NO_3^- + H_2O + 2e^- \qquad E^\ominus_{ox} = -0.01 \, V \qquad (3)$$

Thermodynamics predict that nitrite and nitrous acid will react with a wide variety of the reducing components present in meat. Kinetic aspects are, however, also important and during meat curing the reaction with added ascorbate or erythorbate is quantitatively the most important. Ascorbic acid (H_2Asc) is a bifunctional acid, although weaker than nitrous acid as it has a $pK_{a,1}$ of 4.04 and a $pK_{a,2}$ of 11.34.[9] Owing to different reactivities of the various acid and base forms, the reaction shows a significant dependence on pH (Figure 2).

Stoichiometry and Rate Laws. Under conditions of excess of ascorbate, the stoichiometry of the reaction between nitrite and ascorbate was established by Dahn *et al.*[11] as

$$\text{ascorbic acid} + 2HNO_2 \rightarrow \text{dehydroascorbic acid} + 2NO + 2H_2O \quad (4)$$

In the pH range 0–5, the rate law varies and four distinct regions have been recognized.[12] In solutions of strong acid, the undissociated ascorbic acid reacts with the nitrous acidium ion:

$$H^+ + HNO_2 \rightleftharpoons H_2NO_2^+ \rightleftharpoons H_2O + NO^+ \quad (5)$$

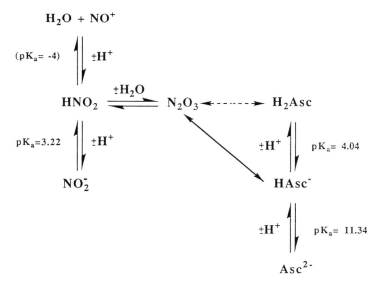

Figure 2 *The pH-dependence of the reaction between nitrite and ascorbate is complex owing to the different reactivities of the species in mutual acid/base equilibria and in hydration equilibria. The reactivity of the nitrite/nitrous acid system increases with decreasing pH, whereas the reactivity of ascorbate as a reductant increases with increasing pH. The reaction between N_2O_3 and hydrogen ascorbate is quantitatively the most important in meat curing followed by reaction of N_2O_3 with ascorbic acid*

Nitrous acidium or the kinetically equivalent nitrosonium ion (NO^+) is a strong electrophile and nitrosating agent. NO^+ or its hydrate is formed only in concentrated acid ($K = 7 \times 10^{-5}$ l mol^{-1}),[13]

$$HNO_2 + H^+ \rightleftharpoons NO^+ + H_2O \qquad (6)$$

and has little if any importance in meat systems. In moderately acidic solution, N_2O_3 in equilibrium with nitrous acid ($K = 9 \times 10^{-3}$ l mol^{-1}),[13]

$$2HNO_2 \rightleftharpoons N_2O_3 + H_2O \qquad (7)$$

reacts with both ascorbic acid and ascorbate, the latter form being the more reactive. At a pH of 4 or above, the reaction between nitrite and ascorbate is zero order with respect to ascorbate, and the rate-determining step is the dehydration of nitrous acid to form N_2O_3. In the pH region of relevance to meat, the most significant reduction of nitrite is the result of the reaction between N_2O_3 and ascorbate ion:[14]

$$2HNO_2 \xrightarrow{k_2} N_2O_3 + H_2O \qquad (8)$$

$$N_2O_3 + Hasc^- \xrightarrow{fast} \text{reaction intermediates} \qquad (9)$$

This reaction sequence agrees with the experimental rate law:[14]

$$-\frac{d[asc]_t}{dt} = k_2[HNO_2]^2 \qquad (10)$$

in which $[HNO_2]$ is the concentration of undissociated nitrous acid and $[asc]_t$ is the total concentration of ascorbic acid. The decrease in rate with increasing pH is dramatic owing to the second-order dependence of *free* nitrous acid.

A Tale of Seven Intermediates. Nitrous acid is reduced by one equivalent to yield the free-radical product NO, and ascorbic acid is oxidized by two equivalents according to the stoichiometry of equation (4). In an excess of nitrite, dehydroascorbic acid is further oxidized via ring-opening to diketogulonic acid to yield polymeric brown products (Figure 3).[15] Up to seven intermediates have been postulated in the non-complementary reaction between nitrite and ascorbate, and efforts have been made to detect radicals derived from ascorbic acid.[15] None of these intermediates have, however, been identified as free radicals, and except for diketogulonic acid no intermediates have been isolated so far. Accordingly, the identities of the reaction intermediates have to be inferred indirectly through their chemical properties. The decrease in free nitrite in a mixture of nitrite and ascorbic acid has been shown[16] to be paralleled by an increasingly high

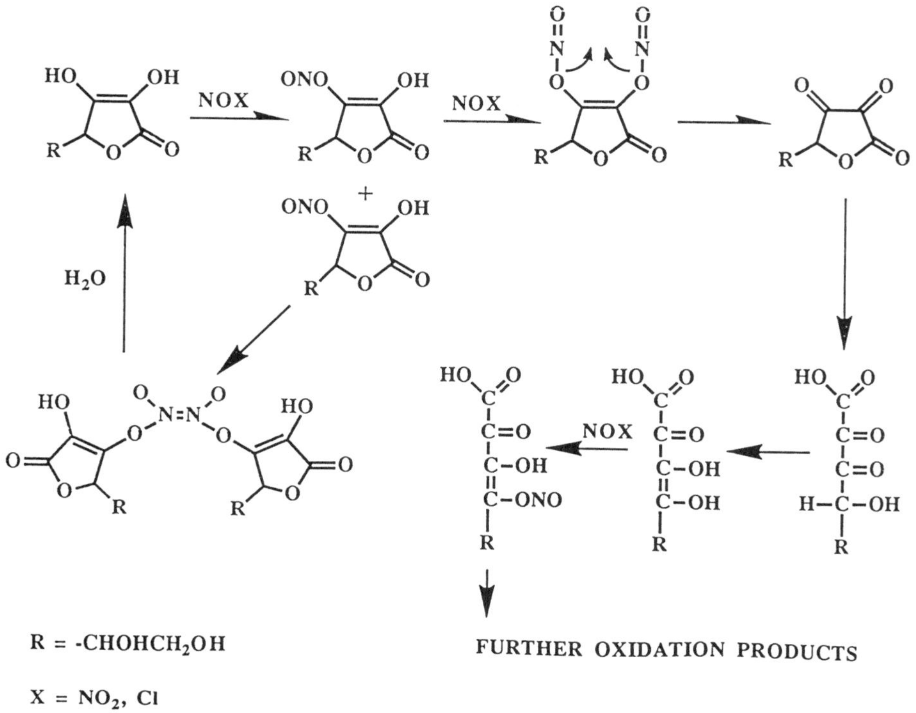

Figure 3 *The reaction between nitrite and ascorbate proceeds through a number of reaction intermediates, some of which are highly reactive as nitrosating agents. The sequence shown has been suggested by Fox* et al.[15]

nitrosating ability toward cytochrome c. At least one of the intermediates is an efficient nitrosating agent and is capable of transferring the NO radical. Aged reaction mixtures gradually lose the nitrosating ability while nitrite is regenerated. Oxygen is not necessary for regeneration of nitrite and, notably, ascorbic acid is not regenerated simultaneously with nitrite. The latter observation is important, as it provides evidence against the reaction mechanism proposed by Fox et al.,[15,17] which entails a dismutation of a dimeric form of 3-nitrosoascorbic acid during which nitrous acid anhydride, N_2O_3, is reformed together with ascorbic acid. The backward reaction was found to be bimolecular and the regeneration of the initial reactants was suggested to occur via the nitroso dimer. This reaction mechanism is, however, not in agreement with the principles of microscopic reversability, which in its most fundamental form states that the forward and reverse of a particular reaction pathway are mirror images.[18]

2,3-Dinitrosoascorbic acid is believed to have a key position in the nitrosation reaction sequence as the nitrosating agent, which generates N_2O_3 and dehydroascorbic acid, or transfers NO to other substrates.[15,19] This nitrous acid ester of an enediol could, however, not be identified with the powerful nitrosating agent detected by Izumi et al.[16] The nature of this latter reaction intermediate is still not clear, and the observation that physical agitation affects the steady-state distribution in the reaction mixture is disturbing, since it suggests the presence of a heterogeneous equilibrium involving either gaseous or solid-state components. The application of new spectroscopic techniques to these important problems should be further encouraged.

Effect of Added Salt. The presence of high chloride concentrations during meat curing may transform nitrous acid into nitrosyl chloride[8]

$$HNO_2 + H^+ + Cl^- \rightarrow NOCl + H_2O \tag{11}$$

NOCl is more reactive than N_2O_3 and less reactive than NO^+. In model systems the rate of formation of nitroso pigments increases with increasing chloride concentration, while in meat systems effects of chloride concentration are small.[20] However, nitrosyl chloride may still be of kinetic importance in the generation of other transnitrosation agents. In this context it should be noted that development of typical cured flavour is a result of the combined effect of salt and nitrite.[8,21]

Nitrosation and Transnitrosation

The fate of nitrite in meat has been investigated in curing procedures in order to minimize the formation of toxic compounds in cured meat. Utilizing nitrite labelled with the stable ^{15}N isotope,[5,7] and in model systems, added nitrite could be completely recovered and accounted for as nitrate, nitrosylmyoglobin, gaseous nitrogen compounds, and residual nitrite.[22] The formation of *N*-nitrosamines has been of special concern:

$$R^1R^2NH + HNO_2 \rightarrow R^1R^2NNO + H_2O \tag{12}$$

At the pH of relevance to meat, ascorbate reacts faster than secondary amines with the nitrosating agent N_2O_3 [*cf.* equation (9)], and this competition is the basis for the inhibition of the formation of carcinogenic *N*-nitroso compounds found in an excess of ascorbate.[23]

In non-haem proteins, tryptophyl residues have been shown to react with nitrite, forming nitroso derivatives. Nitrite-modified lysozyme was separated from free nitrite and was shown to be capable of transferring NO to metmyoglobin in the presence of ascorbate to yield nitrosylmyoglobin.[24] The thermodynamics of this example of transnitrosation have been studied using *N*-acetyl-*N*'-tryptophan as a model substance:[25]

$$N\text{-acetyltryptophan} + HNO_2 \rightleftharpoons N\text{-acetyl-}N'\text{-nitrosotryptophan} + H_2O \tag{13}$$

The reversibility of this nitrosation reaction is the result of a subtle balance between a negative enthalpy of reaction and a negative entropy of reaction ($\Delta H^{\ominus} = -54 \text{ kJ mol}^{-1}$ and $\Delta S^{\ominus} = -140 \text{ J mol}^{-1} \text{ K}^{-1}$, respectively). The relation to oxidative stability is that the protein fraction in cured meat represents a reservoir of nitroso groups available through entropy-driven reactions to regenerate NO-based antioxidants.

The Role of Cytochrome c. During respiration cytochrome *c* alternates between iron(II) and iron(III) in the electron transport chain. Cytochrome *c* in its iron(III) state reacts readily with NO, and the product formed from the two paramagnetic reactants was shown by Ehrenberg and Szczepkowski to be diamagnetic, indicating electron transfer and electron pairing:[26]

$$\text{Cyt } c(\text{Fe}^{\text{III}}) + \text{NO} \rightarrow \underset{(\text{I})}{\text{Cyt } c(\text{Fe}^{\text{II}}-\text{NO}^+)} \qquad (14)$$

The resulting ferrocytochrome *c* nitrosyl compound I appears to play a central role in nitrite 'metabolism' in meat. At the pH of meat, compound I is formed in the presence of ascorbate, most likely through the action of an unknown nitrosating agent.[16] Ascorbate is capable of reducing compound I to ferrocytochrome *c* nitrosyl compound II:

$$\text{Cyt } c(\text{Fe}^{\text{II}}-\text{NO}^+) + e^- \rightarrow \underset{(\text{II})}{\text{Cyt } c(\text{Fe}^{\text{II}}-\text{NO})} \qquad (15)$$

Compound II was found by Izumi *et al.*[27] to be less stable than compound I and was suggested to be involved in transnitrosation or formation of gaseous product through subsequent oxidation. Hydroxide has been found to release NO^+ from compound I in alkaline solution,

$$\text{Cyt } c(\text{Fe}^{\text{II}}-\text{NO}^+) + \text{OH}^- \rightarrow \text{Cyt } c(\text{Fe}^{\text{II}}) + \text{HNO}_2 \qquad (16)$$

in effect regenerating nitrite.[26] Other nucleophiles such as chloride and thiocyanate may be involved in similar reactions at the pH of meat, yielding NOCl and NOSCN respectively, as part of further transnitrosation reactions. Walters *et al.* have suggested that compound I in the presence of a reductant transfers NO to metmyoglobin.[28]

Reactions of Myoglobins in Curing

Myoglobin binds oxygen in a reversible reaction,

$$\text{MbFe}^{\text{II}} + \text{O}_2 \rightleftharpoons \text{MbFe}^{\text{II}}\text{O}_2 \qquad (17)$$

and stores oxygen and facilitates oxygen diffusion. In contrast to the physiological role of the cytochromes, oxygen binding to myoglobin does not involve a formal change in the oxidation state of the haem iron. The change in co-ordination number of Fe^{II} during binding of oxygen, however, changes the spin state of this d^6 chromophore from high to low spin as evidenced by the colour change from purple to cherry red. Oxidation of either oxymyoglobin or myoglobin yields the brown metmyoglobin,[29]

$$MbFe^{II} + H_2O \rightarrow MbFe^{III}H_2O + e^- \qquad E^{\ominus}_{ox} = -0.06 \text{ V} \qquad (18)$$

and oxidants such as nitrite $[E^+_{red} = -0.46 \text{ V}$, equation (2)] are capable of oxidizing the iron(II) forms of myoglobin:

$$MbFe^{II} + NO_2^- + H_2O \rightarrow MbFe^{III}H_2O + NO + 2OH^- \qquad (19)$$

Reaction (19) and a similar reaction for oxymyoglobin are the first steps in a complex sequence of reactions leading to the low-spin iron(II) complex nitrosylmyoglobin and its heat-denaturated form nitrosyl myochrome. The latter provides cooked cured meat with its characteristic pink colour.

Formation of Nitrosylmyoglobin. The major reaction path for the formation of nitrosylmyoglobin involves the transient formation of metmyoglobin, and oxymyoglobin and metmyoglobin are directly involved in the reduction of nitrite to nitric oxide. The ligand substitution

$$MbFe^{II}O_2 + NO \rightleftharpoons MbFe^{II}NO + O_2 \qquad (20)$$

for which an equilibrium constant K of 10^5 may be estimated from available thermodynamic data,[30,31] seems to be of little importance, since the competing reaction

$$MbFe^{II}O_2 + NO + H_2O \rightarrow MbFe^{III}H_2O + NO_3^- \qquad (21)$$

has been found to be very fast ($k_2 = 3.7 \times 10^7 \text{ l mol}^{-1}\text{s}^{-1}$ at 25 °C).[32] However, in meat where oxygen is exhausted, the fast combination reaction[33]

$$MbFe^{II} + NO \rightarrow MbFe^{II}NO \qquad (22)$$

contributes significantly to the formation of nitrosylmyoglobin (Figure 4).
Both nitrite and nitric oxide may react with metmyoglobin[34,35]

$$MbFe^{III}H_2O + NO_2^- \rightarrow MbFe^{III}NO_2 + H_2O \qquad (23)$$

$$MbFe^{III}H_2O + NO \rightarrow MbFe^{III}NO + H_2O \qquad (24)$$

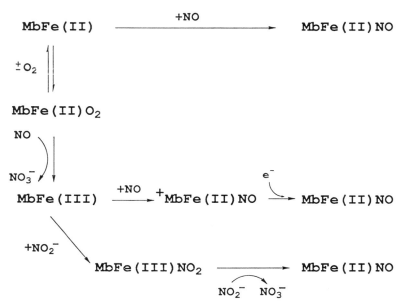

Figure 4 *The formation of nitrosylmyoglobin considered as the result of three parallel reaction paths, which are all thermodynamically favourable. The reduction of MbFe^{III}NO by NO is not energetically favoured*

For a series of model porphyrinate compounds, the Fe^{III}NO$_2$/NO$_2^-$ system has recently been shown to be unstable as it undergoes single atom oxygen transfer:[36]

$$MbFe^{III}NO_2 + NO_2^- \rightarrow MbFe^{II}NO + NO_3^- \qquad (25)$$

For myoglobin this reaction will be thermodynamically favoured, provided that the stability constant for MbFe^{III}NO$_2$ is less than 10^{19} l mol^{-1}, as calculated from available thermodynamic data.[8,29,30] This is a reasonable assumption, and the reaction in which uncoordinated nitrite is a reductant may be a source of nitrate in meat. In meat curing other reductants will compete with nitrite in the reduction of MbFe^{III}NO$_2$, and the pathway to MbFe^{II}NO via MbFe^{III}NO$_2$ was detected kinetically for the reductants cysteine and NADH by Fox and Ackerman in their classical study of nitrosylmyoglobin formation.[35]

The intermediate MbFe^{III}NO is often invoked in kinetic schemes for nitrosylmyoglobin formation.[34,35,37,38] A reaction sequence similar to that for cytochrome *c* [*cf.* equations (14)–(16)] has been suggested by Neto *et al.*[36] in excess NO:

$$MbFe^{III}NO \rightleftharpoons MbFe^{II} + NO^+ \qquad (26)$$

$$NO^+ + OH^- \rightarrow HNO_2 \qquad (27)$$

$$MbFe^{II} + NO \rightarrow MbFe^{II}NO \tag{28}$$

However, calculations based on thermodynamic data[8,29,30] show that the net reaction in acidic solution

$$MbFe^{III}NO + NO + H_2O \rightarrow MbFe^{II}NO + HNO_2 + H^+ \tag{29}$$

is only thermodynamically feasible provided that $MbFe^{II}$ binds NO 10^{29} times more strongly than $MbFe^{III}$ does. The stability constant for binding of NO to $MbFe^{II}$ is $K = 10^{11}$ l mol^{-1}, and thus reaction (29) cannot have any importance in meat, even at high pH. However, NO does reduce iron(III) forms of haem protein,[39] and it was suggested by Killday et al.[40] that $MbFe^{III}NO$ is more adequately described as the imidazole-centred protein radical of Figure 5. This radical undergoes autoreduction yielding $MbFe^{II}NO$, and, in the absence of exogeneous reductants, reducing groups within the protein are apparently capable of donating electrons to the imidazole radical. Very recently quantum mechanical calculations have been reported, which show the existence of such low-lying imidazolate π-cation radical states in compound I of peroxidase models.[41] Similar effects of transient deprotonization of the imidazole moiety of the proximal histidine may stabilize the nitrosylmyoglobin radical and facilitate its reduction in meat by electron transfer from added reductants.

In conclusion, the formation of nitrosylmyoglobin is best considered to be the result of three parallel reaction pathways as shown in Figure 4.

3 Cured Meat Pigments

Methods are available for the synthesis of $MbFe^{II}NO$ of high purity.[42] In solution $MbFe^{II}NO$ is deep red, whereas cooked cured meat is pink. The nitrosylmyoglobin formation in meat is further complicated by heating, which as well as causing protein denaturation may allow further nitrosation of the pigment.[43-45] The colour formation seems to increase sharply at a

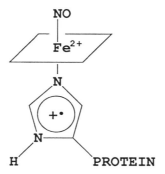

Figure 5 *Imidazolate π-cation radical postulated as intermediates in NO reduction of metmyoglobin*

temperature of 70 °C and correlates with denaturation of certain sarcoplasmic proteins.[46] However, the widely used acetone extraction method for cooked cured meat pigments originally developed by Hornsey[47] extracts both native MbFe^{II}NO and the heat-denaturated pigment.[48] Thus acetone extraction cannot be used to fractionate MbFe^{II}NO and cooked cured-meat pigments.

Structure of Nitrosylmyoglobin

In nitrosylmyoglobin, nitric oxide co-ordinates to the sixth position in the octahedron around iron(II) (Figure 5). NO has a strong *trans*-labilizing effect, and it was suggested in 1962 by Tarladgis[43] that a second nitric oxide replaced the proximal histidine as ligand in nitrosylmyoglobin during heating (Figure 6). Equine metmyoglobin contains 215 nitrogen atoms, and labelling of the nitrosyl ligand with the stable isotope ^{15}N is thus expected to increase the ^{15}N content from the natural level of 0.37% to 0.83% for nitrosylmyoglobin and to 1.29% for dinitrosylmyoglobin. Labelling experiments performed by Lee and Cassens[44] and by Renerre and Rougie[45] confirmed these predictions and are widely quoted as evidence for the co-ordination of two nitric oxide molecules in the heat-denaturated nitrosyl myochrome. However, ESR studies by Bonett *et al.* in 1980 showed that the nitrosyl pigment in uncooked bacon contains one nitric oxide ligand in both pentaco-ordinate and hexaco-ordinate iron(II) complexes.[49] On gentle heating of bacon, the conversion of the hexaco-ordinate iron(II) complex to the pentaco-ordinate iron(II) complex could be directly monitored by ESR. By comparison with the synthetic pigment in co-ordinating and non-co-ordinating solvents, this conversion was shown to be the result of a heat-induced breaking of an imidazole–iron(II) bond. However, this process was not followed by the co-ordination of a second nitric oxide molecule. Bonett *et al.* concluded that the pigment of cooked cured meat is a pentaco-ordinate nitrosylprotohaem trapped in a matrix of denaturated globin, and that reactions of nitrite with amino acid side chains may contribute to denaturation.[49] Killday *et al.* confirmed these results by NMR and mass spectrometry,[40] and these authors suggest that a protein radical reacts with a second nitrite, resulting in the stoichiometry found in the original labelling experiments. However, on prolonged heating, it is possible to increase the labelling to a significantly higher degree than the 1.29% corresponding to the 2:1 stoichiometry,[50] suggesting that further nitrosation reactions of the globin are of importance.

4 Oxidative Processes in Cured Meat

The development of warmed-over flavour in cooked cured meat is very slow compared with non-cured products.[51-55] Warmed-over flavours are the result of lipid oxidation, and lipid oxidation is inhibited in cured meat products. The exact nature of the antioxidants and the mechanism of their

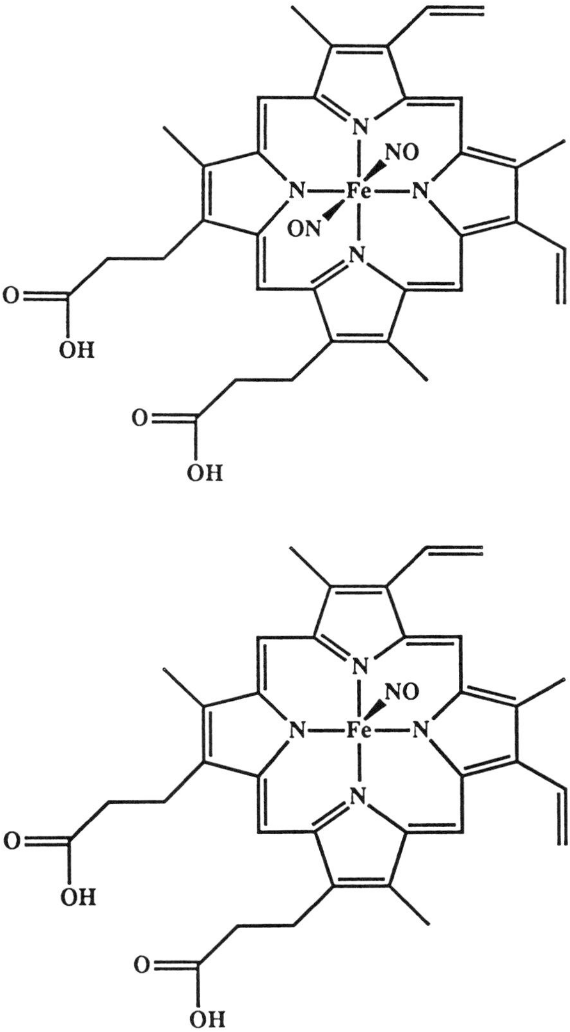

Figure 6 *Pentaco-ordination versus hexaco-ordination in cured meat pigment. A long-standing discussion*

action is, however, still a matter of discussion. On the basis of studies with meat pigment extracts and ground beef, in which lipid oxidation was assessed by the TBA method, Igene *et al.* concluded that at least three mechanisms are responsible for the antioxidative effect of curing: (i) nitrite prevents release of Fe^{II} from haem pigments; (ii) nitrite interacts directly with liberated iron(II) ions; and (iii) nitrite stabilizes the unsaturated lipids within the membranes. The prevention of release of Fe^{II} from the haem pigments during cooking was suggested to be the most important of these antioxidative mechanisms.[54] Kanner *et al.* found evidence for an antioxidative effect of S-nitrosocysteine, a compound generated during curing.[56,57] An antioxidative effect of the cured meat pigments is also well documented, and has been explained by chain breaking in the radical processes associated with lipid oxidation.[52,53]

Initiation and Propagation of Lipid Oxidation

The often addressed question as to whether nitrite *per se* or a product from reaction of nitrite with reactants present in meat is the actual antioxidant in cured meat has proved to be difficult to answer.[51,53,54] However, from a growing body of experimental evidence, it appears to be safe to conclude that not one single mechanism but several mechanisms involving compounds derived from nitrite are responsible for the improved oxidative stability of cured products. Moreover, it is important to note that nitrite-derived compounds affect both the initiation and propagation steps of lipid oxidation. The different pools of nitrite-modified compounds, *i.e.* nitrosated cytochromes and nitrosated tryptophyl moieties on the proteins, the unidentified nitrosating derivatives of ascorbic acid, and the nitrosyl pigments, seem to be rapidly interchanging with each other and with residual nitrite and with N_2O_3 (Figure 7). Exhaustion of one group of compounds due to oxidation or participation in antioxidative reactions will, under optimal conditions, be followed by transnitrosation reactions and reformation of the antioxidant. Evidence in support of this proposal is that, in addition to the results from the study of nitrosation of myoglobins by nitrosated lysozyme,[24] the *reversible* fading of the red colour of vacuum-packed sliced ham during storage after exposure to light, indicating reformation of the nitrosylmyoglobin initially photo-oxidized,[58,59] and the low activation barrier for nitric oxide exchange in nitrosylmyoglobin:

$$MbFe^{II\,15}NO + NO \rightleftharpoons MbFe^{II}NO + {}^{15}NO \qquad (30)$$

Reaction (30) is a radical exchange, for which, by isotopic labelling, the enthalpy of activation has been found to have a value $\Delta H^{\neq}$ of 47 kJ mol^{-1}.[60] The low activation barrier indicates a low-temperature dependence for the rate of this particular transnitrosation reaction. If this result is applicable to other transnitrosation reactions it suggests that restoration of the active antioxidants in cured meat operates during storage at low temperatures.

The ensuing discussion of the protection of lipids against oxidation will focus on the initial stages of oxidation, *i.e* the initiation by metal-ion catalysis, and chain propagation.

Protection against Metal Catalysis. Metal ions and in particular iron(II) are known to be active in the generation of free radicals, as in the Fenton reaction:

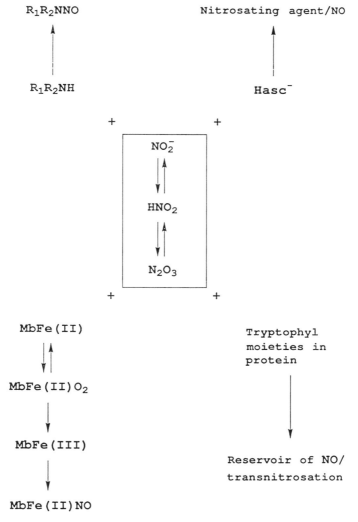

Figure 7 *Different pools of nitrite-modified compounds in meat seem to be in rapid relationship with each other through transnitrosation reactions in which the nitric oxide radical is transferred*

$$Fe^{2+} + H_2O_2 \rightarrow Fe^{3+} + \cdot OH + OH^- \quad (31)$$

The hydroxy radical and corresponding alkoxy radicals formed by reduction of hydroperoxides are all efficient initiators of lipid oxidation, and the antioxidative properties of nitrite have been ascribed to its function as a metal-ion chelator.[51] However, the formation of nitrite iron(II) complexes has not been documented except for mixed-ligand complexes with strong π-accepting ligands such as the pentacyanoiron(II) ion:[61]

$$\text{Fe(CN)}_5\text{H}_2\text{O}^{3-} + \text{NO}_2^- \rightleftharpoons \text{Fe(CN)}_5\text{NO}_2^{4-} + \text{H}_2\text{O} \qquad (32)$$

$\text{Fe(CN)}_5\text{NO}_2^{4-}$, in which nitrite is co-ordinated with a stability constant of 3×10^3 l mol^{-1}, is formally designated as an iron(II) complex, but has, however, an effective charge on the iron significantly greater than two owing to π-back-bonding. Under conditions relevant to cured meat, little evidence has been presented for a similar binding of nitrite to 'inorganic' iron(II). NO is a more likely ligand:

$$\text{Fe(H}_2\text{O)}_6^{2+} + \text{NO} \rightleftharpoons \text{Fe(H}_2\text{O)}_5\text{NO}^{2+} + \text{H}_2\text{O} \qquad (33)$$

The so-called 'ferronitroso-ion' has an EDTA analogue,

$$\text{Fe}^{II}(\text{EDTA}) + \text{NO} \rightleftharpoons \text{Fe}^{II}(\text{EDTA})\text{NO} \qquad (34)$$

and the well studied $\text{Fe}^{II}(\text{EDTA})$ complexes may serve as more realistic models than $\text{Fe(CN)}_5\text{H}_2\text{O}^{3-}$ for the simple iron(II) species. NO binds strongly to iron(II), and reaction (34) has an equilibrium constant of 3.5×10^6 l mol^{-1} at 39 °C.[62] When NO_2^- or HNO_2 is allowed to react with $\text{Fe}^{II}(\text{EDTA})$, the product is $\text{Fe}^{II}(\text{EDTA})\text{NO}$ and an iron(III) species,[63] providing further evidence that nitrite is not, as a metal chelator, capable of preventing metal-ion catalysis of the initiation of lipid oxidation.

The reactions of nitrite with haem pigments may, however, indirectly halt the initiation of lipid oxidation. Metmyoglobin appears to be a more effective catalyst for lipid oxidation than oxymyoglobin and 'inorganic' iron.[64] Nitrite in combination with ascorbic acid prevents the formation and accumulation of metmyoglobin in cooked meat as shown in Figure 4. In effect it protects the lipids against metmyoglobin-catalysed peroxidation. Reaction of metmyoglobin with hydrogen peroxide produces an iron(IV) species,

$$\text{MbFe}^{III} + \text{H}_2\text{O}_2 \rightarrow \cdot\text{MbFe}^{IV}{=}\text{O} + \text{H}_2\text{O} \qquad (35)$$

and the ferrylmyoglobin radical ($\cdot\text{MbFe}^{IV}{=}\text{O}$) and ferrylmyoglobin ($\text{MbFe}^{IV}{=}\text{O}$) are both strong oxidants and have been suggested as initiators of lipid oxidation.[65] Notably, nitrosylmyoglobin yields metmyoglobin and *not* a ferryl species on reaction with hydrogen peroxide:[66,67]

$$\text{MbFe}^{II}\text{NO} + \text{H}_2\text{O}_2 \rightarrow \text{MbFe}^{III} + \text{other products} \qquad (36)$$

The metmyoglobin can be recycled by reduction and transnitrosation to nitrosylmyoglobin. One important aspect of nitrite curing could be the prevention of formation of the higher oxidation states of iron in myoglobin; further kinetic experiments are needed.

Chain Breaking. The reactions involved in the propagation of lipid oxidation,

$$R \cdot + O_2 \rightarrow ROO \cdot \qquad (37)$$

$$ROO \cdot + RH \rightarrow ROOH + R \cdot \qquad (38)$$

$$RO \cdot + RH \rightarrow ROH + R \cdot \qquad (39)$$

differ significantly in rate, and since reaction (37) is almost diffusion controlled, the peroxy radicals (ROO·) are the longer lived of the free radical intermediates. Reactions yielding non-radical products require reaction of the involved radicals with other free radicals, and Morrissey and Tichivangana[53] found evidence that nitrosylmyoglobin can act as a chain terminator, supporting the suggestions of Kanner *et al.*:[52]

$$ROO \cdot + MbFe^{II}NO \rightarrow \text{non-radical products} \qquad (40)$$

Although the nature of the products is uncertain, ROO· is most likely a one-electron oxidizer:

$$ROO \cdot + 3MbFe^{II}NO + 2H_2O \rightarrow ROH + 3MbFe^{II} + 3NO_2^- + 3H^+$$

$$(41)$$

The proposed reaction products are myoglobin and nitrite, both reactants for regeneration of the antioxidant nitrosylmyoglobin, *cf.* Figure 7.

Nitric oxide has recently been found by Kanner *et al.*[67] to prevent hydroxylation of benzoate by hydroxy radicals generated by ultraviolet irradiation of H_2O_2 or by the Fenton reaction [equation (31)]. MbFe^{II}NO was also found to inhibit membrane lipid oxidation by hydrogen peroxide, and it was concluded that nitric oxide modulates the reactivity of both haem and non-haem iron and that nitric oxide is a scavenger of free radicals during peroxidation of lipids. The lifetime of nitric oxide is short in the presence of dioxygen and other oxidants. However, nitric oxide bound in nitrosylmyoglobin represents a free-radical buffer, stable against oxidation by dioxygen, which can be easily mobilized, *cf.* equation (30).

Pigment Oxidation

The function of nitrosylmyoglobin as an antioxidant and as a buffer for nitric oxide radicals capable of terminating chain reactions in lipid oxidation makes protection of the nitrosyl pigments against oxidation of vital importance for the protection of cured meat products against oxidative deterioration. The colour stability of cured meat products has been the subject of numerous studies, mainly to ensure consumer acceptability of products such as ham or bologna and other sausages,[62-72] and improved packaging procedures for such products have been devised.[58,59] It has been recognized for many years that the rate of oxidation of nitrosylmyoglobin and of the analogous haemoglobin compound decreases with decreasing oxygen partial pressure and is strongly accelerated by light.[73-77] The detailed reaction mechanisms of both thermally activated oxidation and photo-oxidation are not, however, fully understood.

Thermal Oxidation. Walsh and Rose[75] established that the transformation of nitrosylmyoglobin into metmyoglobin in air-saturated solutions follows first-order kinetics and suggested the following stoichiometry for the autoxidation:

$$MbFe^{II}NO + \tfrac{1}{2}O_2 \rightarrow MbFe^{III} + NO_2^- \qquad (42)$$

and reported a rate constant of 3×10^{-4} s^{-1} at 26 °C and an energy of activation of 92 kJ mol^{-1}. Whilst the rate constant has been confirmed, the stoichiometry has, by the use of combined spectroscopic and electrochemical techniques, now been shown to be:[50]

$$MbFe^{II}NO + O_2 \rightarrow MbFe^{III} + NO_3^- \qquad (43)$$

The mechanism of this unusual four-electron transfer process and the effect of protein denaturation on the mechanism are the subject of current research. However, heat denaturation of nitrosylmyoglobin increases the activation barrier, and the improved colour stability of cooked cured meat products compared with non-cooked products is believed to be an enthalpy rather than an entropy effect linked to loss of myoglobin organization.[50] The thermal oxidation is slow with a half-life of *ca.* 10 h at 10 °C in air-saturated solution,[50] especially when compared with the 'reverse' oxidation reaction,[37] which occurs when MbFe^{II}O$_2$ is allowed to react with NO, *cf.* equation (21). The significant difference in rate relates to the role of π-back-bonding in nitrosylmyoglobin:

$$MbFe^{II}(NO) \leftrightarrow MbFe^{III}(NO^-) \qquad (44)$$

In nitrosylmyoglobin, the FeNO group is bent,[78] and the net donation of electron density from Fe^{II} to NO and the resulting decrease in radical character may well contribute to the inertness of co-ordinated NO against oxidation although it is still mobile for exchange reactions [compare the high energy of activation for reaction (42) with the low energy of activation for reaction (30)]. However, these observations are not in agreement with the view that thermal oxidation of nitrosylmyoglobin is initiated by a rate-determining ligand dissociation:[6,30]

$$MbFe^{II}NO \rightarrow MbFe^{II} + NO \qquad (45)$$

Further studies must focus on the role of the amino acids in the haem cavity in radical exchange and electron transfer.

Photo-oxidation. Nitrosylmyoglobin is very sensitive to light in the presence of oxygen.[74,76] Light absorption of haem proteins results in ligand dissociation,[79] and for nitrosylmyoglobin the chemical process following electronic excitation is

$$MbFe^{II}NO \xrightarrow{h\nu} MbFe^{II} + NO \qquad (46)$$

Photo-oxidation is expected to be of a more dissociative nature than thermal oxidation, *i.e.* a free nitric oxide radical is oxidized by oxygen in photo-oxidation. Experimental data are, however, not available for a more detailed understanding of the reaction steps following ligand dissociation. The wavelength dependence is less significant for photo-oxidation of nitrosylmyoglobin[74,77] than for oxymyoglobin,[80] and the wavelength dependence of the photo-oxidation quantum yields and the activation parameters for the thermal oxidation enabled Figure 8 to be constructed.[77] For

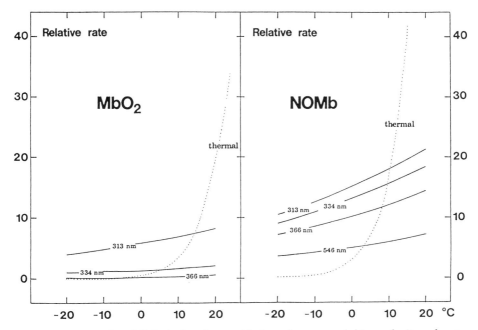

Figure 8 *Thermal and light-induced autoxidation of oxymyoglobin and nitrosylmyoglobin under conditions relevant to meat products during retail display. The rate is determined for air-saturated aqueous solution with pH = 5.5, illuminated by monochromatic light of wavelength of high constant intensity in fluorescent light*
(Reproduced from Ref. 77 with the permission of the Danish Meat Research Institute)

light intensities and temperatures relevant to meat, light-induced oxidation of nitrosylmyoglobin is at least an order of magnitude greater than thermal oxidation at the surface of the product. This is in contrast to oxymyoglobin where photo-oxidation, compared with thermal oxidation, is only significant below a certain temperature. UV-impermeable films thus protect frozen beef during storage in light, but yield little protection for ham[77] where oxygen has to be excluded prior to exposure to light for colour stability.[58,59]

5 Conclusion

The improved oxidative stability of cured meat products is the result of reduction of nitrite to yield the free radical nitric oxide, which is stored in several 'chemical reservoirs' in the product. The cured meat pigment nitrosylmyoglobin is such a free-radical buffer, which may provide nitric oxide to terminate free-radical processes during lipid oxidation. The formation of nitrosylmyoglobin also prevents metmyoglobin-catalysed and ferryl-catalysed lipid oxidation. The colours of cured meat products provide an indication of the oxidative status of the product and it is essential to protect cured meat products against light-induced pigment oxidation.

6 Acknowledgement

The continuous support of the Danish Technical Research Council and the Danish Agricultural and Veterinary Research Council is gratefully acknowledged. The research is part of the FØTEK programme sponsored by the Danish Ministry of Education and Research. Claus Jensen is thanked for his help with the literature search, Astrid Woollen for her help with the preparation of the manuscript, and Kevin Jørgensen for his help with the preparation of the Figures.

7 References

1. R.G. Cassens, 'Nitrite-Cured Meat: A Food Safety Issue in Perspective', Food and Nutrition Press Inc., Trumbull, CT, 1990.
2. R.M.J. Palmer, A.G. Ferrige, and S. Moncada, *Nature (London)*, 1987, **327**, 524.
3. R.M.J. Palmer, D.S. Ashton, and S. Moncada, *Nature (London)*, 1988, **333**, 664.
4. A.R. Butler, *Chem. Br.*, 1990, **26**, 419.
5. R.G. Cassens, J.G. Sebranek, G. Kubberod, and G. Woolford, *Food Prod. Develop.*, 1974, **8**, 50.
6. G.G. Giddings, *CRC Crit. Rev. Food Sci. Nutr.*, 1977, **9**, 81.
7. R.G. Cassens, M.L. Greaser, T. Ito, and M. Lee, *Food Technol.*, 1979, **33**, 46.
8. J.G. Sebranek and J.B. Fox, *J. Sci. Food Agric.*, 1985, **36**, 1169.

9. L.G. Sillén and A.E. Martell, 'Stability Constants of Metal-Ion Complexes', The Chemical Society, London, 1964.
10. A.J. Bard, R. Parsons, and J. Jordan, 'Standard Potentials in Aqueous Solution', Marcel Dekker Inc., New York, 1985.
11. H. Dahn, L. Loewe, E. Lüscher, and R. Menassé, *Helv. Chim. Acta*, 1960, **43**, 285.
12. H. Dahn, L. Loewe, and C.A. Bunton, *Helv. Chim. Acta*, 1960, **43**, 320.
13. T.A. Turney and G.A. Wright, *J. Chem. Soc.*, 1958, 2415.
14. H. Dahn and L. Loewe, *Helv. Chim. Acta*, 1960, **43**, 310.
15. J.B. Fox, R.N. Fiddler, and A.E. Wasserman, *J. Food Protection*, 1981, **44**, 28.
16. K. Izumi, R.G. Cassens, and M.L. Greaser, *Meat Sci.*, 1989, **26**, 141.
17. J.B. Fox and J.S. Thomson, *Biochemistry*, 1963, **2**, 465.
18. J.H. Espenson, 'Chemical Kinetics and Reaction Mechanisms', McGraw-Hill, New York, 1981.
19. R. Izumi, R.G. Cassens, and M.L. Greaser, *J. Food Protection*, 1985, **48**, 346.
20. J.G. Sebranek, J. Park, and J.B. Fox, Proceedings of the 32nd European Meeting of Meat Research Workers, 1986, Vol.2, p.319.
21. D.A. Froehlich, E.A. Gullett, and W.R. Usborne, *J. Food Sci.*, 1983, **48**, 152.
22. M. Fujimaki, M. Emi, and A. Okitani, *Agric. Biol. Chem.*, 1975, **39**, 371.
23. S.S. Mirvish, L. Wallcave, M. Eagen, and S. Shubik, *Science*, 1972, **111**, 65.
24. T. Ito, R.G. Cassens, M.L. Greaser, M. Lee, and K. Izumi, *J. Food Sci.*, 1983, **48**, 1204.
25. P.O. Mellet, P.R. Noel, and R. Goutefongea, *J. Agric. Food Chem.*, 1986, **34**, 892.
26. A. Ehrenberg and T.W. Szczepkowski, *Acta Chem. Scand.*, 1960, **14**, 1684.
27. K. Izumi, R.G. Cassens, and M.L. Greaser, *J. Food Sci.*, 1982, **47**, 1419.
28. C.L. Walters, R.J. Casseden, and A.M. Taylor, *Biochem. Biophys. Acta*, 1967, **143**, 310.
29. M. Brunori, G. Saggese, G.C. Rotilio, E. Antinini, and J. Wyman, *Biochemistry*, 1971, **19**, 1604.
30. E.G. Moore and Q.H. Gibson, *J. Biol. Chem.*, 1976, **256**, 2788.
31. G.D. Armstrong and A.G. Sykes, *Inorg. Chem.*, 1986, **25**, 3135.
32. M.P. Doyle and J.W. Hoekstra, *J. Inorg. Biochem.*, 1981, **14**, 351.
33. R.J. Morris and Q.H. Gibson, *J. Biol. Chem.*, 1980, **255**, 8050.
34. J.B. Fox, *J. Agric. Food Chem.*, 1966, **14**, 207.
35. J.B. Fox and S.A. Ackerman, *J. Food Sci.*, 1968, **33**, 364.
36. L.M. Neto, O.R. Nascimento, M. Tabak, and I. Caracelli, *Biochim. Biophys. Acta*, 1988, **956**, 189.
37. B.M. Watts and B.T. Lehman, *Food Res.*, 1952, **17**, 100.
38. J.F. Reith and M. Szakaly, *J. Food Sci.*, 1967, **32**, 188.
39. J.C.W. Chien, *J. Am. Chem. Soc.*, 1969, **91**, 2166.
40. K.B. Killday, M.S. Tempesta, M.E. Bailey, and C.J. Metral *J. Agric. Food Chem.*, 1968, **36**, 909.
41. G.H. Loew, F.U. Axe, J.R. Collins, and P. Du, *Inorg. Chem.*, 1991, **30**, 2291.
42. A.R. Kamarei and M. Karel, *J. Food Sci.*, 1982, **47**, 682.
43. B.G. Tarladgis, *J. Sci. Food Agric.*, 1962, **13**, 485.
44. S.H. Lee and R.G. Cassens, *J. Food Sci.*, 1976, **41**, 969.
45. M. Rennerre and P. Rougie, *Ann. Technol. Agric.*, 1979, **28**, 423.
46. T. Okayama, M. Fujii, and M. Yamanoue, *Meat Sci.*, 1991, **30**, 49.

47. H.C. Hornsey, *J. Sci. Food Agric.*, 1956, **7**, 534.
48. R. Sakata and Y. Nagata, *Jpn. J. Zootech. Sci.*, 1983, **54**, 667.
49. R. Bonett, S. Chardra, A.A. Charalambides, K.D. Sales, and P.A. Scourides, *J. Chem. Soc. Perkin 1*, 1980, 1706.
50. H.J. Andersen and L.H. Skibsted (unpublished data).
51. B. MacDonald, J.I. Gray, and L.N. Gibbins, *J. Food Sci.*, 1980, **45**, 893.
52. J. Kanner, I. Ben-Gera, and S. Berman, *Lipids*, 1980, **15**, 944.
53. P.A. Morrissey and J.Z. Tichivangana, *Meat Sci.*, 1985, **14**, 175.
54. J.O. Igene, K. Yamauchi, A.M. Pearson, J.I. Gray, and S.D. Aust, *Food Chem.*, 1985, **18**, 1.
55. J. Kanner, S. Harel, J. Shaglovich, and S. Berman, *J. Agric. Food Chem.*, 1984, **32**, 512.
56. J. Kanner, *J. Am. Oil Chem. Soc.*, 1979, **56**, 74.
57. J. Kanner and B. Juven, *J. Food Sci.*, 1980, **45**, 1105.
58. H.J. Andersen, G. Bertelsen, L. Boegh-Soerensen, C.K. Shek, and L.H. Skibsted, *Meat Sci.*, 1988, **22**, 283.
59. H.J. Andersen, G. Bertelsen, A. Ohlen, and L.H. Skibsted, *Meat Sci.*, 1990, **28**, 77.
60. H.J. Andersen, H.S. Johansen, C.K. Shek, and L.H. Skibsted, *Z. Lebensm.-Unters. Forsch.*, 1990, **191**, 293.
61. J.H. Swinehart and P.A. Rock, *Inorg. Chem.*, 1966, **5**, 573.
62. Y. Hishinuma, R. Kaji, H. Akimoto, F. Nakajima, T. Mori, T. Kamo, Y. Arikawa, and S. Nozawa, *Bull. Chem. Soc. Jpn.*, 1979, **52**, 2863.
63. V. Zang and R. van Eldik, *Inorg. Chem.*, 1990, **29**, 4462.
64. D. Ledward, *Food Sci. Technol. Today*, 1987, **1**, 153.
65. J. Kanner, J.B. German, and J.F. Kinsella, *CRC Crit. Rev. Food Sci. Nutr.*, 1987, **25**, 317.
66. L. Grawet, H.J. Andersen, and L.H. Skibsted (unpublished data).
67. J. Kanner, S. Harel, and R. Granit, *Arch. Biochem. Biophys.*, 1991, **289**, 130.
68. H.-S. Lin and J.G. Sebranek, *J. Food Sci.*, 1979, **44**, 1451.
69. H.-S. Lin, J.G. Sebranek, D.E. Galloway, and K.D. Lind, *J. Food Sci.*, 1980, **45**, 115.
70. J.C. Acton, L.B. Ferguson, and R.L. Dick, *Poult. Sci.*, 1986, **65**, 1124.
71. P.G. Klettner, G. Ott, and H. Pöllein, *Mitteilungsbl. Bundesanst. Fleischforsch.*, 1986, **92**, 6956.
72. A.J. Iversen, Proceedings of the European Food Packaging Conference, Vienna, 1984, p.201.
73. W.M. Urbain and L.B. Jensen, *Food Res.*, 1940, **5**, 593.
74. R.F. Kampschmidt, *J. Agric. Food Chem.*, 1955, **3**, 510.
75. K.A. Walsh and D. Rose, *J. Agric. Food Chem.*, 1956, **4**, 352.
76. M.E. Bailey, R.W. Frame, and H.D. Naumann, *J. Agric. Food Chem.*, 1964, **12**, 89.
77. H.J. Andersen, G. Bertelsen, and L.H. Skibsted, Proceedings of the 35th International Congress on Meat Science and Technology, Copenhagen, 1989, p.454.
78. J.C. Maxwell and W.S. Caughey, *Biochemistry*, 1976, **15**, 388.
79. M. Brunori, G.M. Giacometti, E. Antonini, and J. Wyman, *Proc. Natl. Acad. Sci. USA*, 1973, **70**, 3141.
80. G. Bertelsen and L.H. Skibsted, *Meat Sci.*, 1987, **19**, 243.

High-intensity Ultrasonics

R.T. Roberts

LEATHERHEAD FOOD RESEARCH ASSOCIATION, RANDALLS ROAD,
LEATHERHEAD, SURREY KT22 7RY, UK

1 Introduction

The use of low-intensity ultrasonic waves in level measurements, flow detection, and medical imaging is well known. Here, a pulse of ultrasound is used to 'probe' the sample under investigation; comparison of the pulse shape before and after transmission and a measurement of the transit time in the sample can provide information on many physical parameters. For these applications, the intensity of the ultrasonic burst is not sufficient to affect the material, and no chemical or physical changes are involved.

At higher intensities, ultrasound can provoke a number of changes within the propagating medium - mainly through the phenomenon of cavitation. Its ability to enhance chemical reactions ('sonochemistry') and its physical effects ('sonoprocessing') have been known for some time, and manufacturers of ultrasound equipment are now beginning to consider the problems of employing the technique on a large scale. It is therefore an appropriate time to consider the use of high-intensity ultrasound as a processing aid in the food industry.

2 Cavitation - Making Holes in Liquids

As a sound wave passes through a liquid, molecules of liquid in the wave path suffer alternate periods of compression (positive acoustic pressure) and rarefaction (negative acoustic pressure). If the intensity of the wave is large enough, the negative pressure produced during the rarefaction half of the cycle can be sufficient to cause the intermolecular separation to exceed a critical value. In this case the liquid is literally pulled apart, and cavities or voids are formed.

The acoustical pressure required to cause cavitation has been calculated to be of the order of 10 000 atm, compared with a figure of 10^{-5} atm for normal sound levels. If allowances are made for evaporation of the liquid vapour into the cavity, the excess (negative) pressure needed to cause cavitation fails to 1 000 atm,[1] and in practice cavitation can be induced at much lower acoustic pressures. This is because minute bubbles of entrained gas provide weak points in the liquid structure at which cavitation can occur. Suspended particles can also provide suitable sites for cavitation. Lorimer and Mason[1] suggest that this is due to gas nuclei present in crevices on the particle surface, and they describe the mechanism by which cavitation is induced.

Ultrasonic irradiation can produce two types of cavitation; stable and transient.[1] Stable cavities exist for many acoustic cycles and may oscillate about a mean size. Transient cavities exist for less than one cycle and then collapse. Very high local temperatures, pressures, and electrical potentials are produced during transient cavity collapse, and it is believed that these extreme conditions are responsible for the majority of sonochemical effects,[2,3] although Lorimer and Mason suggest[1] that stable cavities also play an important role in sonochemistry. The following description of stable and transient cavitation is from Lorimer and Mason.[1]

Stable Cavitation

Stable cavities are produced at lower intensities (1–3 W cm^{-2}) than transients and contain gas and vapour. They exist for many acoustic cycles and their comparatively long lifetime is sufficient to allow gas diffusion into and out of the cavity. During their lifetime the cavities may grow by a process known as rectified diffusion occurring as a result of different gas diffusion rates during compression and rarefaction. In the compression half of the cycle (high, positive excess pressure) the liquid is unsaturated, and gas diffuses out of the cavity into the liquid. During rarefaction (high, negative excess pressure) the liquid medium surrounding the cavity is supersaturated with vapour and gas diffuses out of the liquid into the cavity. Since the surface area of the cavity is larger during the negative half cycle there is a net flow of gas into the cavity.

Cavities enlarged in this way may coalesce with similar stable cavities until they reach a size where their buoyancy is such that they float to the surface and are expelled. Alternatively, they may collapse, but the cushioning effect of the vapour in the cavity means that this process does not produce such extremes of conditions as collapse of a transient cavity.

Maximum power transfer from the medium to the bubble will occur when the frequency of the applied ultrasound coincides with the resonant frequency of the bubble (which will depend on its size, pressure, temperature, surface tension, *etc.*). It has been estimated that local temperatures of the order of 1500 K and pressures of 150 000 times hydrostatic pressure (pressure exerted by the weight of the liquid above the cavity) can be produced as a result of resonant vibration of stable cavities.

Transient Cavitation

Transient cavities are voids, although they may be filled with vapour, and are produced at relatively high acoustic intensities, *ca.* 10 W cm^{-2}. The lifetime of the bubble is so short that diffusion of gas into and out of the liquid is negligible; however, evaporation and condensation of liquid can take place. As there is no permanent gas present in the cavity to act as a cushion, collapse of a transient cavity is extremely violent and estimates of the local conditions during collapse give temperatures of the order of 40 000 K and pressures around 1000 bar (990 atm).

3 Parameters Affecting Cavitation

It has already been suggested that most sonochemical effects are attributed to the phenomenon of cavitation. The formation or not of cavities is dependent upon many factors including frequency and intensity of ultrasound, external pressure, liquid viscosity, and the temperature.

Frequency

Crawford[3] suggested that the frequency of irradiation has little effect on cavitation and hence on the rate of ultrasound-induced chemical reactions. More recently, Lorimer and Mason[1] observed that the intensity required to produce cavitation in water increases with frequency. Ten times more power is required to induce cavitation in water at 400 kHz than at 10 kHz. Other workers have also reported that the threshold intensity needed for cavitation increases with frequency.

Intensity

Once a threshold intensity has been reached, below which cavitation cannot be induced, the degree of cavitation rises with increasing intensity of ultrasound. Lorimer and Mason[1] point out that the temperatures and pressures generated during cavitation collapse will be greater at increased intensities.

Liquid Properties

A number of liquid properties can have an effect on cavitation. For example, the cavitation threshold intensity increases slightly with liquid viscosity.[1] This is to be expected, since the production of cavity requires that the internal cohesive forces of the liquid be overcome.

Using a liquid with a greater vapour pressure increases the amount of gas present in the cavity. As stated before, this gas acts as a cushion against collapse, so that an increase in gas in the cavity reduces the maximum temperature during collapse, and chemical reaction rates would fall. This effect has been demonstrated by Suslick et al.[4] for non-aqueous solutions.

Cavitation is readily induced in liquids with a high dissolved gas content, and is greatly inhibited by degassing before irradiation takes place. Cavities produced in such liquids will themselves contain large amounts of gas (due to increased diffusion rates under a large concentration gradient) and the conditions during cavity collapse will be less extreme.

Temperature

Raising the temperature of the liquid increases its equilibrium vapour pressure and decreases its surface tension. As a result the shock waves produced during cavitation collapse are partially damped.

However, increasing the temperature also increases the number of internal weak points within the liquid, which act as nucleating centres for cavitation, reducing the threshold intensity required for cavity formation.

In many cases the effect of temperature on processes induced by cavitation can be complex.

4 Applications in the Food Industry
Oxidation and Reduction in Aqueous Solutions

Of particular relevance to the food industry is the effect of ultrasonic irradiation on oxidation/reduction reactions in aqueous solutions. The primary reaction that takes place within cavities that are filled with water vapour is:

$$H_2O \rightarrow \cdot H + \cdot OH \qquad (1)$$

This creation of free radicals is fundamental in explaining the oxidation/reduction effects of ultrasound and is the typical reaction produced by electrical discharges in water vapour and by ionizing radiations in water. Indeed, early workers concluded that these radicals were produced by electrical discharge associated with cavitation collapse. However, more recently it has been suggested[5] that they are generated by thermal dissociation of water molecules in the compression phase of oscillating gas bubbles.

The presence of the $\cdot$H and $\cdot$OH free radicals in sonicated biological systems was confirmed by Edmonds and Sancier[6] and Christmann et al.[7] using electron spin resonance (ESR). Heusinger[8] compared the oxidation products obtained in aerated aqueous α-D-glucose solutions after irradiation with ultrasound, and those obtained by irradiation with X-rays. He discovered that both forms of irradiation gave identical products and concluded that the same primary species were responsible and the same secondary reactions occurred.

Although oxidation/reduction reactions are important in nearly all food systems, for some the process of oxidation is essential - perhaps for the development of flavour or colour, and for others oxidation causes degradation, and considerable efforts are made to avoid the incorporation of air within the food. Food components that are adversely affected by oxidation include oils and fats, colours, and vitamins A and C. However, it should be noted that a large number of workers have subjected a wide range of foods to high-intensity ultrasound, and that relatively few have reported degradation effects due to oxidation.

Ageing of Alcoholic Beverages

An example of an essential oxidation reaction is the oxidation of polyphenols in the maturing of wines and spirits. There are conflicting reports on the ability of ultrasound to enhance this effect. Spect[9] reported that irradiation at 1 MHz changes the alcohol/ester balance and results in a change in taste usually associated with ageing. However, Ensminger[10] refers to a study carried out on a range of wines, which reports that the compositional changes produced had a detrimental effect on their taste.

Chambers and Smith[11] describe a ultrasonic technique for accelerating the ageing of whisky. The document is a patent and contains claims rather than experimental evidence.

The ageing of whisky is a complex series of chemical reactions; the exact process is very subtle and depends on the type of liquor, the storage container, and the storage time, but generally includes an increase in the levels of acids, esters, furfural, solids, and tannins, as well as a deepening of colour. Storage times vary from one to seven years.

Several techniques for accelerating this process are described in the patent. It is claimed that the ageing time can be reduced to less than a year by applying ultrasound to liquor in a standard barrel. The patent also presents a number of systems for more rapid acceleration, which involve passing high-intensity ultrasound through a mixture of liquor and charred wood or sawdust. Cell disruption in biological materials as a consequence of ultrasonic irradiation is a well-known effect, and it is likely that this phenomenon would accelerate the release of compounds from the wood into the liquor.

Enzyme Inhibition

Several workers have reported the ability of ultrasonic irradiation to inhibit enzyme activity. Crawford[3] has reported that the inversion of sucrose could be inhibited by inactivation of enzymes brought about by ultrasonic irradiation. Chambers[12] observed that pure pepsin was inactivated by ultrasonic irradiation and linked the effect to the occurrence of cavitation in the presence of dissolved air. Enzymic activity fell exponentially with time, but this effect only occurred when air was present and cavitation was induced. No inactivation was observed in the presence of nitrogen or hydrogen, or in a degassed system.

The activity of impure enzyme extractions increased initially after sonication, but then followed an exponential decay curve. To explain this it was proposed that less pure enzyme extractions existed as molecular aggregates, and the initial rise in activity was due to the dispersal of these aggregates by the ultrasound. This phenomenon was also found to be dependent on cavitation, but not on the presence of dissolved air, and hence, by saturating with nitrogen or hydrogen (which would not cause inactivation after cavitation) before sonication, it was possible to produce a four-fold increase in the activity of the less pure extraction.[12]

However, Naimark and Mosher[13] also observed a decrease in the activity of pepsin under sonication, but they did not see any initial rise in activity, and suggested that the solutions used by Chambers contained pepsinogen, which was converted into active pepsin.

Naimark and Mosher[13] made the following general observations on the effects of ultrasound on enzymes:

(a) Oxidases are usually inactivated.
(b) Catalases are only affected at low concentrations.
(c) Reductases and amylases are highly resistant to sonication.

Hydrogenation of Oils

This is an example of a catalytic reaction that can be speeded up by sonication. Hydrogenation is an important process in the food industry, which is used to convert unsaturated fatty acids into their corresponding saturated fatty acids by the addition of hydrogen atoms at the double bond. Hydrogen is bubbled through the fat in the presence of a solid catalyst (usually nickel) at 200–300 °C. Since the melting point of the fat is increased by this process it is used to produce fats that are semi-solid at room temperature from vegetable oils. Such fats are used extensively as shortening agents in bakery products. In addition, saturated fats are less prone to oxidation and the consequent development of rancidity.

An improved rate of reaction under ultrasound irradiation has been reported by Moulton et al.,[14] Jordan,[15] and Moulton et al.[16] A number of mechanisms exist by which the observed enhancement may occur.

(a) The strong shock waves produced by cavitation break up the hydrogen bubbles produced by the normal injection system, thus increasing the total area of the oil/hydrogen interface.
(b) If cavitation takes place on the surface of the catalyst, action similar to ultrasonic cleaning is induced. This would increase the rate at which unhydrogenated oil is presented to the surface of the catalyst (Suslick et al.[17]).
(c) The high local pressure associated with cavitation may accelerate the rate of hydrogenation.

Emulsification

The rate of reaction between two immiscible liquids clearly depends upon the degree of emulsification of the two phases. Davidson et al.[18] found that the saponification of glycerides by sodium hydroxide could be enhanced by the application of ultrasound. The hydrolysis of wool waxes was studied in detail; sonication (in an ultrasonic bath) produced greater reaction rates, owing to smaller globule size and better homogenization, and saponification at lower temperatures. Energy requirements were reduced, and product quality was improved as a result.

Formed Meat Products

In recombined meat products such as beef rolls, the pieces of muscle are held together by a protein gel, formed by the myofibrillar proteins released during processing. This release is achieved by tumbling the meat pieces and adding salt. A sticky exudate is formed on the pieces, which binds them together when they are compressed.

Vimini et al.[19] examined the effect of ultrasound on the extraction of proteins, using the ultrasound to disrupt the myofibres of the meat. They compared the effect of salt and ultrasound on the amount of exudate formed, its binding strength, water-holding capacity, product colour, and cooking yields. Samples that received both ultrasonic irradiation and tumbling in salt were superior in all these properties to specimens that had only one treatment. Products that had received only sonication were similar in exudate yield, cooking yield, and water-holding capacity to products produced by the conventional salt treatment, but had much lower binding strengths, since salt is necessary to gel the protein. Product colour was also improved, owing to the reduced destruction of myoglobin.

Similar observations were made on cured ham rolls by Reynolds et al.[20] Ultrasonic treatment enhanced the extraction of myofibrillar proteins, which led to an increase in the mechanical strength of the reformed meat product. However, Takai et al.[21] obtained inconclusive results on the effect of ultrasound on the extraction of water-soluble proteins from fish meal.

Meat Tenderization

Experiments at the Leatherhead Food Research Association have shown that high-powered ultrasound has a tenderizing effect on meat. Sonicating beef muscle at $2\,W\,cm^{-2}$ for 2 h at 40 kHz showed damage to the perimysal connective tissue resulting in improved eating texture.

Extraction of Enzymes

Chymosin (rennin) is an enzyme obtained from the stomachs of calves, which is used to coagulate milk in cheese production. The normal method of extraction is to soak the chopped material in an extracting liquid (usually containing sodium chloride) that has had its pH adjusted for optimum performance. Zayas[22] was able to increase the yield of rennin by the application of ultrasound. Activity of the enzyme extracted was also increased. In improving the extraction of enzymes, care must be taken when choosing the ultrasonic parameters (*e.g.* frequency, intensity, irradiation time) so that the enzymic activity is not inhibited. Enzyme response to ultrasound is determined by their molecular weight, structure, concentration in solution, temperature, and pH, and the process must be carefully assessed to ensure that the enzyme is not inactivated by the ultrasound.

Extraction of Proteins

Studies made of the extraction of protein from defatted soya-bean flakes (Wang[23]) showed that sonication increased the percentage of the available protein extracted under all conditions, compared with conventional stirred systems. This initial investigation was followed up in pilot-scale study by Moulton and Wang.[24] A continuous processing system, capable of handling 12–15 kg of soya-bean flake/solvent slurry per hour was set up. The slurry was irradiated by an ultrasonic probe delivering 550 W at 20 kHz. The authors examined the effect of a number of variables including solvent type, flake/solvent ratio, and ultrasonic irradiation on the yield of protein. In all cases sonication raised the amount of protein extracted, the energy required being of the order of $1-5$ W g^{-1} of protein obtained.

Childs and Forte[25] compared the effects of enzymic treatment and ultrasonic irradiation of the extraction of proteins from cottonseed meal. Sonication increased the yield of protein from the cottonseed flour but it had little effect when used alone on the expressed meal. However, combining ultrasound and enzymic (trypsin) treatment of the meal produced a significantly greater yield of protein.

The quality of the extractions obtained by this procedure is also altered; they have a significantly higher emulsifying capacity, and contain fewer low molecular weight proteins. The authors have attributed this greater emulsifying capability to the higher average molecular radius, but there is some uncertainty whether the hydrolysing behaviour of the enzyme has been changed by the sonication resulting in fewer low molecular weight species, or whether the ultrasound has induced aggregation amongst the smaller molecules. Work by Wang[26] and Wang and Wolf[27] appears to support this latter mechanism. They observed that ultrasonic irradiation of soya-bean proteins specifically affected the 7S proteins, causing them to aggregate.

Crystallization

The crystallization process is an important part of many production operations in the food industry. For example, the manufacture of many confectionery products involves the controlled crystallization of sugar solutions - as indeed does the manufacture of sugar itself. The hardening of fat is a complex process, but is basically one of crystallization, although from the melt rather than from a solution. Controlling the crystallization of the fat phase is of vital importance in the production of many foods including chocolate, margarine, and shortenings. The relevant characteristics of the crystals formed are their size and their polymorphic form. Similarly, the formation of ice crystals in frozen foods is an important consideration. Control of crystal size is essential to prevent damage in certain foods, and can have an effect on the thaw behaviour of frozen foods, as well as having a bearing on the organoleptic properties of, say ice cream.

The process of crystallization, whether from a melt or from a solution, can be divided into two stages - nucleation and growth. In the nucleation stage microscopic crystal nuclei are formed, which develop into larger macroscopic crystals during the growth stage. The total mass of solids that will crystallize from a given melt or solution is determined by the final temperature and pressure of the system, and hence the size of individual crystals produced will be dependent upon the total number of nuclei formed during the initial nucleation stage. Efficient nucleation leads to a large number of small crystals, whereas poor nucleation results in fewer, larger crystals. In most food systems, large crystals, which tend to give food a gritty texture, are undesirable, and a large number of small crystals is preferred.

Control of the nucleation stage then is crucial in the food industry. Unfortunately, many of the parameters that govern this stage also affect the growth rate, and it is difficult to enhance one independently of the other. Subjecting the specimen to high shear, for example in a scraped-surface heat exchanger, can improve nucleation to some degree, and this also distributes the nuclei so formed evenly throughout the system.

Several workers have reported that ultrasound can induce nucleation in crystallizing systems. It has been used to reduce the grain size in solidifying metals, and has induced nucleation in aqueous salt solutions (Crawford[3]) and sugar solutions. There are several possible mechanisms by which this could occur:

(a) It is possible that stresses produced during cavitation could break up existing crystals and nuclei. The fragments resulting from this action would then act as nucleating centres. The net effect would be to produce a large number of small crystals in the final product.

(b) It is known that ultrasound can cause nucleation at very low levels of supersaturation. Van Hook and Frulla[28] claim that nucleation is induced at 102% of saturation. At these low concentrations there are no existing crystals that can disintegrate, and it is believed that in these cases the tiny cavitation bubbles themselves act as nucleating sites, in the same way as other impurities such as dust particles.

(c) Nucleation may also occur in the regions of high/low pressure associated with cavitation. The behaviour of the crystallization process in sucrose under ultrasonic irradiation has been studied in detail. Honig[29] reported that sonication had no effect on the growth rate, but did speed up the rate of nucleation. He suggests an optimum frequency of ca. 10 kHz and a minimum power level of 2 W cm^{-2}. Turner et al.[30] found the best frequency to be in the range 8–20 kHz and suggested that powers of 100 W cm^{-2} were necessary to enhance nucleation significantly. They achieved a reduction in the width of the crystal size distribution by irradiating sucrose syrup for 30 seconds at 8.25 kHz and a power of 300 W cm^{-2}.

Sonication has also been found to aid nucleation in some sugars that are normally reluctant to crystallize, such as d-fructose and sorbitol.[28]

5 Conclusions

This paper has attempted to explain briefly the effects that high-intensity sound waves have on materials and to review some of the work that has been carried out on food systems. It is a subject that is attracting a great deal of excitement at present for many applications. It is a purely physical technique that offers an alternative to additives. Equipment manufacturers are also becoming aware of the potential offered by the processing industry and are taking a keen interest in the scale-up of many of the bench-scale trials now under way.

6 References

1. J.P. Lorimer and T.J. Mason, *Chem. Soc. Rev.*, 1987, **16**, 239.
2. A. Weissler, *J. Acoust. Soc. Am.*, 1953, **25**, 651.
3. A.E. Crawford, 'Ultrasonic Engineering, with Particular Reference to High Power Applications', Butterworths Scientific Publications, London, 1955.
4. K.S. Suslick, J.J. Gawienowski, P.F. Schubert, and H.H. Wang, *Ultrasonics*, 1984, **22**, 33.
5. A. Henglein, *Ultrasonics*, 1987, **25**, 6.
6. P.D. Edmonds and K.M. Sancier, *Ultrasound Med. Biol.*, 1983, **9**, 635.
7. C.L. Christmann, A.J. Carmichael, N.M. Mopsoba, and P. Riez, *Ultrasonics*, 1987, **25**, 31.
8. H. Heusinger, *Z. Lebensm.-Unters. Forsch.*, 1967, **185**, 106.
9. W. Spect, *Nuovo, Chim. Ser.*, 1951.
10. D. Ensminger, 'Ultrasonics. The Low and High Intensity Applications', Marcel Dekker, Inc., New York, 1973.
11. L.A. Chambers and E.W. Smith, 'Method of Ageing Alcoholic Liquors', US Patent No. 2 088 585, 1937.
12. L.A. Chambers, *J. Biol. Chem.*, 1937, **117**, 639.
13. G.M. Naimark and W.A. Mosher, *J. Acoust. Soc. Am.*, 1953, **25**, 289.

14. K.J. Moulton, S. Koritala, and E.N. Frankel. *J. Am. Oil Chem. Soc.*, 1983, **60**, 1257.
15. M. Jordan, *Leatherhead Food Res. Assoc. Res. Rep.*, 1986, No. 544.
16. K.J. Moulton, S. Koritala, K. Warner, and E.N. Frankel. *J. Am. Oil Chem. Soc.*, 1987, **64**, 542.
17. K.S. Suslick, D.J. Casadonte, M.L.H. Green, and M.E. Thompson, *Ultrasonics*, 1987, **25**, 56.
18. R.S. Davidson, A. Safdar, J.P. Spencer, and B. Robinson, *Ultrasonics*, 1987, **25**, 35.
19. R.J. Vimini, J.D. Kemp, and J.D. Fox, *J. Food Sci.*, 1983, **48**, 1572.
20. J.B. Reynolds, D.B. Anderson, G.R. Schmidt, D.M. Theno, and D. G. Siegel, *J. Food Sci.*, 1978, **43**, 866.
21. R. Takai, H. Watanabe, S. Mizusawa, and H. Hasegawa, *Bull. Jpn. Soc. Sci. Fish.*, 1984, **50**, 907.
22. J.F. Zayas, *J. Dairy Sci.*, 1986, **69**, 1767.
23. L.C. Wang, *J. Food Sci.*, 1975, **40**, 549.
24. K.J. Moulton and L.C. Wang, *J. Food Sci.*, 1982, **47**, 1127.
25. E.A. Childs and J.F. Forte, *J. Food Sci.*, 1976, **41**, 652.
26. L.C. Wang, *J. Agric. Food Chem.*, 1981, **29**, 177.
27. L.C. Wang and W.J. Wolf, *J. Food Sci.*, 1983, **48**, 1260.
28. A. Van Hook and A. Frulla, *Ind. Eng. Chem.*, 1952, **44**, 1305.
29. P. Honig, 'Principles of Sugar Technology, Part II', Elsevier Publishing, 1959, p.136.
30. C.F. Turner, T.T. Galkowski, W.F. Radle, and A. Van Hook, *Int. Sugar J.*, 1950, 298.

High Pressure

D.E. Johnston

DEPARTMENT OF AGRICULTURE FOR NORTHERN IRELAND AND THE QUEEN'S UNIVERSITY OF BELFAST, NEWFORGE LANE, BELFAST BT9 5PX, UK

1 Introduction

The investigation of the effects of applying high hydrostatic pressures to food was first reported about a century ago. Hite, a chemist working at the Agricultural Experimental Station of the West Virginia University, investigated the application of high pressure as a means of preserving milk, meats, and fruit juices.[1] Despite formidable problems with the construction and operation of their apparatus, his group was able to demonstrate increased storage life of liquid milk samples treated at ambient temperatures. In addition meat samples treated at 126 °F (52 °C) were found to be in very good condition when opened three months later. This early work was extended to apple juice, peaches, pears, blackberries, raspberries, tomatoes, peas, beans, and beets with varying degrees of success in preserving the qualities of the fresh material.[2] As part of these early studies more detailed studies were carried out exploring the kinetics of the death of micro-organisms as a result of pressurization.

Another pioneer of high-pressure science, P. W. Bridgman, reported in 1914 that egg albumen could be coagulated under suitable conditions, demonstrating that apart from its ability to kill certain micro-organisms high-pressure treatment could alter protein reactivity.[3]

The technological problems associated with the routine handling of materials at high pressure was an inhibiting factor in the exploration of the technique for many years but in more recent times the availability of suitable equipment has improved. Many engineering processes in the fields of plastics, ceramics, and metal forming now employ such pressures on a routine basis. As a consequence the interest of food scientists in the technique has been greatly stimulated. Consumer demand for 'natural' foods and processing techniques has provided an additional spur for new investigations.

Because of the time-scale over which developments in the field have taken place and the diversity of scientific disciplines of those involved, a variety of units are found in the literature. Typically the pressures involved in treating food samples are of the order of 100 MPa and above. Other units equivalent to this bench mark are 1019.7 kgf cm^{-2}, 1.0 kbar, 986.9 atm, and 14503.8 lbf in^{-2}. A few examples provide some perspective on the comparison of such pressure to more common experience. A lady weighing 60 kg, placing her entire weight on one stiletto heel (area 0.66 cm^2) would exert a pressure of 8.8 MPa. A typical maximum operating pressure for a 13 mm diameter die used to prepare KBr discs for IR spectroscopy would be *ca.* 750 MPa.

2 The Effects of Pressure on Structure and Reactivity

From a food science viewpoint the effects of high pressure are the destruction of micro-organisms, the denaturation of protein, and the alteration of enzymatic reactions. These effects can be understood by considering the changes which high pressure causes to molecules and reactions. Applying high pressure will enhance those reactions which give rise to volume decrease and retard those reactions which give rise to a volume increase.[4,5]

The performance and properties of molecules in biological systems often depend critically on conformation or specific changes in conformation. In turn conformation is determined by solute–solute and solute–solvent interactions. The main types of interaction are all subject to volume change and therefore are influenced by pressure change.

Electrostriction

In an aqueous system an electrically charged ion causes the surrounding water molecules to align themselves according to its coulombic field. The alignment results in a more compact arrangement of water molecules and the overall reaction is accompanied by a volume decrease. The phenomenon is known as electrostriction. The ionization of acidic or basic groups found in biological molecules will involve a volume contraction and therefore ionization will be enhanced by pressure increase. The ionization of buffer systems and indeed the ionization of water itself can introduce complications into biochemical studies at high pressure. Careful consideration and choice of buffer can help overcome the problems.[4]

Hydrophobic Interactions

Large regions of biological molecules can be non-polar in character. In aqueous surroundings these non-polar or hydrophobic regions cause a structuring effect on the layer of water molecules immediately adjacent to them. This structuring restricts rotational and translational freedom of the molecules in the layer relative to the surrounding water and hence the overall entropy of the system is lowered.[6] This is an unstable state and the system attempts to achieve a more favourable state by forcing two or more such hydrophobic regions together so that they are surrounded by only a single hydration shell. The water molecules released to the bulk by the merging of the individual shells give an overall entropy increase which drives the process. Overall, the movement of molecules from an ordered to a disordered state causes a small volume increase to the system.[4,5] As a result hydrophobic interactions are disrupted by the application of high pressure below 100 MPa. However, above 100 MPa hydrophobic interactions tend to be stabilized by high pressure owing to an associated volume decrease.

Hydrogen Bonds

A third interaction accompanied by volume change is hydrogen bond formation. In general this causes a small volume decrease as the interatomic distance is shortened. However, in circumstances where hydrogen bonding is strongly directed, an open structure may result, as in the α-helix or β-pleated structure of proteins. This can partially compensate the volume decrease.[4,5] Applying high pressure would therefore be expected to disrupt structured regions of molecules maintained by hydrogen bonds.

Enzyme Kinetics

Enzymes are specialized proteins which catalyse a diversity of biological reactions. Because of their importance in living systems, they have been studied by biochemists interested in how humans and other organisms respond to high pressures such as those encountered in a deep sea dive.[5] The kinetics of enzyme-catalysed reactions can be altered by the influence of high pressure on their mechanism, especially the binding step and the catalytic step. These changes can result in either increased or decreased enzyme activity. In cases where high pressure causes major structural changes to the enzyme, through altered balance of various interacting forces, it may lead to irreversible denaturation.[5]

3 Effects of High Pressure on Micro-organisms

Controlling the activity of micro-organisms was the first reported use of high pressure in connection with food processing.[1,2] Applied at room temperature high hydrostatic pressure is able to kill vegetative micro-organisms.[7] Spores are more resistant but with appropriate choice of conditions[8] they too may be inactivated by the technique. The application of pressure in conjunction with heat is felt to be a particularly promising method from a point of extending the technique.[5,9] A great deal of development still remains to be done in this area. Just as micro-organisms vary in their susceptibility to heat so too there is variability in their susceptibility to high pressure, and pressure resistance in an organism appears to be independent of heat resistance.[5]

High pressure can potentially affect a micro-organism at a number of different levels to inhibit its activity or kill it. The cell membrane has an important role in transport processes and separates the intracellular material from the environment. Serious alterations in the permeability of the cell membrane will result in cell death. The membrane is composed mainly of phospholipid and protein. Application of high pressure will cause changes in electrostrictive effects, hydrophobic interactions, and hydrogen bonding which will alter the structure of these molecules and hence the permeability of the cell membrane. This is felt to be the initial site of damage in micro-organisms subjected to high pressure.[5]

Within the cell the essential biochemical reactions catalysed by enzymes can be disrupted, causing the various cycles either to become out of step or to cease altogether, leading once again to death of the organism.[5]

Changes in pressure can also lead to complete mechanical disruption of the cell wall, releasing the cell contents and killing the organism.[5]

4 Application of High Pressure to Meat

Since the original report by Hite[1] of the preservative effect of high pressure on meat the most significant development has been the report by Macfarlane that high-pressure treatment of meat could have a tenderizing effect.[10] This area of work has been the subject of much research in view of its important commercial potential and has been thoroughly reviewed.[11] In building an understanding of the effects of high-pressure treatment on meat it is helpful to consider first its effects on the main meat proteins. The ability of high pressure to alter ionization, hydrophobic interactions, and hydrogen bonds is the basis of the effects and they can vary in degree from slight conformational change, through changes in aggregation–disaggregation equilibrium, to complete denaturation.

Effects on Individual Meat Proteins

Actin. It has been reported by Ikkai and Ooi that F actin is denatured by a pressure of 147 MPa in the absence of ATP and by 245 MPa in its presence.[12] This suggests that the ATP exerts a protective effect. The disaggregation of F actin is thought to occur in the presence of ATP as adding EDTA, known to denature G actin readily but not F actin, gives rise to a pressure–denaturation curve which is almost the same as that obtained when ATP is absent, *i.e.* irreversible denaturation of F actin begins at 147 MPa.

Myosin. At pressures up to 35 MPa a shift in the monomer–polymer equilibrium of myosin can be observed in favour of the monomeric form of myosin.[13] The equilibrium constant appears to be temperature independent and this would be consistent with association being due to charged groups interacting. On removal of the pressure the equilibrium shifts back.

At higher pressures irreversible changes take place. There are irreversible increases in molecular weight,[14] and the reactivity of SH groups[15] and the exposure of hydrophobic regions both increase.[16] When pressure-treated myosin is subsequently heat-denatured its setting properties are found to be improved.[17]

Actomyosin. The aggregation of actomyosin is also influenced by the application of high pressure.[14] In the absence of ATP there is still some association between the actin and myosin[16] but in its presence results suggest a dissociation to actin and myosin.[18]

Tenderization of Meat by High–pressure Treatment

There is great commercial potential for a process capable of providing meat with improved tenderness cost effectively. Macfarlane and co-workers have pioneered the development of high-pressure application to meat to provide increased tenderness. This body of work has included high-pressure application to *pre-rigor* meat,[10] *post-rigor* meat,[19] and a combined pressure and heat treatment.[20]

Pre-rigor Meat. Various combinations of high pressure, temperature, and holding time have been examined by Macfarlane and co-workers for the ability to tenderize *pre-rigor* meat.[10] *Biceps femoris* muscles pressurized *pre-rigor* to 103 MPa at 35 °C for 2 min were found to have improved eating quality reflected in lower cooking loss, increased moisture content, and lower Warner–Bratzler shear values. The improved acceptability was confirmed by taste panel scores. Similar improvements were also reported by Kennick and Elgasim who suggested four potential mechanisms for the effects,[21] *i.e.* breakdown of myofibrillar structure, F to G transformation of actin or myosin or both, early release of lysosomal enzymes, or creation of breaks in fibre structure due to massive contraction.

Observations with the light microscope and transmission and scanning electron microscopes all reveal that severe disruption of the muscle structure results from *pre-rigor* pressure treatment.[11] It is likely that the first two suggested mechanisms are interrelated. Strips of *pre-rigor* muscle under tension at 30 °C contract immediately when pressure is applied but after 1–5 min they begin to lengthen again.[22] This lengthening was presumed to occur because of weakening of myofilaments under pressure. This weakening as a result of contraction is proposed to account for the lower shear values found for pressure-treated *pre-rigor* muscle compared with muscle contracted to the same extent by cold-shortening.[23]

Post-rigor Meat. Being able to process *post-rigor* muscle to increase tenderness would avoid the time constraints associated with *pre-rigor* treatment and provide greater flexibility of operations in a meat plant. Working in the temperature range 0–20 °C pressure treatment of *post-rigor* meat provides no apparent improvement in the Warner–Bratzler shear values of the cooked meat.[19] Nevertheless ultrastructure studies do show that disruption occurs as a result of the treatment and further evidence from differential scanning calorimetry suggests that major changes to the actin result from the pressure treatment.[19]

Combined High Pressure–Heat Treatment. By contrast to treatment of *post-rigor* muscle in the temperature range 0–20 °C it has been found that increasing the temperature during pressure treatment to 45–60 °C produces marked improvements in cooked tenderness.[24] This pressure–heat treatment can be successfully applied to muscles which have entered *rigor* in a wide range of states of contraction to produce similar Warner–Bratzler shear values.[20] In addition the pressure–heat treatment is able to overcome the toughness associated with cold-shortening. Compression, adhesion, and Warner–Bratzler shear measurements of the properties of meat subjected to pressure–heat treatment indicate that while the myofibrillar component of the toughness is decreased the connective tissue component is not affected.[24,25] Based on examination of SDS gel electrophoresis patterns of pressure–heat-treated meat samples and electron microscopy it has been suggested that most of the tenderization is achieved by an irreversible disaggregation of the myosin of the thick filaments.[26]

Effects of High Pressure on Comminuted Meats

Gel formation by proteins is a phenomenon which involves a variety of protein–protein interactions. Under appropriate conditions a three-dimensionally cross-linked network is formed throughout the system and considerable mechanical strength may be developed. Pressure-induced gelation of egg albumen was originally reported in 1914[3] and gelation of horse serum albumen in 1933.[27] It was established in 1941 that protein denaturation was involved in pressure-induced gelation and that SH groups were reactive after high-pressure treatment.[28]

In comminuted meat products the protein–protein interactions caused by high-pressure treatment provide improved characteristics in meat patties. Macfarlane has reported increased amounts of soluble protein expressable from homogenates of meat pressure treated at 150 MPa for 5 min and improvements in water-holding capacity.[29] Binding between meat particles was found to be improved in cooked meat patties which had been high-pressure treated prior to cooking.[30]

In Japan the potential of pressure-induced meat gels is being explored. Pressure-induced gels have been prepared from crude carp actomyosin and rabbit meat paste and their properties compared with those of heat-induced gels.[31] Applied pressures of 100–300 MPa for 30 min gave gels which increased in rigidity with increasing pressure but which were glossier in appearance and softer than the corresponding heat-induced gels. Studies with rabbit myosin in 0.1 M KCl and 20 mM phosphate buffer of pH 6.0 showed that the pressure threshold for gel formation was concentration dependent.[32] The strength of these gels was almost proportional to protein concentration and was comparable to that of heat-induced gels. Electron microscopy of the pressure- and heat-induced gels showed that they had similar structures. At pressurization times above 10 min particles could be observed on the surface of the myosin filaments in scanning electron micrographs of the pressure-induced gels. Conformational changes in the tail region of the molecule were thought to be involved as well as aggregation of the myosin heads.

5 Future Prospects

If high-pressure treatment in the food industry is to be taken up commercially then it is essential that it is competitive with existing technologies or can establish a unique position. A number of applications of high-pressure technology are already under investigation in Japan, USA, and Australia.[33] Broadly speaking, these potential developments can be grouped around the abilities of the process to kill micro-organisms, to change the structure and behaviour of food biopolymers, and to regulate the activity of enzymes.

Applications to Meat and Meat Products

Preservation. Killing micro-organisms by the application of heat frequently alters the food from raw to cooked by a variety of chemical reactions. Thermal gradients are established in techniques which depend on heat conduction and this in turn can lead to over-cooking of some regions in order to achieve the required temperature at the slowest heating point. By comparison high hydrostatic pressure provides a uniform treatment of the material independent of the size of the portion, pack, or pieces within the pack. Volatile flavour molecules and heat-sensitive vitamins are conserved. Many of the applications of irradiation currently being considered could also be achieved by high-pressure treatment but without the associated problems of consumer resistance. Commercial interest in preservation by high pressure or a combination of heat and high pressure is high.[5,9,33]

Tenderization. Despite many scientific aspects of the phenomenon having been established, the use of high-pressure tenderization has not been taken up commercially. Remaining barriers could be economic or logistical or both. It may be that future developments in technology could make the technique more economically attractive.

Re-structuring. The basic phenomenon of improvement in properties of restructured meat products has been established. There has been some investigation of the ability of the technique to permit lower levels of salt and phosphates in formulation[30] but again the gains have not attracted commercial interest. The use of pressure as a means of inducing gelation may prove more attractive but this is still at an early stage.

Upgrading of Waste Protein. With increasing consumer demand for 'natural' ingredients to gel, emulsify, stabilize, and foam many proteins are being examined for their potential as functional ingredients. Provided that the economics are sound, there could be scope for pressurization to be used to create new or improved functional properties of lower-grade proteins by controlled unfolding of their structure.

Future Technological Development

Technology Transfer. If high-pressure processing is to become an established food-processing method the technology must be successfully transferred from existing sources in the metal-forming, ceramics, and plastics industries. An early objective of food scientists must therefore be to establish the working parameters needed for a particular application so that engineers can modify existing equipment or re-design to meet the requirements. Consideration will also need to be given to the various materials involved and their potential interactions with food in the factory situation.

Working Volumes. Because of the various problems involved in the construction and operation of high-pressure equipment, laboratory research machines usually have small working volumes of the order of a litre. This is sufficient to investigate fundamental aspects and provide basic information. In development larger volumes are required and use of chambers of *ca.* 42 l volume for meat tenderization, capable of holding several large boneless cuts, such as those used in commercial practice, has been reported.[21] Prospects for the development of larger-capacity equipment for every day use in the food industry would appear to be good, as vessels are already available for the cold isostatic pressing process with capacities up to 3000 l and operating pressures of 400 MPa.[33]

Packaging and Handling. At present, foods to be treated by high pressure are placed in a flexible plastic container with good burst strength and low permeability and sealed without air. The sealed packs are then placed in the working fluid inside the apparatus and the apparatus is closed and brought up to working pressure. After the required time–pressure treatment the apparatus is de-pressurized and opened to allow removal of the food. The operation of the apparatus in the batch mode is the only mode possible at present but current high hydrostatic pressure machinery can be cycled rapidly and this should offset the inconvenience of batch operation.[33]

Working Fluid. Existing research apparatus reported in the literature employs hydrocarbons, specialist hydraulic oils, or water as the fluid medium to transmit the pressure to the food. Water is cheap, non-toxic, non-flammable, and readily available. These are significant advantages and it could well be that this is the best long-term prospect for future food-processing developments. Inert gases such as nitrogen may offer an alternative but working with gases as the pressure transmission medium as opposed to liquids poses additional problems.[11,33]

6 Conclusions

High-pressure treatment of foods continues to attract commercial and academic interest. The major research investment is in Japan where the Research Institute for Food Science at Kyoto University set up a special unit in 1989 to bring together the Japanese Ministry of Agriculture, Forestry and Fisheries in collaboration with twenty-one food and engineering companies to develop high pressure for use in the food industry. The unit is to have a four year life span and its first annual budget was *ca.* US$1 million.[33] This demonstrates a high degree of confidence in the technique. Given the achievements of Japanese industry in heavy engineering, electronic consumer goods, cameras, motor cars, and motorcycles can the UK food industry afford to ignore high pressure?

7 References

1. B.H. Hite, *Bull. W. Va. Univ. Agric. Exp. Stn.*, 1899, **58**, 15.
2. B.H. Hite, N.J.Giddings, and C.E. Weakly, *Bull. W. Va. Univ. Agric. Exp. Stn.*, 1914, **146**, 3.
3. P.W. Bridgman, *J. Biol. Chem.*, 1914, **19**, 511.
4. E. Morild, *Adv. Protein Chem.*, 1981, **34**, 93.
5. D.G. Hoover, C. Metrick, A.M. Papineau, D.F. Farkas, and D. Knorr, *Food Technol.*, 1989, **43**, 99.
6. F. Franks, in 'Water Relations of Foods', ed. R.B. Duckworth, Academic Press, London, 1975, p.3.
7. W. J. Timson and A.J. Short, *Biotechnol. Bioeng.*, 1965, **7**, 139.
8. A.J.H. Sale, G.W. Gould, and W.A. Hamilton, *J. Gen. Microbiol.*, 1970, **60**, 323.
9. D.F. Farkas and D.G. Hoover, *Act. Rep. Res. Dev. Assoc.*, 1990, **42**, 37.
10. J.J. Macfarlane, *J. Food Sci.*, 1973, **38**, 294.
11. J.J. Macfarlane, in 'Developments in Meat Science', ed. R. Lawrie, Elsevier Applied Science, London, 1985, Vol.3, p.155.
12. T. Ikkai and T. Ooi, *Biochemistry*, 1966, **5**, 1551.
13. R. Josephs and W.F. Harrington, *Biochemistry*, 1968, **7**, 2834.

14. J.M. O'Shea, D.J. Horgan, and J.J. Macfarlane, *Aust. J. Biol. Sci.*, 1976, **29**, 197.
15. Y.N. Berg, N.A. Lebedeva, E.A. Markina, and L.L. Ivanov, *Biokhimiya*, 1965, **30**, 277.
16. J.M. O'Shea and R.K. Tume, *Aust. J. Biol. Sci.*, 1979, **32**, 415.
17. T. Suzuki and J.J. Macfarlane, *Meat Sci.*, 1984, **11**, 263.
18. T. Ikkai and T. Ooi, *Biochemistry*, 1969, **8**, 2615.
19. J.J. Macfarlane, I.J. McKenzie, R.H. Turner, and P.N. Jones, *Meat Sci.*, 1981, **5**, 307.
20. P.E. Bouton, A.L. Ford, P.V. Harris, and J.J. Macfarlane, *J. Food Sci.*, 1977, **42**, 132.
21. W.H. Kennick and E.A. Elgasim, Proceedings of the 34th Annual Reciprocal Meat Conference, American Meat Science Association, Chicago, 1981, p.68.
22. J.J. Macfarlane, I.J. McKenzie, and R.H. Turner, *Meat Sci.*, 1982, **7**, 169.
23. J.J. Macfarlane, R.H. Turner, and D. Ratcliff, *J. Food Sci.*, 1976, **41**, 1447.
24. D. Ratcliff, P.E. Bouton, A.L. Ford, P.V. Harris, J.J Macfarlane, and J.M. O'Shea, *J. Food Sci.*, 1977, **42**, 857.
25. P.E. Bouton, P.V. Harris, J.J Macfarlane, and J.McK. Snowden, *J. Texture Stud.*, 1977, **8**, 297.
26. J.J. Macfarlane, I.J. McKenzie, and R.H. Turner, *Meat Sci.*, 1986, **17**, 161.
27. J. Basset, M. Machboeuf, and G. Sandor, *C. R. Acad. Sci.*, 1933, **197**, 796.
28. E.A. Grant, R.B. Dow, and W.R. Franks, *Science*, 1941, **94**, 616.
29. J.J. Macfarlane, *J. Food Sci.*, 1974, **39**, 542.
30. J.J. Macfarlane, I.J. McKenzie, R.H. Turner, and P.N. Jones, *Meat Sci.*, 1984, **10**, 307.
31. M. Okamoto, T. Kawamura, and R. Hayashi, *Agric. Biol. Chem.*, 1990, **54**, 183.
32. K. Yamamoto, T. Miura, and T. Yasui, *Food Struct.*, 1990, **9**, 269.
33. D. Farr, *Trends Food Sci. Technol.*, 1990, **1**, 14.

Irradiation of Meat and Poultry

M.H. Stevenson

DEPARTMENT OF AGRICULTURE FOR NORTHERN IRELAND AND THE
QUEEN'S UNIVERSITY OF BELFAST, NEWFORGE LANE, BELFAST BT9 5PX, UK

1 Introduction

The use of ionizing radiation for the preservation of food has been extensively studied for many years. The irradiation sources which are permitted[1] for use with food are γ-photons from the radionuclides ^{60}Co and ^{137}Cs, high-energy electrons generated by machines, maximum energy 10MeV, and X-rays, maximum energy 5 MeV.

The major difference between the sources is their penetrating power. The γ- and X-rays are highly penetrating and can treat food in bulk whereas the high-energy electrons are only useful for surface irradiation or for the treatment of thin packages.

In the case of red meat and poultry, irradiation can be used for sterilization of the product (radappertization), enhancement of food safety through the inactivation of pathogenic micro-organisms, such as *Salmonella* or *Campylobacter* (radicidation), and extension of shelf-life by the elimination of the micro-organisms responsible for normal spoilage (radurization).

The doses needed to achieve the above objectives vary. Thus for sterilization, doses of *ca.* 25–50 kGy* (depending on the product) are necessary whereas for extension of shelf-life and the enhancement of food safety doses below 10 kGy are sufficient.

Following the report of the Food and Agriculture Organization (FAO)/International Atomic Energy Agency (IAEA)/World Health Organization (WHO) Joint Expert Committee on the Wholesomeness of Irradiated Food[2] which concluded that

> 'irradiation of food up to an overall average dose of 10 kGy produced no toxicological hazard and introduced no special nutritional or microbiological problems',

greater emphasis has been directed towards the lower dose applications. Nevertheless, the initial work on irradiation sterilization of meat and other protein foods carried out by the US Army at their Natick Laboratories

---

*1 gray (Gy) = 1 J of absorbed energy per kg.

from the 1950s to about 1980 provided invaluable information on the chemical changes induced in irradiated food. The research also resulted in development of the practical application of the technology.

This chapter describes the various applications of the technology to meat and poultry, outlines the changes occurring in the major chemical components, indicates their effect on various quality attributes, and introduces the methods being developed to detect irradiation in these foods.

2 Radappertization

This involves the treatment of pre-cooked (enzyme inactivated) foods that are hermetically sealed in metal cans, flexible pouches, or metal or plastic trays with sterilizing doses of γ-rays or electrons.

The key information for radappertization is the minimum radiation dose (MRD) which will provide both product stability and safety. This MRD is based on the most radiation-resistant micro-organism associated with the product, namely *Clostridium botulinum*. Measurement of the radiation resistance of the spores of various strains of *C. botulinum*, types A and B, suggested that it was *ca.* 46 kGy. This value was obtained by employing a 12 logarithm reduction of the number of organisms present and was calculated from the experimentally determined D_{10} values for *C. botulinum* spores. The D_{10} value is the dose required to obtain a 90% reduction of the organisms present. The use of a 12-D factor placed irradiation sterilization on a similar footing to thermal sterilization. Alternatively, it has been recommended that the MRD should be determined experimentally, and when this is done the measured values range from 24 kGy for corned beef to 41 kGy for beef.[3] The resulting radappertized products are free from all food spoilage micro-organisms and organisms of public health significance including pathogens such as *C. botulinum* and *Salmonella*.

Most products do not need special preparation and the process has been applied to bacon, roasts of beef, pork, lamb, beef steaks, and frankfurters.[4–9] In uncured meats, addition of small amounts of sodium chloride (0.5–1%) along with 0.3% condensed phosphates is beneficial in improving flavour, texture, juiciness, overall acceptance, and the yield of the products. The addition of phosphates improves water holding capacity,[6,7] and also controls lipid oxidation and has some antibacterial effects.[10]

Irradiation alone will not inactivate proteolytic enzymes which during storage will result in deterioration of the product, so in order to attain long-term storage enzyme inactivation is achieved through pre-cooking the food to an internal temperature of 70–75 °C.[4]

An integral part of the whole processing procedure is the combination of the treatment with high-vacuum packaging which will ensure the exclusion of oxygen and so minimize the lipid oxidation which can occur on irradiation.

It is essential that the vacuum-packaged product is frozen to −30 to −40 °C and irradiated in the frozen state. This prevents off-flavour development by reducing production of radiolysis products generated by the high radiation doses.[11] The formation of these products begins at *ca.* $-20\,°C$[11] and thus it is important that during irradiation processing the centre of the container does not exceed this temperature. Some products such as ham and corned beef are not so susceptible to temperature but chicken and turkey show detectable loss in quality if irradiation sterilization is not carried out at temperatures below $-20\,°C$.[4]

Following irradiation with sterilizing doses, the foods can be stored and distributed without refrigeration. The length of storage depends on the enzyme inactivation temperature and the storage temperature. A product, shelf-stable for 2 years, can be obtained if it has been pre-cooked at 70–75 °C and subsequently stored at 21 °C.[7,12] If the storage temperature is lowered, the shelf-life can be extended even further. The high quality of some irradiation-sterilized products (ham, beef steaks, corned beef, turkey slices) has been confirmed by both American and Soviet astronauts during their space flights in the 1970s.[4]

3 Radicidation

The incidence of food-borne disease is increasing and according to Kampelmacher[13] these diseases will soon become the largest cause of morbidity in Europe after respiratory disease. Raw red meats and poultry are both carriers of pathogenic micro-organisms such as *Salmonellae* and it has been reported that up to 64% of poultry carcasses may be contaminated with this organism. In order to enhance food safety, the numbers of these harmful micro-organisms need to be reduced. Despite attempts to control the problem it still persists, and, because these micro-organisms are sensitive to irradiation, this process offers potential for the enhancement of food safety. This does not mean that the process should be used as an alternative to good manufacturing practice and, in fact, it is essential that only high-quality products are treated with irradiation if the optimum benefit is to be derived from the process.

The sensitivity of the pathogen (D_{10} value) to irradiation depends on the strain of the micro-organism, the medium in which it is present, the temperature of irradiation, and the gas atmosphere in the packaging.[14,15] In order essentially to free a food of *Salmonellae*, it was suggested that a seven-fold reduction of the numbers present was needed, and this would require a dose of *ca.* 3.5–4 kGy. Subsequent views consider that a five-fold decrease would be sufficient and thus the required dose is decreased to *ca.* 2.5 kGy.

Although much of the initial concern about pathogenic organisms was directed towards *Salmonellae*, other organisms such as *Campylobacter jejuni, Staphylococcus aureus*, and more recently *Listeria monocytogenes* are also important. The D_{10} values for these organisms vary but their sensitivities to irradiation are similar to that of *Salmonellae*.[14]

As well as the pathogenic micro-organisms, there are also parasites found in fresh meats which present a health hazard. In the case of pork infected with *Trichinella spiralis*, a dose of 300 Gy has been shown[16] to be effective in making the fresh pork non-infectious and this low dose is well below the threshold level of 1.75 kGy, the dose above which sensory changes are detectable.[17]

4 Radurization

The purpose of radurization of meats and poultry is to obtain an extension of their shelf-life through reduction of the initial microbial load.

The bacteria which most commonly spoil fresh meats and poultry are *Pseudomonas* and *Achromobacter*. *Pseudomonas* is very sensitive to radiation and thus doses used to control *Salmonellae* will wipe out these normal spoilage bacteria.[14] Although *Achromobacter* are also radiation sensitive, they may survive low doses of radiation and thus the spoilage which will occur on prolonged storage will arise from an outgrowth of these and other bacteria.

Although low-dose irradiation is effective in reducing microbial load, other pathways of spoilage can still occur and need to be controlled. In the case of fresh red meats the following procedure has been developed.[18]

The product is dipped in 0.5% solution of sodium tripolyphosphate and individual portions are wrapped in an oxygen-permeable, moisture-impermeable film. A number of portions are placed in a bulk container, vacuum packed, and irradiated in the range 1–2 kGy at a temperature between 0 and 10 °C using γ-rays. The product is stored and transported in bulk at temperatures not exceeding 5 °C. The bulk packages are opened 30 min prior to displaying the retail cuts at temperatures between 0 and 5 °C for sale within 72 hours.

The phosphate solution controls drip and helps to maintain colour while the double packaging system provides anaerobic conditions until the individual portions are removed, thus limiting lipid oxidation. The individual packs are oxygen permeable so oxygen can diffuse into the pack and ensure a good red colour. Vacuum-packaged beef treated in this way has been found to have acceptable sensory qualities for up to 24 days after irradiation.[18]

There are important differences between the different red meats and poultry. For example, discoloration is a problem with beef[19] but pork, veal, and chicken, which are less pigmented, are generally not so susceptible to these problems.[20,21] Nevertheless, irradiation of fresh poultry at doses as low as 2.5 kGy produced a pink coloration which was detectable both by a sensory panel and spectrophotometrically (Table 1).

Although it is well established that irradiation can reduce the microbial load and thus slow down the rate at which microbial spoilage takes place, the influence of irradiation on the organoleptic qualities of red meat and poultry is very important. Threshold levels of irradiation have been determined for different species, and at doses above these odour and

Table 1 *Effects of irradiation dose on the colour of chicken meat assessed using a trained sensory panel and spectrophotometrically*

Dose (kGy)	Sensory score Pinkness	$^a a^*$ (redness)	
		Breast	Leg
0.0	4.4	7.5	14.3
2.5	5.8	12.8	17.6
5.0	5.8	14.8	21.2
7.5	6.3	14.4	21.7
10.0	6.3	14.5	21.5
SEM	0.28	0.71	0.66
SIG	***	***	***

^{a} CIE a^* value, illuminant D_{65}.
SEM = standard error of the mean; SIG = significance.

flavour changes can be detected which render the products unacceptable.[17] By combining low-dose irradiation with other processes such as vacuum packaging or modified atmosphere packaging it is possible to minimize the changes which occur in products such as corned beef,[22] pork loins,[23] minced pork,[21] and sheep carcasses.[24]

These odour and flavour changes together with the effects on colour arise as a result of chemical changes occurring in the food. Thus an understanding of the fundamental chemistry involved helps to explain why these changes occur and has provided evidence of the wholesomeness of irradiated food.

5 Interaction of Ionizing Radiation with Matter

When γ- or X-rays interact with matter, different types of energy transfer can occur but in the case of food irradiation the Compton effect predominates. In this case, the incident photon interacts with the absorbing medium in such a way that an orbital electron is ejected. The direction of the incident photon changes after the collision and it loses some of its original energy but it may go on and react with other atoms to form secondary electrons. It is these ejected electrons which ionize and excite the components of the system producing, for example, free radicals. These in turn are highly reactive and attack other compounds in the medium before forming stable end products. This series of reactions may take only fractions of a second.

Because the Compton electrons produced from either γ- or X-rays produce ionizations and excitations in the same way as high-energy electrons, the chemical changes induced in food following irradiation are similar irrespective of the source of the ionizing radiation. The extent of any change produced by irradiation is dependent on the amount of radiation deposited in the food and this is known as the absorbed dose, the unit of which is the gray (Gy).

M.H. Stevenson

6 Chemical Changes Occurring in Foods

Water

Meat and poultry contain substantial amounts of water as well as fat and protein and lesser amounts of vitamins. In the case of food irradiation, the radiolysis of water is particularly important and the following highly reactive entities are formed:[25]

$$H_2O \rightarrow \cdot OH + e_{aq}^- + H \cdot + H_2 + H_2O_2 + H_3O^+ \quad (1)$$

Since the hydroxyl radical is a powerful oxidizing agent while the hydrated electron and hydrogen atom are reducing agents, all water-containing foods are likely to undergo oxidation and reduction reactions.

The presence of oxygen during irradiation can influence the course of radiolysis and the hydroperoxyl radical, $\cdot OOH$, and the superoxide radical, $\cdot O_2^-$, which are formed are both oxidizing agents. In addition, oxygen can also add to other radicals in the food to give peroxy radicals, $\cdot OOR$.

The temperature during irradiation is also important because in a frozen product the reactive intermediates of water radiolysis are trapped and thus are not free to interact with each other or with other food components. Thus freezing is effective in minimizing some of the adverse organoleptic changes which may occur in irradiated meat[4] and chicken.[26]

When a food is irradiated the chemical changes which occur may arise as a result of direct action on the fats, proteins, vitamins, *etc.* or by indirect action mediated through the reactive intermediates formed on the radiolysis of water. Since meat and poultry contain substantial amounts of water, the indirect effects are important. This type of terminology also applies to mixtures of substances since there can be direct action on individual substances as well as indirect action caused by reactive species produced by other components of the system. Thus in a multicomponent matrix such as a food, the constituent components exert a certain degree of mutual protection and therefore the extent of change in any one component is minimized. This emphasizes the need to extrapolate carefully the results obtained with, for example, pure aqueous solutions to complex food systems.[25]

In the case of red meats and poultry, the effects of irradiation on the protein and fat components may influence the appearance, odour, flavour, and texture of the products.

Proteins

Considerable research has been undertaken to elucidate the chemical changes which occur in irradiated proteins and this has been reviewed in detail elsewhere.[26,27] An understanding of the basic changes which occur in irradiated amino acids has been particularly important. The reactions which occur may be different in aqueous and dry systems but in the case of food the aqueous system is more important. In the absence of O_2, simple amino acids undergo reductive deamination and decarboxylation. If O_2 is present, the hydrated electron and the hydrogen atom are removed and this blocks the reductive deamination. Instead oxidative deamination takes place through reaction with the ·OH radical.

A number of radiolytic products are formed, the exact nature of which depends on the amino acid. Aqueous solutions of alanine, for example, yield pyruvic and propionic acids, acetaldehyde, ethylamine, ammonia, carbon dioxide, and hydrogen.[26,27]

In the case of the sulfur amino acids, irradiation results in the formation of compounds such as hydrogen sulfide and methyl mercaptan.

When proteins are irradiated in the presence of water all the reactions which could occur with amino acids are possible. Proteins may contain up to 20 amino acids and these together with the highly reactive water intermediates give the potential for a large number of interactions. Additional effects are also exerted by the tertiary structure of the proteins and in practice many of the constituent amino acids are protected from attack because they are inaccessible to radical reactions. Some degradation and aggregation of proteins can occur on irradiation and this may alter the viscosity of the proteins. Despite the potential to attack individual amino acids and thus affect the nutritional quality of the protein, irradiation, even at high doses (up to 50 kGy), does not significantly alter protein quality.[28]

Many of the changes induced in proteins are minor but some types of proteins merit special consideration, namely the enzymes and the chromoproteins. A large proportion of the energy deposited goes into protein denaturation, that is changes in secondary and tertiary structure, but this denaturation is much less extensive than that caused by heating.[25] Enzymes are generally not denatured by irradiation and thus, in irradiation sterilization, a pre-heat treatment is needed to inactivate the enzymes which are not affected by the high-dose irradiation, but which on long-term storage could result in proteolytic breakdown of the food.

Lipids

The lipid portion of foods consists primarily of triglycerides and as for proteins an understanding of the basic mechanisms involved in the chemical changes occurring in irradiated lipids has been obtained using model systems, e.g. pure triglycerides.[29,30] In contrast to proteins, where the indirect effects mediated through water play an important role, in the case of lipids the direct effects of electrons on the lipid fraction are more important.

Cleavage occurs preferentially at bonds in the vicinity of the carbonyl group (Figure 1) but can also occur at other locations. Sixteen different free radicals have been postulated as being formed and these free radicals then lead to the formation of end-products through a number of different

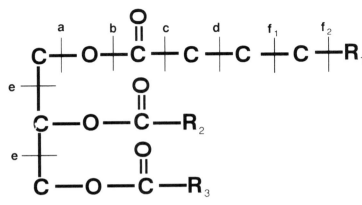

Figure 1 *Cleavage of triglycerides following irradiation:* a, b, c, d, e, f_1, f_2, *cleavage points;* R_1, R_2, R_3, *fatty acids*

pathways involving abstraction, dissociation, recombination, disproportionation, and radical–molecule interaction. Thus a diverse range of products could be formed and much detailed research has been carried out to identify many of these compounds.[29,30]

If oxygen is present during or after irradiation, normal autoxidation is accelerated and the compounds formed parallel those which are formed during normal oxidative breakdown of lipids.[31]

Off-flavours and Odours in Irradiated Meats

A characteristic odour develops in irradiated meat products, the intensity of which is dependent on the radiation dose applied.[32-34] This odour has been described as being metallic, sulfide, wet dog, goaty, or burnt.

Batzer and Doty[34] were among the first to try to identify the compounds responsible for the irradiation odour. Initially they identified hydrogen sulfide and methyl mercaptan and assumed that these sulfur compounds arose from protein degradation during irradiation. A few years later a significant increase in volatile amines, mainly ammonia, was reported in meats following irradiation.[35] There was also an increase in methyl- and ethyl-amine and four other amines were detected but could not be identified. It was hypothesized that these compounds arose from the non-protein nitrogen constituents since they were not formed when pure proteins were irradiated.

Using a low-temperature vacuum distillation technique followed by gas–liquid chromatographic analysis the first data on the total volatiles present in irradiated meats were produced.[36] The compounds identified included methyl mercaptan, acetaldehyde, dimethyl sulfide, acetone, methanol, ethanol, methyl ethyl ketone, dimethyl disulfide, ethyl mercaptan, and isobutyl mercaptan. All the compounds except the last two increased with irradiation dose.

Somewhat later Wick et al.[37] employed a sniffing technique and found two components which resembled the irradiation odour. These compounds were not detected in non-irradiated beef. One of these compounds was identified as methional, and on the basis of its low odour threshold, characteristic odour, and presence in irradiated meat they concluded that it was a major contributor to irradiation odour.

Since meat is a mixture of protein and fat, it is perhaps not unexpected that the irradiation odour may arise from an interaction of fat and protein radicals yielding new odour compounds which are not present in either fraction alone. In fact, the irradiation odour was not detected when protein and fat fractions were treated separately but was present when a meat lipoprotein fraction was irradiated.[38]

The major radiolytic products formed when beef and pork fat were irradiated were the hydrocarbons and, in order to establish the significance of these compounds in the generation of the radiation odour, the odour thresholds of several alkanes, alkenes, alkynes, and alkadienes have been studied.[39] The quantities of hept-1-ene and oct-1-ene in irradiated beef fat exceeded their odour threshold, hex-1-ene was present near its threshold, and the others were below their odour threshold. Contrary to these results, Wick et al.[37] reported that the main classes of compounds in irradiated meats were aldehydes and alcohols with a much smaller proportion of hydrocarbons. However, the difference between these results may be due to the fact that one group used isolates from partially cooked meats held in air[37] rather than vacuum-packed meat. Wick et al.[37] following identification of the compounds in the odour isolates then subsequently added the key compounds to meat and evaluated the samples organoleptically. They concluded that methional, n-nonanal, and phenylacetaldehyde were responsible for the characteristic odour of irradiated beef and that hydrocarbons were not important. Although this early research has indicated that sulfur and carbonyl compounds may contribute to irradiation odour in beef, the actual compounds responsible are not clearly defined. Work is needed to determine the significance of each class of compound and why odour apparently does not change in character due to variations in oxygen content or meat composition. This initial work was carried out with beef and it is likely, although not confirmed, that these same groups of compounds will be important with lamb, pork, and also poultry.

Colour Changes

As well as the possible contribution of the proteins to the flavour changes which may be present in irradiated meat and poultry, it has also been well established that irradiation affects the colour of meat and poultry.[40] During irradiation, meat becomes pink and the mechanism of this colour change has been investigated in different species[19,41-46] but the precise nature of this pink pigment has yet to be conclusively resolved. A number of different mechanisms have been proposed, some of which attributed the pink colour to oxymyoglobin[44] whereas others suggested that other compounds may be responsible for the enhanced colour.[47,48] The colour changes may affect the acceptability of the product and so have important implications in consumer acceptance of irradiated meats.

7 Wholesomeness of Irradiated Meat

Questions about the safety of irradiated food have been and are still being debated despite the very extensive studies which have been carried on for many years. The results of safety studies including animal feeding studies, *in vitro* studies, and chemical investigations carried out in various laboratories throughout the world have been evaluated many times by expert groups, such as the Joint FAO/IAEA Expert Committee (JECFI)[2] and the Advisory Committee on Irradiated and Novel Foods, UK[49] and in every case support the wholesomeness of these products. Nevertheless, consumers are still very cautious about accepting irradiated food but this is perhaps not surprising when they are subjected to biased and inaccurate information. In the UK, most consumer studies have involved questionnaires and it is not unexpected that the response to irradiated foods is unfavourable since individuals are being asked to assess products which they have never seen.[50] In other countries, where properly conducted market trials have been undertaken and where the consumer has the opportunity to see and compare irradiated and unirradiated food, the response is more positive and consumers have indicated that they are prepared to try the irradiated product.[51]

Although on health grounds there is no reason why irradiated food should be labelled, consumers are demanding that this should be so and recently in the UK the Food Labelling Regulations[52] have stipulated that food treated with ionizing radiation will have the words 'irradiated' or 'treated with ionizing radiation' on the label. This will give the consumer the opportunity to choose either to purchase or to avoid irradiated food.

In order to help enforce these labelling regulations, the availability of a test or tests to identify irradiated food is now recognized as desirable. At a recent meeting in Geneva, it was recommended that national Governments should encourage research into the development of methods for the detection of irradiated food as this would help to promote international trade in irradiated food and enhance consumer confidence in existing control procedures exercised at irradiation facilities.[53] Thus in the past few years considerable progress has been made towards the development of identification methods and these have been reviewed recently.[54] Research efforts in this area are being actively encouraged by the EC in their Community Bureau of Reference (BCR) programme on detection methods and by the IAEA through their Analytical Detection Methods for Irradiation Treatment of Food (ADMIT) programme.

8 Detection of Irradiated Food

It is unlikely that one single detection method will be available for all foods but rather a number of methods will be developed to cover the range of foods which are likely to be irradiated. For meat and poultry, the methods which show most promise at present are electron spin resonance (ESR) spectroscopy[55-57] and the detection of volatile compounds formed from irradiated lipids.[58,59]

ESR detects unpaired electrons and so can be used to identify free radicals. Normally, these free radicals persist for only fractions of a second, but if they can be trapped within the food then this technique can be used for their detection. For food containing bone, free radicals are trapped in the crystalline fraction of the bone and these give a very characteristic ESR signal (Figure 2) which is quite different from that present in unirradiated bone. To date, no other processing technique has been shown to generate a similar ESR signal and thus this characteristic signal shape is indicative of irradiation treatment. Qualitative identification of irradiation will be all that will be required in many instances but in some cases it may also be desirable to obtain an estimate of the irradiation dose received by the food. The facts that the intensity of the ESR signal increases with irradiation dose,[55] that it is not destroyed by cooking,[60] is present in frozen irradiated chicken,[61] and is similar irrespective of the source of bone within

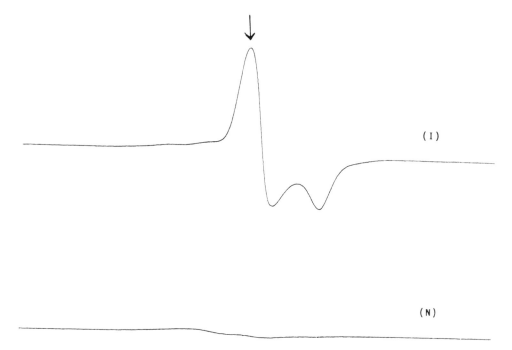

Figure 2 *ESR spectra of irradiated* (I) *and non-irradiated* (N) *chicken bone:* ↓ *centre field* 350 mT; *scan width* 10 ± 5 mT

the carcass[62] lend support to the feasibility of using the technique to quantify the dose obtained. Collaborative blind trials organized by the Ministry of Agriculture Fisheries and Food, UK[63] and BCR[64] have indicated that certainly from a qualitative point of view there is no difficulty in detecting irradiated meat containing bone. The technique can even be applied to mechanically recovered meat[65] since only a few milligrams of bone are needed for detection of the characteristic radiation-induced signal.

The disadvantage of this type of approach is that it depends on the presence of bone and not all meats will contain bone fragments. Thus the other methods which are based on the volatile compounds formed from lipids will have wider application.

Two groups of compounds have been investigated as markers of irradiation. These are the hydrocarbons and the 2-alkylcyclobutanones. As long ago as 1969, it was suggested that the long-chain hydrocarbons could be used as markers of irradiation.[66] More recently, this work has been extended[58,67–69] and it appears that at least three long-chain hydrocarbons, hexadecatriene, tetradecene, and heptadecene, show potential as indicators of irradiated meat. Further research will continue to assess the impact of irradiation and processing variables on the formation of these compounds in irradiated meats.

In the early 1970s, Nawar and co-workers[70,71] also isolated another group of compounds, the 2-alkylcyclobutanones, from pure triglycerides irradiated in vacuum at high doses (60 kGy). The compounds are four-membered ring ketones with an alkyl group located at the 2-position and have the same number of carbon atoms as the parent fatty acid from which they are formed. Thus it is postulated that a series of these compounds will exist in irradiated meat. Using chicken as a model of a fat-containing food, a solvent extraction procedure has been developed for the extraction of one of these compounds, 2-dodecylcyclobutanone, and it has been detected using gas chromatography–mass spectrometry.[59,72,73] The compound is not detectable in non-irradiated or cooked chicken meat and is present in chicken meat irradiated at 5 kGy (Figure 3). It is also not destroyed by cooking and is not detectable in chicken meat which has undergone normal spoilage.[73] Thus the compound has potential as a specific marker for irradiation. The amount of compound formed also increases as irradiation dose increases and does not disappear on storage,[58,73] so it has potential for estimating the irradiation dose applied. Although most of the systematic evaluation of this approach has concentrated on the use of 2-dodecylcyclobutanone, it has also been shown using an infra-red detector in line with the mass spectrometer that a series of these compounds is present in irradiated triglycerides (Figure 4). The 2-alkylcyclobutanones show an infra-red absorption at 1798 cm^{-1} which is characteristic of a ketone group attached to a four-membered ring. Qualitative evidence of the presence of a number of these compounds has also been confirmed in irradiated sterilized chicken meat (M.H. Stevenson, unpublished). Thus in the future,

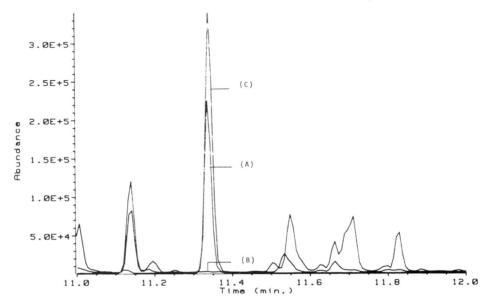

Figure 3 *Selected ion monitoring of ions* m/z 98 *and* m/z 112 *from* (A) *standard 2-dodecylcyclobutanone*: (B) cooked, unirradiated, and (C) irradiated (5 kGy) chicken meat

it should be possible to isolate a number of these compounds and so base the identification of irradiation on several cyclobutanones rather than solely on one of these compounds.

Other approaches are also being investigated such as changes in DNA[74] and the presence of *o*-tyrosine[75,76] in irradiated meats but these methods still require considerable development.

9 Conclusions

The effect of ionizing radiation on red meat and poultry has been investigated for many years and there is an extensive literature on the changes which occur in irradiated meat as well as on the potential applications of the technology for the preservation of these products. As yet, the commercial application of the process to meat and poultry is fairly limited and this is probably due at least in part to the perceived consumer reluctance to accept irradiated food. One of the consumers' concerns has been that it is not possible to tell whether food has been irradiated but the significant progress made in recent years indicates that this is no longer the case. Given time, it is expected that irradiation preservation of meat and poultry will find its place among the other existing technologies which are used for the preservation of food. It is not a panacea for all food processing problems but has the potential to enhance the safety of meat and poultry products.

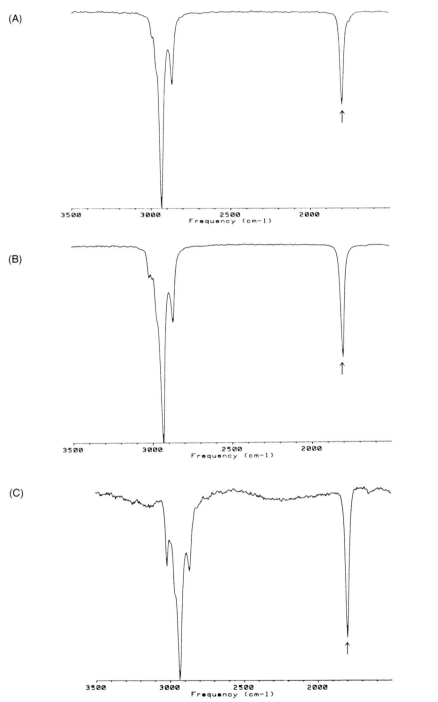

Figure 4 *IR spectra of irradiated* (A) *tristearin,* (B) *triolein, and* (C) *trilinolein*: ↑ *characteristic absorption peak at* 1798 cm^{-1}

10 References

1. Anon., Codex Alimentarius Commission, Rome, 1984, Vol. XV, Edition 1.
2. Anon., Report of a Joint FAO/IAEA/WHO Expert Committee, World Health Organization, Geneva, 1981, Technical Report Series 659.
3. W.M. Urbain, *Food Irradiat. Inf.*, 1978, **8**, 14.
4. E. Wierbicki, Proceedings of the 26th European Meeting of Meat Research Workers, 1980, Vol.1, p.196.
5. G.W. Shults, J.S. Cohen, V.C. Mason, and E. Wierbicki, *J. Food Sci.*, 1977, **42**, 885.
6. G.W. Shults, D.R. Russell, and E. Wierbicki, *J. Food Sci.*, 1972, **37**, 860.
7. G.W. Shults and E. Wierbicki, Ref.4, p.202.
8. E. Wierbicki and F. Heiligman, Ref.4, p.198.
9. R.N. Terrell, G.C. Smith, Z.L. Carpenter, F. Heiligman, and E. Wierbicki, Ref.4, p.206.
10. R. Firstenburg-Eden, D.B. Rowley, and E. Shattuck, *Appl. Envir. Microbiol.*, 1980, **39**, 159.
11. C. Merritt, Jr., P. Angelini, and R.A. Graham, *J. Agric. Food Chem.*, 1978, **26**, 29.
12. F. Heiligman, *Food Technol.*, 1965, **19**, 114.
13. E.H. Kampelmacher, *Br. Poult. Sci.*, 1987, **28**, 3.
14. I.R. Grant and M.F. Patterson, *Int. J. Food Sci. Technol.*, 1991, **26**, 521.
15. I.R. Grant, 'The Microbiology of Irradiated Pork', Ph.D. Thesis, The Queen's University of Belfast, 1990.
16. R.J. Brake, K.D. Murrell, E.E. Ray, J.D. Thomas, B.A. Muggenburg, and J.S. Sivinski, *J. Food Safety*, 1985, **7**, 127.
17. S. Sudarmadji and W.M. Urbain, *J. Food Sci.*, 1972, **37**, 671.
18. W.M. Urbain, 'Proceedings of Bombay Symposium on Radiation Preservation of Food', International Atomic Energy Agency, Vienna, 1973, p.505.
19. A.R. Kamarei, M.Karel, and E. Wierbicki, *J. Food Sci.*, 1979, **44**, 25.
20. K.M. Blythe, 'The Effect of Irradiation on the Organoleptic Quality of Fresh Chicken Carcasses', M.Sc. Thesis, The Queen's University of Belfast, 1990.
21. I.R. Grant and M.F. Patterson, *Int. J. Food Sci. Technol.*, 1991, **26**, 507.
22. P.A. Wills, J.J. Macfarlane, B.J. Shay, and A.F. Egar, *Int. J. Food Microbiol.*, 1987, **4**, 313.
23. M.L. Mattison, A.A. Kraft, D.G. Olson, H.W. Walker, R.E. Rust, and D.B. James, *J. Food Sci.*, 1986, **51**, 284.
24. S.L. Beilker, B. Bill, F.H. Grau, I. Griffiths, J.J. Macfarlane, and P. Vanderlinde, Proceedings of the 33rd European Meeting of Meat Research Workers, 1987, p.177.
25. J.F. Diehl, 'Safety of Irradiated Foods', Marcel Dekker Inc., New York, 1990, p.41.
26. R.S. Hannan and H.J. Shepherd, *J. Sci. Food Agric.*, 1959, **10**, 286.
27. M.G. Simic, in 'Preservation of Food by Ionizing Radiation', ed. E.S. Josephson and M.S. Peterson, CRC Press, Florida, 1983, p.18.
28. B.O. Eggum, in 'Decontamination of Animal Feeds by Irradiation', IAEA, Vienna, 1979, p.55.
29. M.F. Dubravic and W.W. Nawar, *J. Am. Oil Chem. Soc.*, 1968, **45**, 656.
30. W.W. Nawar, in 'Radiation Chemistry of Major Food Components', ed. P.S. Elias and A.J. Cohen, Elsevier, Amsterdam, 1977, p.21.

31. M. Vajdi, W.W. Nawar, and C. Merritt, Jr., *J. Am. Oil Chem. Soc.*, 1982, **59**, 38.
32. W. Huber, A. Brasch, and A. Waly, *Food Technol.*, 1953, **7**, 109.
33. F.P. Mehrlich, Proceedings of Symposium on Food Irradiation, IAEA, Vienna, 1966, p.673.
34. O.F. Batzer and D.M. Doty, *J. Agric. Food Chem.*, 1955, **3**, 64.
35. R.E. Burks, E. Baker, P. Clark, J.E. Esslingety, and J. Lacey, Jr., *J. Agric. Food Chem.*, 1959, **7**, 778.
36. C. Merritt Jr., S.R. Brosnick, M.L. Bazinet, J.T. Walsh, and P. Angelini, *J. Agric. Food Chem.*, 1959, **7**, 784.
37. E.L. Wick, E. Murray, J. Mizutani, and M. Koshika, 'Radiation Preservation of Foods', Advances in Chemistry Series, American Chemical Society, Washington, DC, 1967, p.12.
38. C. Merritt, Jr., Ref.33, p.197.
39. J.R. Champagne and W.W. Nawar, *J. Food Sci.*, 1969, **34**, 335.
40. I.A. Taub, F.M. Robbins, M.G. Simic, J.E. Walker, and E. Wierbicki, *Food Technol.*, 1979, **33**, 184.
41. G.G. Giddings and P. Markakis, *J. Food Sci.*, 1972, **37**, 361.
42. I.D. Ginger, U.J. Lewis, and B.S. Schweigert, *J. Agric. Food Chem.*, 1955, **3**, 156.
43. R. Clark and J.F. Richards, *J. Agric. Food Chem.*, 1971, **19**, 170.
44. A.L. Tappel, *Food Res.*, 1956, **21**, 650.
45. J.D. Selman, *Food Manuf.*, 1982, **57**, 29.
46. K.D. Whitburn, M.Z. Hoffman, and I.A. Taub, *J. Food Sci.*, 1982, **46**, 1814.
47. C. Bernofsky, J.B. Fox, and B.S. Schweigert, *Arch. Biochem. Biophys.*, 1959, **80**, 9.
48. W.D. Brown and J.A. Akoyunoglou, *Arch. Biochem. Biophys.*, 1964, **107**, 239.
49. Advisory Committee on Irradiated and Novel Foods (ACINF), Report on the Safety and Wholesomeness of Irradiated Foods, HMSO, London, 1986.
50. Anon, *Which Magazine*, 1989, p.541.
51. M. Marcotte, 'Consumer Acceptance of Irradiated Food', Atomic Energy of Canada Ltd., Ontario, 1989.
52. The Food Labelling (Amendment) (Irradiated Food) Regulations, 1990, SI 2489, HMSO, London.
53. Anon, *Food Irradiat. Newslett.*, 1989, **13**, 7.
54. IAEA, 'Analytical Detection Methods for Irradiated Foods', IAEA, Vienna, 1991.
55. M.H. Stevenson and R. Gray, *J. Sci. Food Agric.*, 1989, **48**, 269.
56. M.F. Desrosiers and M.G. Simic, *J. Agric. Food Chem.*, 1988, **36**, 601.
57. J.S. Lea, N.J.F. Dodd, and A.J. Swallow, *Int. J. Food Sci. Technol.*, 1988, **23**, 625.
58. W.W. Nawar, Z.R. Zhu, and Y.J. Yoo, in: 'Food Irradiation and the Chemist', ed. D.E. Johnston and M.H. Stevenson, Special Publication No. 86, The Royal Society of Chemistry, Cambridge, 1990, p.13.
59. D.R. Boyd, A.V.J. Crone, J.T.G. Hamilton, M.V. Hand, M.H. Stevenson, and P.J. Stevenson, *J. Agric. Food Chem.*, 1991, **39**, 789.
60. R. Gray and M.H. Stevenson, *Int. J. Food Sci. Technol.*, 1989, **24**, 447.
61. M.H. Stevenson and R. Gray, Ref.58, p.80.
62. R. Gray and M.H. Stevenson, *Int. J. Food Sci. Technol.*, 1990, **25**, 506.

63. S.L. Scotter, P. Holley, and R. Wood, *Int. J. Food Sci. Technol.*, 1990, **25**, 512.
64. J.J. Raffi, 'A European Intercomparison on Electron Spin Resonance Identification of Irradiated Foodstuffs', EUR 13630 EN, 1991.
65. R. Gray and M.H. Stevenson, *Radiat. Phys. Chem.*, 1989, **34**, 899.
66. W.W. Nawar, in 'Health Impact Identification and Dosimetry of Irradiated Foods', ed. K.W. Bögl, D.F. Regulla, and M.J. Suess, WHO, Copenhagen, 1988, p.287.
67. W. Meier, in 'Potential New Methods of Detection of Irradiated Food', ed. J.J. Raffi and J.-J. Belliardo, CEC EUR No. 133331, Brussels, 1991, p.194.
68. A. Splegelberg, L. Heide, and K.W. Bögl, Ref.66, p.177.
69. J. Touminen, J. Klutamo, A.-M. Sjöberg, and S. Leinonen, Ref.66, p.197.
70. P.R. LeTellier and W.W. Nawar, *Lipids*, 1972, **7**, 75.
71. A.P. Handel and W.W. Nawar, *Radiat. Res.*, 1981, **86**, 437.
72. M.H. Stevenson, A.V.J. Crone, and J.T.G. Hamilton, *Nature (London)*, 1990, **344**, 202.
73. A.V.J. Crone, J.T.G. Hamilton, and M.H. Stevenson, *J. Sci. Food Agric.*, 1992, **58**, 249.
74. D.J. Deeble, A.W. Jabir, B.T. Parsons, C.J. Smith, and P. Wheatley, Ref.58, p.57.
75. W. Meier, R. Bürgin, and D. Fröhlich, *Beta Gamma*, 1988, **1**, 34.
76. N. Chuaqui-Offermanns and T. McDougall, *J. Agric. Food Chem.*, 1991, **39**, 300.

Subject Index

Alkyl cyclobutanones, detection, 319, 321
Alkyl cyclobutanones, formation, 319
Accelerated lipid autoxidation, 315
Actin, molecular weight, 213
Alaska pollack, 207
Amino acids, reactions with sugars, 171, 176
Ammonia, 33
Amylase, 58
Anabolic agents, 47
Anaerobic glycolysis, 46
Animal factors, beef production, 17
Animal welfare, stress, 63
Antioxidants, 158, 199, 277
Ascorbates, 159, 267
Autolytic action, 44

Beef
  consumption, 16
  dark-cutting, 53
  fat, 261
  industry, lack of integration, 17
  production, 16
  production, lean beef strategies, 24
Blood, 188
Boar meat quality, 70
Boar taint, 11
Bovine muscles, shear force 107
Bovine serum albumen, 190
Breed, carcass evaluation, 19
Breed, differences, 18
Breeding programmes, 36
Broilers, 29

Canned crustaceans, colour changes, 142
Canned fish products, green discoloration, 142
Canned meat and fish colour, 142
Canned meat/fish, 201
Carbon dioxide, 33
Carbonyl compounds, 152
Carcass
  characteristics, bulls and steers, 21
  composition, 17
  conformation, breed differences, 19
  fat, 18
  quality, DFD risk, 22
Catabolic changes, 57
Cathepsin-L, 55
Cavitation, 288
Cell disruption, 291
Chicken fat, 233
Chikuwa, 219
Cholesterol, 161, 200, 228
Chymosin, 293
Co-precipitation of proteins, 190
Coccidiostat, 29
Cold-shortening, 47
Collagen, 261
  aldimine cross-link, 116, 117
  cleavage, 124
  dimers, 112
  endomysium, 79, 80
  epimysium, 79, 80
  fibre gelatinization, 119, 120
  fibre size, 115
  fibre, residual strength, 120, 121, 122
  genetic type, 116
  hydrothermal shrinkage, 119
  intra-muscular, 113
  keto-imine cross-link, 116, 117
  molecules, microfibre, 111
  perimysial fibres, role in texture, 119, 122
  perimysium, 79, 80
  solubility, 115
  structure, 88
  total collagen content, 115

Collagen (*continued*)
  types, classification, 108
  types, distribution, 110, 114
  types, supramolecular structures, 108
Collagens, non-fibrous, 110
Colour stability
  age effects, 135
  oxygen pressure, 136
  storage conditions, 136
  temperature effects, 137
Comminuted meat, 54
Compensatory growth, 24
Compression testing, 234, 241
Compressive force, collagen content, 109
Compton effect, 312
Conditioning, 123
Conditioning, and flavour, 176
Conditioning, and texture, 97–101
Connective tissue, 79, 230, 238 (see also Collagen)
Consumer acceptance, 145
Consumer choice, 63
Contractile apparatus of muscle, 81, 82
Contractile proteins, 43
Cooked meat
  colour, 139
  fracture mechanics, 122
  fracture propagation, 122
  haemoproteins, 140
  haemoproteins, assessment of 'doneness', 141
  tensile properties, 121
Cooking of meat
  protein denaturation, 106
  shear force changes, 106, 108
  structural changes, 106
Cryoprotectants, 214, 223, 252, 254
Crystal size, 295
Crystallization, 294
Curing, 266
  antioxidant effects, 277
  chemistry of, 267
  flavour, 271
  oxidative processes, 276
  pigments, 275
  role of cytochrome *c*, 272
Cysteine, 191
Cytochromes, 34, 153
Cytoskeletal framework, 83, 84

Dark, firm, and dry (DFD) meat, 65, 175
  classification, 67, 73
  colour, 129
  prediction, 67, 73
  preslaughter handling, 72
  quality problems, 72
Deamination, 44
Decanter in surimi preparation, 225, 227, 232
Denaturation, 213, 300
Depot fats, 193
Desmin, 87
Detection of irradiated meat, 318
Dewatering of surimi, 217, 226
Diet type, carcass gain, 22
Differential scanning calorimetry, 262
Dynamic viscosity, 235

Eating quality, bull beef, 22
Eating quality, lean pork, 5
Elastin, 88, 91, 114
Electrical stimulation, colour effects, 132
Electron spin resonance spectroscopy, 318
Electrostriction, 299
Emulsification, 293
Endocrine function, meat quality, 68
Enzyme activity, 56, 291
Enzyme denaturation, irradiation, 309, 314
Enzyme kinetics, 300
Extensive breeding, 27
Extrinsic factors and meat quality, 45

Fat thickness, carcass quality, 6
Fattening period, 27
Fatty acid composition, pigmeat, 10
Fatty acid compositions, 195
Fatty acids, 147
Feed efficiency, bulls and steers, 20
Feed efficiency, diet composition, 22
Feed restriction, qualitative effects, 24
Fenton reaction, 278
Fibril-associated collagens, 113
Fibril-forming collagens, 110
Filamentous collagens, 112
Fish muscle, 210
Fish oils, 193

Index

Flavour
  chemistry of, 170
  definition, 169
  deterioration, 163
  diet effect of, 174
  fatness effects of, 175
  heat-induced reactions in, 170
  key components in, 177
  species effect of, 173
  stress and pH effects of, 175
Folding test for surimi, 227, 228
Free radicals in bone, 318
Freeze denaturation, 256
Freezer storage, 150
Frequency optimum, 296
Frequency spectrum, 246

Gadoid species, 213
Gap filaments, 85, 86
Gastrointestinal tract, 189
Gelatin, 238, 241
Gel formation, 303
Genetic potential, meat quality, 68
Genetics, 145
Glycogen, depletion, 52
Glycogen, residual, 52
Glycolysis, 30
Growth promoting agents, 47, 123
Growth rates, collagen cross-linking, 123

Haem, 184
  iron, 153
  pigments, 154, 241
Haemoproteins, 185
Halothane, 46
Halothane, sensitivity, meat quality, 8
Heat-stressed birds, 34
High-oleic acid diet, pork chop eating quality, 11
Hormone and β-agonist use, 25, 63
Hydrocarbons in irradiated meat, 316
Hydrogenation, 292
Hydrogen bonds, 300
Hydrogen peroxide, 154, 280, 281
Hydrophobic interactions, 220, 299
Hydroxyl radical, 156

Interaction of ionizing radiation with matter, 312
Intestinal proteases, 216

Intramuscular collagen, role in meat texture, 107
Intrinsic factors and meat quality, 45
Iodine value, 195
Iron
  absorption, 189
  bioavailability, 183
  complexation, 191
  deficiency, 183
  digestion, 188
  ligands, 186
  release, heated meat, 141
Irradiated foods, consumer acceptance, 317
Irradiated meat
  colour changes in, 316
  labelling, 317
  marker compounds, 319
  odour, 315
  safety, 317
Irradiation, 289
  application in food processing, 308
  doses, 308
  odour, compounds identified, 315, 316
  of meat, outlook, 320
  sources, 308

Kamaboko, products, 207, 222

Lean meat
  animal welfare, 62
  consumer selection, 62
  production, quality implications, 73
  stress sensitivity, 70
Lean pork, taste panel assessment, 5
Linolenic acid, pig diet, 10
Lipid breakdown, irradiation, 315
Lipid, composition, 34
Lipid oxidation, 36, 171, 198, 278
Lipid oxidation, and Maillard reaction, 172
Lipids, 146
Lipoxygenase, 157
Long-chain hydrocarbons, 319
Lumen, 189
Lysosomal proteases, 57
Lysosomes, 55

Maillard reaction, 58, 141, 147

Maillard reaction, and lipid oxidation, 172
Marbling fat
  Duroc pigs, 7
  eating quality, 6
  measurement, 7
Meat, absorption/scattering spectra, 133
Meat 'blooming', 132
Meat colour, storage effects, 129
Meat, effect, 187
Meat, factor, 187
Meat quality
  attributes classification, 62
  parameters, 67
  preslaughter factors, 64
  problems, cattle, 71
  sex differences, 70
Meat radappertization, 309
Meat tensile strength, longitudinal, 124
Meat tensile strength, transverse, 124
Mechanical properties, of fish, 95
Mechanical properties, of raw meat, 92
Mechanically recovered meat, 222
Metmyoglobin, 156, 272
Metmyoglobin, equilibrium level, 137
Metmyoglobin formation, 133
  electrical stimulation, 135
  inhibition, 138
  oxygen consumption rate, 134
  oxygen pressure, 134
  photo-oxidation, 138
  resistance, 134
Metmyoglobin, reduction, 134
Microbial spoilage, 50
Microscopy, 236, 238
Modari phenomenon, 215
Monomeric complexes, 186
Muscle cell, 80
Muscle contraction, 83
Muscle fibre type, sex differences, 73
Myobind, 231
Myofibril, 80
Myofibrillar protein denaturation, 121
Myofibrillar proteins, 223
Myoglobin, 293
  conformational change, 140
  differences, 128, 129
  heat stability, 140
  structure, 130, 131
  structure, species differences, 131
Myopathy, 31
Myosin, content, lightness, 130
Myosin, molecular weight, 212
Myotomes, 79, 97

Neopyrithamin, 53
Nitric oxide, cured meats, 266
Nitrile, 158
Nitrite, curing, 266
Nitrosation, 271
Nitrosylmyoglobin, 267, 273, 276
Nucleation, 295

Off-flavours, 174, 198
Oleic acid, pig diet, 11
Oscillatory deformation, 245
Oxidative deterioration, 145
Oxidizing agents, 313
Oxygen consumption rate, pH effects, 137
Oxygen, pressure, 49
Oxyhaemoglobin, 155
Oxymyoglobin, formation, 131
Oxymyoglobin, structure, 132

Pale, soft, and exudative (PSE) meat, 176
  classification, 66
  colour, 129
  eating quality, 7
  prediction, 66
  thyroid function, 69
Paramagnetic oxygen, 155
Pelagic species, 220
Phosphates, 159, 240
Phospholipids, 149
Photo-oxidation, 283
Phytic acid, 187
Picro-sirius red, 240
Pigmeat
  carcass classification, 3
  carcass compositional changes, 4
  carcass fat composition, 4
  carcass fat content, 4
  cooked, cured, 276
  integrated production systems, 13
  quality, 3
  quality, assurance schemes, 13
  tenderness, 8
  tenderness, *ad libitum* feeding, 8, 9

Index

tenderness, β-agonists, 9
tenderness, growth rate, 9
Pithing, 47
Plane of nutrition
  carcass quality, 23
  live weight gain, 23
  restricted feeding, 23, 24
Polydextrose, 254, 260
Polyunsaturated fatty acids, 35, 162
Polyunsaturated/saturated (P/S) ratio, 195–196, 233
Porcine muscle, pH decline, 66
Porphyrin ring, 185
Post mortem
  ageing, 56
  glycolysis, 43
  glycolysis, meat quality, 64, 65
Potato starch, 259
Poultry
  muscles, 30
  quality, 28
  yield, 28
Prerigor chilling, 49
Preslaughter handling, 32
Preslaughter stress, 53
  meat quality, 73
  sex differences in cattle, 72
Pressure, enzymes, 299
Pressure, structure, 299
Pressure–heat treatment, 303
Processed meat, 203
Protein irradiation, 313
Proteoglycans, 114
Proteolysis, 54

Radappertization
  destruction of micro-organisms, 309
  minimum radiation dose, 309
  product quality, 310
Radicidation, pathogenic
  micro-organisms, 310
Radiolytic products from protein, 314
Radurization, spoilage
  micro-organisms, 311
Radurized meat
  product appearance, 311
  product manufacture, 311
  products, flavour changes, 312
  products, sensory panel assessment, 312
Rancidity, 162, 171, 194, 228, 232, 258

Rancimat, 199
Raw meat
  colour, 128
  colour, haemoproteins, 128
  green discolorations, 138
Rectified diffusion, 288
Restructuring, 305
Rigor mortis, 92

Saleable meat yield, 21
Sarcomere shortening, 48
Sarcoplasmic proteins, 216
Sarcoplasmic reticulum, 87, 88
Satsumaage, 219
Scalding, 33
Selection of leanness, stress sensitivity, 68
Separator, 226
Serum ferritin, 184
Sex differences, carcass composition, 20
  20
Shear values, 32
Shortening, cold, 92, 97
Shortening, *rigor*, 92
Silage digestibility, carcass fat, 23
Skatole, adipose tissue, 12
Slaughter weight, 18
Slaughter weight, feeding period, 20
Soya bean flakes, 294
Soya isolate, 259
Storage modulus, 235, 245
Stress
  behavioural studies, 63
  cycle, 64
  definition, 63
Stress-sensitive animals, 65
Stress sensitivity
  breed differences, 69
  corticosteroid levels, 69
  thyroid function, 70
Stress-susceptible steers, 51
Sugar beet feed, meat quality, 12
Sulfmyoglobin formation, 138, 139
Surimi, 222
  beef, 230
  Surimi, colour, 229
  composition, 237
  fish, 207
  folding test, 227, 228
  gels, 172, 208, 227
  grading, 228

Surimi (*continued*)
  industry, 208
  mutton, 231
  pork, 229
  poultry, 232
  process, 215
  production, 210
  products, 281
  quality, 233
  rope, 219
  setting process, 218
  structure, 239
Suwari, 214

Tannins, 187
Tenderization, 304
Tensile adhesive strength, 234
Thermal breakdown of meat components, 171
Thermal oxidation, 282
Thermocoagulation, 218
Thermorheological profile, 248
Thick filaments, 81–83

Thin filaments, 81–83
Thiobarbituric acid value, 258
Thyroid function, PSE in pig meat, 71
Thyroid function, sex and breed differences in pigs, 71
Tocopherol, 35
Triacylglycerols, 146
Trivalent cross-links in collagen, 117, 118

Ultrasonic irradiation, 290
Ultrasound, 287

Vacuum packaging, 161
Vitamin E, 160

Warmed-over flavour, 148, 153, 171, 276
Waste protein, 305
Water radiolysis, 313
Water radiolysis, temperature effects, 313
Working volumes, 305